SUPERSYMMETRY IN PARTICLE PHYSICS

Supersymmetry has been a central topic in particle physics since the early 1980s, and represents the culmination of the search for fundamental symmetries that has dominated particle physics for the last 50 years. Traditionally, the constituents of matter (fermions) have been regarded as essentially different from the particles (bosons) that transmit the forces between them. In supersymmetry, however, fermions and bosons are unified.

This is the first textbook to provide a simple pedagogical introduction to what has been a formidably technical field. The elementary and practical treatment brings readers to the frontier of contemporary research, in particular, to the confrontation with experiments at the Large Hadron Collider. Intended primarily for first-year graduate students in particle physics, both experimental and theoretical, this volume will also be of value to researchers in experimental and phenomenological supersymmetry. Supersymmetric theories are constructed through an intuitive 'trial and error' approach, rather than being formal and deductive. The basic elements of spinor formalism and superfields are introduced, allowing readers to access more advanced treatments. Emphasis is placed on physical understanding, and on detailed, explicit derivations of all important steps. Many short exercises are included making for a valuable and accessible self-study tool.

IAN AITCHISON is Emeritus Professor of Physics at the University of Oxford. His research interests include time-dependent effective theories of superconductors, field theories at finite temperature and topological aspects of gauge theories.

SUPERSYMMETRY IN PARTICLE PHYSICS

An Elementary Introduction

IAN J. R. AITCHISON

Department of Physics, University of Oxford,
The Rudolf Peierls Centre for Theoretical Physics

and

Stanford Linear Accelerator Center,
Stanford University

CAMBRIDGE
UNIVERSITY PRESS

CAMBRIDGE UNIVERSITY PRESS
Cambridge, New York, Melbourne, Madrid, Cape Town,
Singapore, São Paulo, Delhi, Mexico City

Cambridge University Press
The Edinburgh Building, Cambridge CB2 8RU, UK

Published in the United States of America by Cambridge University Press, New York

www.cambridge.org
Information on this title: www.cambridge.org/9780521880237

First published 2007

A catalogue record for this publication is available from the British Library

ISBN 978-0-521-88023-7 Hardback

For Danny

Contents

Preface

This book is intended to be an elementary and practical introduction to supersymmetry in particle physics. More precisely, I aim to provide an accessible, self-contained account of the basic theory required for a working understanding of the 'Minimal Supersymmetric Standard Model' (MSSM), including 'soft' symmetry breaking. Some simple phenomenological applications of the model are also developed in the later chapters.

The study of supersymmetry (SUSY) began in the early 1970s, and there is now a very large, and still growing, research literature on the subject, as well as many books and review articles. However, in my experience the existing sources are generally suitable only for professional (or intending) theorists. Yet searches for SUSY have been pursued in experimental programmes for some time, and are prominent in experiments planned for the Large Hadron Collider at CERN. No direct evidence for SUSY has yet been found. Nevertheless, for the reasons outlined in Chapter 1, supersymmetry at the TeV scale has become the most highly developed framework for guiding and informing the exploration of physics beyond the Standard Model. This dominant role of supersymmetry, both conceptual and phenomenological, suggests a need for an entry-level introduction to supersymmetry, which is accessible to the wider community of particle physicists.

The first difficulty presented by conventional texts on supersymmetry – and it deters many students – is one of notation. Right from the start, discussions tend to be couched in terms of a spinor notation that is generally not familiar from standard courses on the Dirac equation – namely, that of either 'dotted and undotted 2-component Weyl spinors', or '4-component Majorana spinors'. This creates something of a conceptual discontinuity between what most students already know, and what they are trying to learn; it becomes a pedagogical barrier. By contrast, my approach builds directly on knowledge of Dirac spinors in a conventional representation, using 2-component ('half-Dirac') spinors, without necessarily requiring the more sophisticated dotted and undotted formalism. The latter is, however, introduced early on (in Section 2.3), but it can be treated as an optional extra; the essential elements of SUSY and the MSSM (contained in Sections 3.1, 3.2, 4.2, 4.4, 5.1 and Chapters 7 and 8) can be understood quite reasonably without it.

Apart from its simple connection to standard Dirac theory, a second advantage of the 2-component formalism is, I think, that it is simpler to use than the Majorana one for motivating and establishing the forms of simple SUSY-invariant Lagrangians. Again, a more powerful route is available via the superfield formalism, to which I provide access in Chapter 6, but the essentials do not depend on it.

On the other hand, I don't think it is wise to eschew the Majorana formalism altogether. For one thing, there are some important sources which adopt it exclusively, and which students might profitably consult. Furthermore, the Majorana formalism appears to be the one generally used in SUSY calculations, since, with some modifications, it allows the use of short-cuts familiar from the Dirac case. So I provide an early introduction to Majorana spinors as well, in Section 2.5; and at various places subsequently I point out the Majorana equivalents for what is going on. I make use of Majorana forms in Section 8.2, where I recover the Standard Model interactions in the MSSM, and also in the calculations of Section 5.2 and of Chapter 12. I believe that the indicated arguments justify the added burden, to the interested reader, of having to acquire some familiarity with a second language.

Moving on from notation, my approach is generally intuitive and constructive, rather than formal and deductive. It is very much a do-it-yourself treatment. Thus in Sections 2.1 and 2.2 I provide a gentle and detailed introduction to the use of Weyl spinors in the 'half-Dirac' notation. Care is taken to introduce a simple (free) SUSY theory very slowly and intuitively in Section 3.1, and this is followed by an appetite-whetting preview of the MSSM, as a relief from the diet of formalism. The simple SUSY transformations learned in Section 3.1 are used to motivate the SUSY algebra in Section 4.2 (rather than just postulating it), and simple consequences for supermultiplet structure are explained in Section 4.4. The more technical matter of the necessity for auxiliary fields (even in such a simple case) is discussed at the end of Chapter 4.

The introduction of interactions in a chiral multiplet follows reasonably straightforwardly in Section 5.1 (the Wess–Zumino model). The more technical – but theoretically crucial – property of cancellation of quadratic divergences is illustrated for some simple cases in Section 5.2.

After the optional detour into chiral superfields, the main thread is taken up again in Chapter 7, where supersymmetric gauge theories are introduced via vector supermultiplets, which are then combined with chiral supermultiplets. Here the superfield formalism has been avoided in favour of a more direct try-it-and-see approach similar to that of Section 3.1.

At roughly the half-way stage in the book, all the elements necessary for understanding the construction of the MSSM (or variants thereof) are now in place. The model is defined in Chapter 8, and immediately applied to exhibit gauge-coupling unification. Elementary ideas of SUSY breaking are introduced in Chapter 9, together with the phenomenolgically important notion of 'soft' supersymmetry-breaking parameters. The remainder of the book is devoted to simple applications: Higgs physics (Chapter 10), sparticle masses (Chapter 11) and sparticle production processes (Chapter 12).

Throughout, emphasis is placed on providing elementary, explicit and detailed derivations of important formal steps wherever possible. Many short exercises are included, which are designed to help the reader to engage actively with the text, and to keep abreast of the formal development through practice at every stage.

In keeping with the stated aim, the scope of this book is strictly limited. A list of omitted topics would be long indeed. It includes, for example: the superfield formalism for vector supermultiplets; Feynman rules in super-space; wider phenomenological implications of the MSSM; local supersymmetry (supergravity); more detail on SUSY searches; SUSY and cosmology; non-perturbative aspects of SUSY; SUSY in dimensions other than 4, and for values of N other than $N = 1$. Fortunately, a number of excellent and comprehensive monographs are now available; readers interested in pursuing matters beyond where I leave them, or in learning about topics I omit, can confidently turn to these professional treatments.

I am very conscious that the list of references is neither definitive nor comprehensive. In a few instances (for example, in reviewing the beginnings of SUSY and the MSSM) I have tried to identify the relevant original contributions, although I have probably missed some. Usually, I have not attempted to trace priorities carefully, but have referred to more comprehensive reviews, or have simply quoted such references as came to hand as I worked my own way into the subject. I apologize to the many researchers whose work, as a consequence, has not been referenced here.

The book has grown out of lectures to graduate students at Oxford working in both experimental and theoretical particle physics. In this, its genesis is very similar to my book with Tony Hey, *Gauge Theories in Particle Physics*, first published in 1982 and now in its third (two-volume) edition. The present book aims to reach a similar readership: in particular, I have tried to design the level so that it follows smoothly on from the earlier one. Indeed, as the title suggests, this book may be seen as 'volume 3' in the series.

However, I would expect theorists and experimentalists to use the book differently. For theorists, it should be a relatively easy read, setting them up for immediate access to the professional literature and more advanced monographs. On the other hand, many experimentalists are likely to find some of the formal parts indigestible, even with the support provided. They should be able to find a reasonably friendly route to the physics they want to learn via the 'essential elements' mentioned earlier (that is, Sections 2.1, 2.2, 3.1, 3.2, 4.2, 4.4, 5.1 and Chapters 7 and 8), to be followed by whatever applications they are most interested in. Much of this material should not be beyond final year maths or physics undergraduates who have taken courses in relativistic quantum mechanics, introductory quantum field theory, and gauge theories. By the same token, the book may also be useful to a wide range of physicists in other areas, who wish to gain a first-hand appreciation of the excitement and anticipation which surround the possible discovery of supersymmetry at the TeV scale.

Ian J. R. Aitchison
February 2007

Acknowledgements

Questions raised by three successive generations of Oxford students have led to many improvements in my original notes. I am grateful to John March-Russell for many clear and patient explanations when I was trying (not for the first time) to learn supersymmetry while on leave at CERN in 2001–2, and subsequently at Oxford. I have also benefited from being able to consult Graham Ross as occasion demanded. I thank Michael Peskin and Stan Brodsky for welcoming me as a visitor to the Stanford Linear Accelerator Center Theory Group, where work on the book continued, supported by the Department of Energy under contract DE-AC02-76SF00515. As regards written sources, I owe most to Stephen Martin's '*A Supersymmetry Primer*' (reference 46), and to '*Weak Scale Supersymmetry*' by Howard Baer and Xerxes Tata (reference 49); I hope I have owned up to all my borrowings. In developing the material for Chapter 6 I was greatly helped by unpublished notes of lectures on superfields by the late Caroline Fraser, given at Annecy in 1981–2.

Above all, and once again, I owe a special debt to my good friend George Emmons, who has been an essential part of this project from the beginning: his comments and queries revealed misconceptions on my part, as well as obscurities in the presentation, and led to many improvements in the developing text; he read carefully through several successive LaTex drafts, spotting many errors, and he corrected the final proofs. His encouragement and support have been invaluable.

1

Introduction and motivation

Supersymmetry (SUSY) – a symmetry relating bosonic and fermionic degrees of freedom – is a remarkable and exciting idea, but its implementation is technically rather complicated. It can be discouraging to find that after standard courses on, say, the Dirac equation and quantum field theory, one has almost to start afresh and master a new formalism, and moreover one that is not fully standardized. On the other hand, 30 years have passed since the first explorations of SUSY in the early 1970s, without any direct evidence of its relevance to physics having been discovered. The Standard Model (SM) of particle physics (suitably extended to include an adequate neutrino phenomenology) works extremely well. So the hard-nosed seeker after truth may well wonder: why spend the time learning all this intricate SUSY formalism? Indeed, why speculate at all about how to go 'beyond' the SM, unless or until experiment forces us to? If it's not broken, why try and fix it?

As regards the formalism, most standard sources on SUSY use either the 'dotted and undotted' 2-component (Weyl) spinor notation found in the theory of representations of the Lorentz group, or 4-component Majorana spinors. Neither of these is commonly included in introductory courses on the Dirac equation (although perhaps they should be), but it is perfectly possible to present simple aspects of SUSY using a notation which joins smoothly on to standard 4-component Dirac equation courses, and a brute force, 'try-it-and-see' approach to constructing SUSY-invariant theories. That is the approach to be followed in this book, at least to start with. However, as we go along the more compact Weyl spinor formalism will be introduced, and also (more briefly) the Majorana formalism. Later, we shall include an introduction to the powerful superfield formalism. All this formal concentration is partly because the simple-minded approach becomes too cumbersome after a while, but mainly because discussions of the phenomenology of the Minimal Supersymmetric Standard Model (MSSM) generally make use of one or other of these more sophisticated notations.

What of the need to go beyond the Standard Model? Within the SM itself, there is a plausible historical answer to that question. The V–A current–current (four-fermion) theory of weak interactions worked very well for many years, when used at lowest order in perturbation theory. Yet Heisenberg [1] had noted as early as 1939 that problems arose if one tried to compute higher-order effects, perturbation theory apparently breaking down completely at the then unimaginably high energy of some 300 GeV (the scale of $G_{\text{F}}^{-1/2}$). Later, this became linked to the non-renormalizability of the four-fermion theory, a purely theoretical problem in the years before experiments attained the precision required for sensitivity to electroweak radiative corrections. This perceived disease was alleviated but not cured in the 'Intermediate Vector Boson' model, which envisaged the weak force between two fermions as being mediated by massive vector bosons. The non-renormalizability of such a theory was recognized, but not addressed, by Glashow [2] in his 1961 paper proposing the $\text{SU}(2) \times \text{U}(1)$ structure. Weinberg [3] and Salam [4], in their gauge-theory models, employed the hypothesis of spontaneous symmetry breaking to generate masses for the gauge bosons and the fermions, conjecturing that this form of symmetry breaking would not spoil the renormalizability possessed by the massless (unbroken) theory. When 't Hooft [5] demonstrated this in 1971, the Glashow–Salam–Weinberg theory achieved a theoretical status comparable to that of quantum electrodynamics (QED). In due course the precision electroweak experiments spectacularly confirmed the calculated radiative corrections, even yielding a remarkably accurate prediction of the top quark mass, based on its effect as a virtual particle . . . but note that even this part of the story is not yet over, since we have still not obtained experimental access to the proposed symmetry-breaking (Higgs [6]) sector. If and when we do, it will surely be a remarkable vindication of theoretical preoccupations dating back to the early 1960s.

It seems fair to conclude that worrying about perceived imperfections of a theory, even a phenomenologically very successful one, can pay off. In the case of the SM, a quite serious imperfection (for many theorists) is the 'SM fine-tuning problem', which we shall discuss in a moment. SUSY can suggest a solution to this perceived problem, provided that supersymmetric partners to known particles have masses no larger than a few TeV (roughly).

In addition to the 'fine-tuning' motivation for SUSY – to which, as we shall see, there are other possible responses – there are some quantitative results (Section 1.2), and theoretical considerations (Section 1.3) , which have inclined many physicists to take SUSY and the MSSM (or something like it) very seriously. As always, experiment will decide whether these intuitions were correct or not. A lot of work has been done on the phenomenology of such theories, which has influenced the Large Hadron Collider (LHC) detector design. Once again, it will surely be extraordinary if, in fact, the world turns out to be this way.

1.1 The SM fine-tuning problem

The electroweak sector of the SM contains within it a parameter with the dimensions of energy (i.e. a 'weak scale'), namely

$$v \approx 246\,\text{GeV}, \tag{1.1}$$

where $v/\sqrt{2}$ is the vacuum expectation value (or 'vev') of the neutral Higgs field, $\langle 0|\phi^0|0\rangle = v/\sqrt{2}$. The occurrence of the vev signals the 'spontaneous' breaking of electroweak gauge symmetry (see, for example [7], Chapter 19), and the associated parameter v sets the scale, in principle, of all masses in the theory. For example, the mass of the $W^{\pm}$ (neglecting radiative corrections) is given by

$$M_W = gv/2 \sim 80\,\text{GeV}, \tag{1.2}$$

and the mass of the Higgs boson is

$$M_H = v\sqrt{\frac{\lambda}{2}}, \tag{1.3}$$

where g is the SU(2) gauge coupling constant, and λ is the strength of the Higgs self-interaction in the Higgs potential

$$V = -\mu^2\phi^\dagger\phi + \frac{\lambda}{4}(\phi^\dagger\phi)^2, \tag{1.4}$$

where $\lambda > 0$ and $\mu^2 > 0$. Here ϕ is the SU(2) doublet field

$$\phi = \begin{pmatrix} \phi^+ \\ \phi^0 \end{pmatrix}, \tag{1.5}$$

and all fields are understood to be quantum, no 'hat' being used.

Recall now that the *negative* sign of the 'mass^2' term $-\mu^2$ in (1.4) is essential for the spontaneous symmetry-breaking mechanism to work. With the sign as in (1.4), the minimum of V interpreted as a classical potential is at the non-zero value

$$|\phi| = \sqrt{2}\mu/\sqrt{\lambda} \equiv v/\sqrt{2}, \tag{1.6}$$

where $\mu \equiv \sqrt{\mu^2}$. This classical minimum (equilibrium value) is conventionally interpreted as the expectation value of the quantum field in the quantum vacuum (i.e. the vev), at least at tree level. If '$-\mu^2$' in (1.4) is replaced by the positive quantity 'μ^2', the classical equilibrium value is at the origin in field space, which would imply $v = 0$, in which case all particles would be massless. Hence it is vital to preserve the sign, and indeed magnitude, of the coefficient of $\phi^\dagger\phi$ in (1.4).

The discussion so far has been at tree level (no loops). What happens when we include loops? The SM is renormalizable, which means that finite results are obtained for all higher-order (loop) corrections even if we extend the virtual momenta

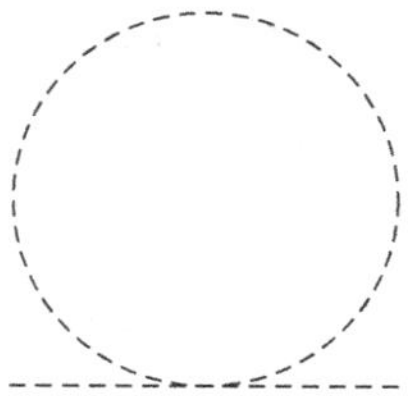

Figure 1.1 One-loop self-energy graph in ϕ^4 theory.

in the loop integrals all the way to infinity; but although this certainly implies that the theory is well defined and calculable up to infinite energies, in practice no one seriously believes that the SM is really all there is, however high we go in energy. That is to say, in loop integrals of the form

$$\int^{\Lambda} \mathrm{d}^4 k \; f(k, \text{external momenta}) \tag{1.7}$$

we do not think that the cut-off Λ *should* go to infinity, physically, even though the reormalizability of the theory assures us that no inconsistency will arise if it does. More reasonably, we regard the SM as part of a larger theory which includes as yet unknown 'new physics' at high energy, Λ representing the scale at which this new physics appears, and where the SM must be modified. At the very least, for instance, there surely must be some kind of new physics at the scale when quantum gravity becomes important, which is believed to be indicated by the Planck mass

$$M_{\mathrm{P}} = (G_{\mathrm{N}})^{-1/2} \simeq 1.2 \times 10^{19}\ \mathrm{GeV}. \tag{1.8}$$

If this is indeed the scale of the new physics beyond the SM or, in fact, if there is *any* scale of 'new physics' even several orders of magnitude different from the scale set by v, then we shall see that we meet a problem with the SM, once we go beyond tree level.

The 4-boson self-interaction in (1.4) generates, at one-loop order, a contribution to the $\phi^\dagger\phi$ term, corresponding to the self-energy diagram of Figure 1.1, which is proportional to

$$\lambda \int^{\Lambda} \mathrm{d}^4 k \, \frac{1}{k^2 - M_{\mathrm{H}}^2}. \tag{1.9}$$

This integral clearly diverges quadratically (there are four powers of k in the numerator, and two in the denominator), and it turns out to be *positive*, producing a correction

$$\sim \lambda \Lambda^2 \phi^\dagger \phi \tag{1.10}$$

to the 'bare' $-\mu^2\phi^\dagger\phi$ term in V. (The '$\sim$' represents a numerical factor, such as $1/4\pi^2$, which is unimportant for the argument here: we shall include such factors explicitly in a later calculation, in Section 5.2.) The coefficient $-\mu^2$ of $\phi^\dagger\phi$ is then replaced by the one-loop corrected 'physical' value $-\mu^2_{\text{phys}}$, where (ignoring the numerical factor) $-\mu^2_{\text{phys}} = -\mu^2 + \lambda\Lambda^2$, or equivalently

$$\mu^2_{\text{phys}} = \mu^2 - \lambda\Lambda^2. \tag{1.11}$$

Re-minimizing V, we obtain (1.6) but with μ replaced by $\mu_{\text{phys}} \equiv \sqrt{\mu^2_{\text{phys}}}$. Consider now what is the likely value of μ_{phys}. With v fixed phenomenologically by (1.1), equation (1.6), as corrected to involve μ_{phys}, provides a relation between the two unknown parameters μ_{phys} and λ: $\mu_{\text{phys}} \approx \sqrt{\lambda}\,123\,\text{GeV}$. It follows that if we want to be able to treat the Higgs coupling λ perturbatively, μ_{phys} can hardly be much greater than a few hundred GeV at most. (A value considerably greater than this would imply that λ is very much greater than unity, and the Higgs sector would be 'strongly interacting'; while not logically excluded, this possibility is generally not favoured, because of the practical difficulty of making reliable non-perturbative calculations.) On the other hand, if $\Lambda \sim M_{\text{P}} \sim 10^{19}$ GeV, the one-loop correction in (1.11) is then vastly greater than $\sim(100\text{ GeV})^2$, so that to arrive at a value $\sim(100\text{ Gev})^2$ *after* inclusion of this loop correction would seem to require that we start with an equally huge value of the Lagrangian parameter μ^2, relying on a remarkable cancellation, or *fine-tuning*, to get us from $\sim(10^{19}\text{ GeV})^2$ down to $\sim(10^2\text{ GeV})^2$.

In the SM, this fine-tuning problem involving the parameter μ_{phys} affects not only the mass of the Higgs particle, which is given in terms of μ_{phys} (combining (1.3) and (1.6)) by

$$M_{\text{H}} = \sqrt{2}\mu_{\text{phys}}, \tag{1.12}$$

but also the mass of the W,

$$M_{\text{W}} = g\mu_{\text{phys}}/\sqrt{\lambda}, \tag{1.13}$$

and ultimately all masses in the SM, which derive from v and hence μ_{phys}. The serious problem posed for the SM by this 'unnatural' situation, which is caused by quadratic mass divergences in the scalar sector, was pointed out by K. G. Wilson in a private communication to L. Susskind [8].[1]

[1] From a slightly different perspective, 't Hooft [9] also drew attention to difficulties posed by theories with 'unnaturally' light scalars. In the context of Grand Unified gauge theories, Weinberg [10] emphasized the difficulty of finding a natural theory (i.e. one that is not fine-tuned) in which scalar fields associated with symmetry breaking are elementary, and some symmetries are broken at the GUT scale $\sim 10^{16}$ GeV whereas others are broken at the very much lower weak scale; this is usually referred to as the 'gauge hierarchy problem'.

This fine-tuning problem would, of course, be much less severe if, in fact, 'new physics' appeared at a scale Λ which was much smaller than M_{P}. How much tuning is acceptable is partly a subjective matter, but for many physicists the only completely 'natural' situation is that in which the scale of new physics is within an order of magnitude of the weak scale, as defined by the quantity v of equation (1.1), i.e. no higher than a few TeV. The question then is: what might this new physics be?

Within the framework of the discussion so far, the aim of an improved theory must be somehow to eliminate the quadratic dependence on the (assumed high) cut-off scale, present in theories with fundamental (or 'elementary') scalar fields. In the SM, such fields were introduced to provide a simple model of spontaneous electroweak symmetry breaking. Hence one response – the first, historically – to the fine-tuning problem is to propose [8] (see also [11]) that symmetry breaking occurs 'dynamically'; that is, as the result of a new strongly interacting sector with a mass scale in the TeV region. In such theories, generically called 'technicolour', the scalar states are not elementary, but rather fermion–antifermion bound states. The dynamical picture is analogous to that in the BCS theory of superconductivity (see, for example, Chapters 17, 18 and 19 of [7]). In this case, the Lagrangian for the Higgs sector is only an effective theory, valid for energies significantly below the scale at which the bound state structure would be revealed, say 1–10 TeV. The integral in (1.9) can then only properly be extended to this scale, certainly not to a hierarchically different scale such as M_{P}, or the GUT scale. This scheme works very nicely as far as generating masses for the weak bosons is concerned. However, in the SM the fermion masses also are due to the coupling of fermions to the Higgs field, and hence, if the Higgs field is to be completely banished from the 'fundamental' Lagrangian, the proposed new dynamics must also be capable of generating the fermion mass spectrum. This has turned out to require increasingly complicated forms of dynamics, to meet the various experimental constraints. Still, technicolour theories are not conclusively ruled out. Reviews are provided by Fahri and Susskind [12], and more recently by Lane [13]; see also the somewhat broader review by Hill and Simmons [14].

If, on the other hand, fundamental scalars are to be included in the theory, how might the quadratic divergences be controlled? A clue is provided by considering why such divergences only seem to affect the scalar sector. In QED the photon self-energy diagram of Figure 1.2 is apparently quadratically divergent (there are two fermion propagators, each of which depends linearly on the integrated 4-momentum). As in the scalar case, such a quadratic divergence would imply an enormous quantum correction to the photon mass. In fact this divergence is absent, provided the theory is regularized in a gauge-invariant way (see, for example [15], Section 11.3). In other words, the symmetry of gauge invariance guarantees that no

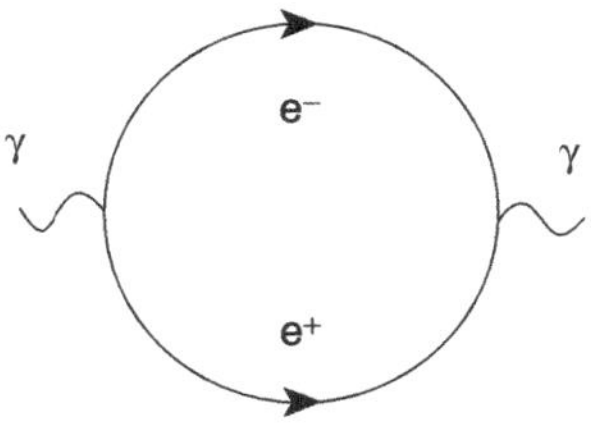

Figure 1.2 One-loop photon self-energy diagram in QED.

term of the form

$$m_\gamma^2 A^\mu A_\mu \tag{1.14}$$

can be radiatively generated in an unbroken gauge theory: the photon is massless. The diagram of Figure 1.2 is divergent, but only logarithmically; the divergence is absorbed in a field strength renormalization constant, and is ultimately associated with the running of the fine structure constant (see [7], Section 15.2).

We may also consider the electron self-energy in QED, generated by a one-loop process in which an electron emits and then re-absorbs a photon. This produces a correction δm to the fermion mass m in the Lagrangian, which seems to vary linearly with the cut-off:

$$\delta m \sim \alpha \int^\Lambda \frac{\mathrm{d}^4 k}{\not k k^2} \sim \alpha\Lambda. \tag{1.15}$$

(Here we have neglected both the external momentum and the fermion mass, in the fermion propagator, since we are interested in the large k behaviour.) Although perhaps not so bad as a quadratic divergence, such a linear one would still lead to unacceptable fine-tuning in order to arrive at the physical electron mass. In fact, however, when the calculation is done in detail one finds

$$\delta m \sim \alpha m \ln \Lambda, \tag{1.16}$$

so that even if $\Lambda \sim 10^{19}$ GeV, we have $\delta m \sim m$ and no unpleasant fine-tuning is necessary after all.

Why does it happen in this case that $\delta m \sim m$? It is because the Lagrangian for QED (and the SM for that matter) has a special symmetry as the fermion masses go to zero, namely chiral symmetry. This is the symmetry under transformations (on fermion fields) of the form

$$\psi \to \mathrm{e}^{\mathrm{i}\alpha\gamma_5}\psi \tag{1.17}$$

in the U(1) case, or

$$\psi \to \mathrm{e}^{\mathrm{i}\boldsymbol{\alpha}\cdot\boldsymbol{\tau}/2\gamma_5}\psi \tag{1.18}$$

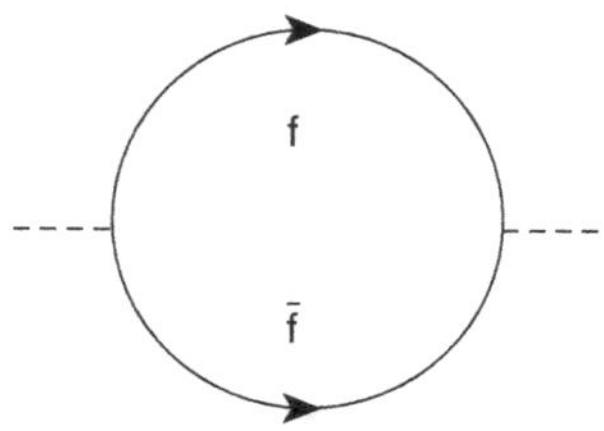

Figure 1.3 Fermion loop contribution to the Higgs self-energy.

in the SU(2) case. This symmetry guarantees that all radiative corrections to m, computed in perturbation theory, will vanish as $m \to 0$. Hence δm must be proportional to m, and the dependence on Λ is therefore (from dimensional analysis) only logarithmic.

In these two examples from QED, we have seen how unbroken gauge and chiral symmetries keep vector mesons and fermions massless, and remove 'dangerous' quadratic and linear divergences from the theory. If we could find a symmetry which grouped scalar particles with either massless fermions or massless vector bosons, then the scalars would enjoy the same 'protection' from dangerous divergences as their symmetry partners. Supersymmetry is precisely such a symmetry: as we shall see, it groups scalars together with fermions (and vector bosons with fermions also). The idea that supersymmetry might provide a solution to the SM fine-tuning problem was proposed by Witten [16], Veltman [17] and Kaul [18].

We can understand qualitatively how supersymmetry might get rid of the quadratic divergences in the scalar self-energy by considering a possible fermion loop correction to the $-\mu^2\phi^\dagger\phi$ term, as shown in Figure 1.3. At zero external momentum, such a contribution behaves as

$$\left(-g_{\rm f}^2 \int^\Lambda {\rm d}^4k\, {\rm Tr}\left[\frac{1}{(\not{k} - m_{\rm f})^2}\right]\right)\phi^\dagger\phi = \left(-4g_{\rm f}^4 \int^\Lambda {\rm d}^4k\, \frac{k^2 + m_{\rm f}^2}{\left(k^2 - m_{\rm f}^2\right)^2}\right)\phi^\dagger\phi. \tag{1.19}$$

The sign here is crucial, and comes from the closed fermion loop. The term with the k^2 in the numerator in (1.19) is quadratically divergent, and of opposite sign to the quadratic divergence (1.10) due to the Higgs loop. Ignoring numerical factors, these two contributions together have the form

$$\left(\lambda - g_{\rm f}^2\right)\Lambda^2\phi^\dagger\phi. \tag{1.20}$$

The possibility now arises that *if* for some reason there existed a boson–fermion coupling $g_{\rm f}$ related to the Higgs coupling by

$$g_{\rm f}^2 = \lambda \tag{1.21}$$

then this quadratic sensitivity to Λ would not occur.

A relation between coupling constants, such as (1.21), is characteristic of a symmetry, but in this case it must evidently be a symmetry which relates a purely bosonic vertex to a boson–fermion (Yukawa) one. Relations of the form (1.21) are indeed just what occur in a SUSY theory, as we shall see in Chapter 5. In addition, the masses of bosons and fermions belonging to the same SUSY multiplet are equal, if SUSY is unbroken; in this simplified model, then, we would have $m_{\rm f} = M_{\rm H}$. Note, however, that the cancellation of the quadratic divergence occurs whatever the values of $m_{\rm f}$ and $M_{\rm H}$, since these masses do not enter the expression (1.20). We shall show this explicitly for the Wess–Zumino model [19] in Chapter 5. It is a general result in any SUSY theory, and has the important consequence that SUSY-breaking mass terms (as are certainly required phenomenologically) can be introduced 'by hand' without spoiling the cancellation of quadratic divergences. As we shall see in Chapter 9, other SUSY-breaking terms which do not compromise this cancellation are also possible; they are referred to generically as 'soft SUSY-breaking terms'.

To implement this idea in the context of the (MS)SM, it will be necessary to postulate the existence of new fermionic 'superpartners' of the Higgs field – 'Higgsinos' – as discussed in Chapters 3 and 8. But this will by no means deal with all the quadratic divergences present in the $-\mu^2\phi^\dagger\phi$ term. In principle, every SM fermion can play the role of 'f' in (1.19), since they all have a Yukawa coupling to the Higgs field. To cancel all these quadratic divergences will require the introduction of scalar superpartners for all the SM fermions, that is, an appropriate set of squarks and sleptons. There are also quadratic divergences associated with the contribution of gauge boson loops to the '$-\mu^2$' term, and these too will have to be cancelled by fermionic superpartners, 'gauginos'. In this way, the outlines of a supersymmetrized version of the SM are beginning to emerge.

After cancellation of the Λ^2 terms via (1.21), the next most divergent contributions to the '$-\mu^2$' term grow logarithmically with Λ, but even terms logarithmic in the cut-off can be unacceptably large. Consider a simple 'one Higgs – one new fermion' model. The $\ln\Lambda$ contribution to the '$-\mu^2$' term has the form

$$\sim \lambda\left(aM_{\rm H}^2 - bm_{\rm f}^2\right)\ln\Lambda, \tag{1.22}$$

where a and b are numerical factors. Even though the dependence on Λ is now tamed, a fine-tuning problem will arise in the case of any fermion (coupling to the Higgs field) whose mass $m_{\rm f}$ is very much larger than the weak scale. In general, if the Higgs sector has any coupling, even indirect via loops, to very massive states (as happens in Grand Unified Theories for example), the masses of these states will dominate radiative corrections to the '$-\mu^2$' term, requiring large cancellations once again.

This situation is dramatically improved by SUSY. Roughly speaking, in a supersymmetric version of our 'one Higgs – one new fermion' model, the boson and fermion masses would be equal ($M_H = m_f$), and so would the coefficients a and b in (1.22), with the result that the correction (1.22) would vanish! Similarly, other contributions to the self-energy from SM particles and their superpartners would all cancel out, if SUSY were exact. More generally, in supersymmetric theories only wavefunction renormalizations are infinite as $\Lambda \to \infty$, as we shall discuss further in the context of the Wess–Zumino model in Section 5.2; these will induce corresponding logarithmic divergences in the values of physical (renormalized) masses (see, for example, Section 10.4.2 of [15]). However, no superpartners for the SM particles have yet been discovered, so SUSY – to be realistic in this context – must be a (softly) broken symmetry (see Chapter 9), with the masses of the superpartners presumably lying at too high values to have been detected yet. In our simple model, this means that $M_{\tilde{H}}^2 \neq m_{\tilde{f}}^2$. In this case, the quadratic divergences still cancel, as previously noted, and the remaining correction to the physical '$-\mu^2$' term will be of order $\lambda(M_{\tilde{H}}^2 - m_{\tilde{f}}^2) \ln \Lambda$. We conclude that (softly) broken SUSY may solve the SM fine-tuning problem, provided that the new SUSY superpartners are not too much heavier than the scale of v (or M_H), or else we are back to some form of fine-tuning.[2] Of course, how much fine-tuning we are prepared to tolerate is a matter of taste, but the argument strongly suggests that the discovery of SUSY should be within the reach of the LHC – if not, as it now seems, of either LEP or the Tevatron. Hence the vast amount of work that has gone into constructing viable theories, and analysing their expected phenomenologies.

In summary, SUSY can *stabilize* the hierarchy $M_{H,W} \ll M_P$, in the sense that radiative corrections will not drag $M_{H,W}$ up to the high scale Λ; and the argument implies that, for the desired stabilization to occur, SUSY should be visible at a scale not much greater than a few TeV. The origin of this latter scale (that of SUSY-breaking – see Chapter 9) is a separate problem. It is worth emphasizing that a theory of the MSSM type, with superpartner masses no larger than a few TeV, is a consistent effective field theory which is perturbatively calculable for all energies up to, say, the Planck, or a Grand Unification, scale without requiring fine-tuning (but see Section 10.3 for further discussion of this issue, within the MSSM specifically). Whether such a post-SUSY 'desert' exists or not is, of course, for experiment to decide.

Notwithstanding the foregoing motivation for seeking a supersymmetric version of the SM (a view that became widely accepted from the early 1980s), the reader should be aware that, historically, supersymmetry was not invented as a response to

[2] The application of the argument to motivate a supersymmetric SU(5) grand unified theory (in which Λ is now the unification scale), which is softly broken at the TeV mass scale, was made by Dimopoulos and Georgi [20] and Sakai [21]. Well below the unification scale, the effective field content of these models is that of the MSSM.

the SM fine-tuning problem. Supersymmetric field theories, and the supersymmetry algebra (see Section 1.3 and Chapter 4), had been in existence since the early 1970s: in two dimensions, in the context of string theory [22–24]; as a graded Lie algebra in four dimensions [25, 26]; in a non-linear realization [27]; and as four-dimensional quantum field theories [19, 28, 29]. Indeed, Fayet [30–33] had pioneered SUSY extensions of the SM before the fine-tuning problem came to be regarded as so central, and before the phenomenological importance of soft SUSY breaking was appreciated; and Farrar and Fayet had begun to explore the phenomenology of the superpartners [34–36].

It may be that, if experiment fails to discover SUSY at the TeV scale, supersymmetry itself may still turn out to have physical relevance. At all events, this book is concerned with the SUSY response to the SM fine-tuning problem, in the specific form of the MSSM. We should however note that, in addition to technicolour, other possibilities have been proposed more recently, in particular the radical idea that the gravitational (or string) scale is actually very much lower than (1.8), perhaps even as low as a few TeV [37]. The fine-tuning problem then evaporates since the ultraviolet cut-off Λ is not much higher than the weak scale itself. This miracle is worked by appealing to the notion of 'large' hidden extra dimensions, perhaps as large as sub-millimetre scales. This and other related ideas are discussed by Lykken [38], for example. Nevertheless, it is fair to say that SUSY, in the form of the MSSM, is at present the most highly developed framework for guiding and informing explorations of physics 'beyond the SM'.

1.2 Three quantitative indications

Here we state briefly three quantitative results of the MSSM, which together have inclined many physicists to take the model seriously; as indicated, we shall explore each in more detail in later chapters.

(a) The precision fits to electroweak data show that $M_{\rm H}$ is less than about 200 GeV, at the 99% confidence level. The 'Minimal Supersymmetric Standard Model' (MSSM) (see Chapter 8), which has two Higgs doublets, predicts (see Chapter 10) that the lightest Higgs particle should be no heavier than about 140 GeV. In the SM, by contrast, we have *no* constraint on $M_{\rm H}$.[3]
(b) At one-loop order, the inverse gauge couplings $\alpha_1^{-1}(Q^2), \alpha_2^{-1}(Q^2), \alpha_3^{-1}(Q^2)$ of the SM run linearly with $\ln Q^2$. Although α_1^{-1} decreases with Q^2, and α_2^{-1} and α_3^{-1} increase, all three tending to meet at high $Q^2 \sim (10^{16}\ \text{GeV})^2$, they do not in fact meet convincingly

[3] Not in quite the same sense (i.e. of a mathematical bound), at any rate. One can certainly say, from (1.3), that if λ is not much greater than unity, so that perturbation theory has a hope of being applicable, then $M_{\rm H}$ can't be much greater than a few hundred GeV. For more sophisticated versions of this sort of argument, see [7], Section 22.10.2.

in the SM. On the other hand, provided the superpartner masses are in the range 100 GeV –10 TeV, in the MSSM they do meet, thus encouraging ideas of unification: see Section 8.3, and Figure 8.1. It is notable that this estimate of the SUSY scale is essentially the same as that coming from 'fine-tuning' considerations.

(c) In any renormalizable theory, the mass parameters in the Lagrangian are also scale-dependent (they 'run'), just as the coupling parameters do. In the MSSM, the evolution of a Higgs (mass)^2 parameter from a typical positive value of order v^2 at a scale of the order of 10^{16} GeV, takes it to a negative value of the correct order of magnitude at scales of order 100 GeV, thus providing a possible explanation for the origin of electroweak symmetry breaking, specifically at those much lower scales. Actually, however, this happens because the Yukawa coupling of the top quark is large (being proportional to its mass), and this has a dominant effect on the evolution. You might ask whether, in that case, the same result would be obtained without SUSY. The answer is that it would, but the initial conditions for the evolution are more naturally motivated within a SUSY theory, as discussed in Section 9.3 (see Figure 9.1). Once again, this result requires that the superpartner masses are no larger than a few TeV. There is therefore a remarkable consistency between all these quite different ways of estimating the SUSY scale.

1.3 Theoretical considerations

It can certainly be plausibly argued that a dominant theme in twentieth-century physics was that of symmetry, the pursuit of which was heuristically very successful. It is natural to ask if our current quantum field theories exploit all the kinds of symmetry which could exist, consistent with Lorentz invariance. Consider the symmetry 'charges' that we are familiar with in the SM, for example an electromagnetic charge of the form

$$Q = e \int \mathrm{d}^3\boldsymbol{x}\ \psi^\dagger \psi, \tag{1.23}$$

or an SU(2) charge (isospin operator) of the form

$$\boldsymbol{T} = g \int \mathrm{d}^3\boldsymbol{x}\ \psi^\dagger (\boldsymbol{\tau}/2) \psi, \tag{1.24}$$

where in (1.24) ψ is an SU(2) doublet, and in both (1.23) and (1.24) ψ is a fermionic field. All such symmetry operators are themselves Lorentz scalars (they carry no uncontracted Lorentz indices of any kind, for example vector or spinor). This implies that when they act on a state of definite spin j, they cannot alter that spin:

$$Q|j\rangle = |\text{ same } j, \text{ possibly different member of symmetry multiplet }\rangle. \tag{1.25}$$

Need this be the case?

We certainly know of one vector 'charge', namely the 4-momentum operators P_μ which generate space-time displacements, and whose eigenvalues are conserved 4-momenta. There are also the angular momentum operators, which belong inside an antisymmetric tensor $M_{\mu\nu}$. Could we, perhaps, have a conserved symmetric tensor charge $Q_{\mu\nu}$? We shall provide a highly simplified version (taken from Ellis [39]) of an argument due to Coleman and Mandula [40] which shows that we cannot. Consider letting such a charge act on a single-particle state with 4-momentum p:

$$Q_{\mu\nu}|p\rangle = (\alpha p_\mu p_\nu + \beta g_{\mu\nu})|p\rangle, \tag{1.26}$$

where the right-hand side has been written down by 'covariance' arguments (i.e. the most general expression with the indicated tensor transformation character, built from the tensors at our disposal). Now consider a two-particle state $|p^{(1)}, p^{(2)}\rangle$, and assume the $Q_{\mu\nu}$ values are additive, conserved, and act on only one particle at a time, like other known charges. Then

$$Q_{\mu\nu}\big|p^{(1)}, p^{(2)}\big\rangle = \big(\alpha\big(p_\mu^{(1)}p_\nu^{(1)} + p_\mu^{(2)}p_\nu^{(2)}\big) + 2\beta g_{\mu\nu}\big)\big|p^{(1)}, p^{(2)}\big\rangle. \tag{1.27}$$

In an elastic scattering process of the form $1 + 2 \to 3 + 4$ we will then need (from conservation of the eigenvalue)

$$p_\mu^{(1)}p_\nu^{(1)} + p_\mu^{(2)}p_\nu^{(2)} = p_\mu^{(3)}p_\nu^{(3)} + p_\mu^{(4)}p_\nu^{(4)}. \tag{1.28}$$

But we also have 4-momentum conservation:

$$p_\mu^{(1)} + p_\mu^{(2)} = p_\mu^{(3)} + p_\mu^{(4)}. \tag{1.29}$$

The only common solution to (1.28) and (1.29) is

$$p_\mu^{(1)} = p_\mu^{(3)},\ p_\mu^{(2)} = p_\mu^{(4)},\ \text{or}\ p_\mu^{(1)} = p_\mu^{(4)},\ p_\mu^{(2)} = p_\mu^{(3)}, \tag{1.30}$$

which means that only forward or backward scattering can occur, which is obviously unacceptable.

The general message here is that there seems to be no room for further conserved operators with non-trivial Lorentz transformation character (i.e. not Lorentz scalars). The existing such operators P_μ and $M_{\mu\nu}$ do allow proper scattering processes to occur, but imposing any more conservation laws over-restricts the possible configurations. Such was the conclusion of the Coleman–Mandula theorem [40], but in fact their argument turns out not to exclude 'charges' which transform under Lorentz transformations as *spinors*: that is to say, things transforming like a fermionic field ψ. We may denote such a charge by Q_a, the subscript a indicating the spinor component (we will see that we'll be dealing with 2-component spinors, rather than 4-component ones, for the most part). For such a charge, equation (1.25) will clearly not hold; rather,

$$Q_a|j\rangle = |j \pm 1/2\rangle. \tag{1.31}$$

Such an operator will not contribute to a matrix element for a 2-particle $\rightarrow$ 2-particle elastic scattering process (in which the particle spins remain the same), and consequently the above kind of 'no-go' argument can not get started.

The question then arises: is it possible to include such spinorial operators in a consistent algebraic scheme, along with the known conserved operators P_μ and $M_{\mu\nu}$? The affirmative answer was first given by Gol'fand and Likhtman [25], and the most general such 'supersymmetry algebra' was obtained by Haag *et al.* [26]. By 'algebra' here we mean (as usual) the set of commutation relations among the 'charges' – which, we recall, are also the generators of the appropriate symmetry transformations. The SU(2) algebra of the angular momentum operators, which are generators of rotations, is a familiar example. The essential new feature here, however, is that the charges that have a spinor character will have *anticommutation* relations among themselves, rather than commutation relations. So such algebras involve some commutation relations and some anticommutation relations.

What will such algebras look like? Since our generic spinorial charge Q_a is a symmetry operator, it must commute with the Hamiltonian of the system, whatever it is:

$$[Q_a, H] = 0, \tag{1.32}$$

and so must the anticommutator of two different components:

$$[\{Q_a, Q_b\}, H] = 0. \tag{1.33}$$

As noted above, the spinorial Q terms have two components, so as a and b vary the symmetric object $\{Q_a, Q_b\} = Q_a Q_b + Q_b Q_a$ has three independent components, and we suspect that it must transform as a spin-1 object (just like the symmetric combinations of two spin-1/2 wavefunctions). However, as usual in a relativistic theory, this spin-1 object should be described by a 4-vector, not a 3-vector. Further, this 4-vector is conserved, from (1.33). There is only one such conserved 4-vector operator (from the Coleman–Mandula theorem), namely P_μ. So the Q_a terms must satisfy an algebra of the form, roughly,

$$\{Q_a, Q_b\} \sim P_\mu. \tag{1.34}$$

Clearly (1.34) is sloppy: the indices on each side do not balance. With more than a little hindsight, we might think of absorbing the '$_\mu$' by multiplying by γ^μ, the γ-matrix itself conveniently having two matrix indices, which might correspond to a, b. This is in fact more or less right, as we shall see in Chapter 4, but the precise details are finicky.

Accepting that (1.34) captures the essence of the matter, we can now begin to see what a radical idea supersymmetry really is. Equation (1.34) says, roughly speaking, that if you do two SUSY transformations generated by the Q terms, one after the other, you get the energy-momentum operator. Or, to put it even more strikingly (but quite equivalently), you get the space–time translation operator, i.e. a derivative. Turning it around, the SUSY spinorial Q's are like square roots of 4-momentum, or square roots of derivatives! It is rather like going one better than the Dirac equation, which can be viewed as providing the square root of the Klein–Gordon equation: how would we take the square root of the Dirac equation?

It is worth pausing to take this in properly. Four-dimensional derivatives are firmly locked to our notions of a four-dimensional space–time. In now entertaining the possibility that we can take square roots of them, we are effectively extending our concept of space–time itself, just as, when the square root of -1 is introduced, we enlarge the real axis to the complex (Argand) plane. That is to say, if we take seriously an algebra involving both P_μ and the Q's we shall have to say that the space–time co-ordinates are being extended to include further degrees of freedom, which are acted on by the Q's, and that these degrees of freedom are connected to the standard ones by means of transformations generated by the Q's. These further degrees of freedom are, in fact, fermionic. So we may say that SUSY invites us to contemplate 'fermionic dimensions', and enlarge space–time to 'superspace'. SUSY is often thought of in terms of (approximately) degenerate multiplets of bosons and fermions. Of course, that aspect is certainly true, phenomenologically important, and our main concern in this book; nevertheless, the fermionic enlargement of space–time is arguably a more striking concept, and we shall provide an introduction to it in Chapter 6.

One final remark on motivations: if you believe in String Theory (and it still seems to be the most promising framework for a consistent quantum theory of gravity), then the phenomenologically most attractive versions incorporate supersymmetry, some trace of which might remain in the theories that effectively describe physics at presently accessible energies.

2

Spinors: Weyl, Dirac and Majorana

Let us begin our Long March to the MSSM by recalling in outline how symmetries, such as SU(2), are described in quantum field theory (see, for example, Chapter 12 of [7]). The Lagrangian involves a set of fields ψ_r – they could be bosons or fermions – and it is taken to be invariant under an infinitesimal transformation on the fields of the form

$$\delta_\epsilon \psi_r = -\mathrm{i}\epsilon\lambda_{rs}\psi_s, \tag{2.1}$$

where a summation is understood on the repeated index s, the λ_{rs} are certain constant coefficients (for instance, the elements of the Pauli matrices), and ϵ is an infinitesimal parameter. Supersymmetry transformations will look something like this, but they will transform bosonic fields into fermionic ones, for example

$$\delta_\xi \phi \sim \xi\psi, \tag{2.2}$$

where ϕ is a bosonic (say spin-0) field, ψ is a fermionic (say spin-1/2) one, and ξ is an infinitesimal parameter. The alert reader will immediately figure out that in this case the parameter ξ has to be a *spinor*. In due course we shall spell out the details of the '$\sim$' here, but one thing should already be clear at this stage: the number of (field) degrees of freedom, as between the bosonic ϕ fields and the fermionic ψ fields, had better be the same in an equation of the form (2.2), just as the number of fields $r = 1, 2, \ldots, N$ on the left-hand side of (2.1) is the same as the number $s = 1, 2, \ldots, N$ on the right-hand side. We can not have some fields being 'left out'. Now the simplest kind of bosonic field is of course a neutral scalar field, which has only one component, which is real: $\phi = \phi^\dagger$ (see [15], Chapter 5); there is only one degree of freedom. On the other hand, there is *no* fermionic field with just one degree of freedom: being a spinor, it has at least two ('spin-up' and 'spin-down', in simple terms). So that means that we must consider, at the very least, a two-degree-of-freedom bosonic field, to go with the spinor field, and that takes us to a complex (charged) scalar field (see Chapter 7 of [15]).

But exactly what kind of a fermionic field could we 'match' the complex scalar field with? When we learn the Dirac equation, among the first results we arrive at is that Dirac wave functions, or fields, have *four* degrees of freedom, not two: in physical terms, spin-up and spin-down particle, and spin-up and spin-down antiparticle. Thus we must somehow halve the number of spinor degrees of freedom. There are two ways of doing this. One is to employ *2-component spinor fields*, called Weyl spinors in contrast to the four-component Dirac ones. The other is to use *Majorana fields*, for which particle and antiparticle are identical. Both formulations are used in the SUSY literature, and it helps to be familiar with both. Nevertheless, it is desirable to opt for one or the other as the dominant language, and we shall mainly use the Weyl spinor formulation, which we shall develop in the next three sections. However, we shall also introduce some Majorana formalism in Section 2.5. The reader is encouraged, through various exercises, to learn some equivalences between quantities expressed in the Weyl and in the Majorana language. As we proceed, we shall from time to time give the equivalent Majorana forms for various results (for example, in Sections 4.2, 4.5 and 5.1). These will eventually be required when we perform some simple SUSY calculations in Section 5.2 and in Chapter 12; for these the Majorana formalism is preferred, because it is close enough to the Dirac formalism to allow familiar calculational tricks to be used, with some modifications.

We have been somewhat slipshod, so far, not distinguishing clearly between 'components' and 'degrees of freedom'. In fact, each component of a 2-component (Weyl) spinor is complex, so there are actually four degrees of freedom present; there are also four in a Majorana spinor. If the spinor is assumed to be on-shell – i.e. obeying the appropriate equation of motion – then the number of degrees of freedom is reduced to two, the same as in a complex scalar field. Generally in quantum field theory we need to go 'off-shell', so that to match the minimal number (four) of spinor degrees of freedom will require two more bosonic degrees of freedom than just the two in a complex scalar field. We shall ignore this complication in our first foray into SUSY in Chapter 3, but will return to it in Chapter 4.

The familiar Dirac field uses two 2-component fields, which is twice too many. Our first, and absolutely inescapable, task is therefore to 'deconstruct' the Dirac field and understand the nature of the two different 2-component Weyl fields which together constitute it. This difference has to do with the different ways the two 'halves' of the 4-component Dirac field transform under Lorentz transformations. Understanding how this works, in detail, is vital to being able to write down SUSY transformations which are consistent with Lorentz invariance. For example, the left-hand side of (2.2) refers to a scalar (spin-0) field ϕ; admittedly it's complex, but that just means that it has a real part and an imaginary part, both of which have spin-0. Hence it is an invariant under Lorentz transformations. On the right-hand side, however, we have the 2-component spinor (spin-1/2) field ψ, which is certainly not invariant under Lorentz transformations. But the parameter ξ is also a 2-component spinor, in fact, and so we shall have to understand how to put the 2-component objects ξ and ψ together properly so as to form a Lorentz invariant, in order to be consistent with the Lorentz transformation character of the left-hand side. While we may be familiar with how to do this sort of thing for 4-component Dirac spinors, we need to learn the corresponding tricks for 2-component ones. The next two sections are therefore devoted to this essential groundwork.

2.1 Spinors and Lorentz transformations

We begin with the Dirac equation in momentum space, which we write as

$$E\Psi = (\boldsymbol{\alpha} \cdot \boldsymbol{p} + \beta m)\Psi, \tag{2.3}$$

where of course we are taking $c = \hbar = 1$. We shall choose the particular representation

$$\boldsymbol{\alpha} = \begin{pmatrix} \boldsymbol{\sigma} & 0 \\ 0 & -\boldsymbol{\sigma} \end{pmatrix} \quad \beta = \begin{pmatrix} 0 & 1 \\ 1 & 0 \end{pmatrix}, \tag{2.4}$$

which implies that

$$\boldsymbol{\gamma} = \begin{pmatrix} 0 & -\boldsymbol{\sigma} \\ \boldsymbol{\sigma} & 0 \end{pmatrix}, \quad \text{and} \quad \gamma_5 = \begin{pmatrix} 1 & 0 \\ 0 & -1 \end{pmatrix}. \tag{2.5}$$

This is one of the standard representations of the Dirac matrices (see for example [15] page 91, and [7] pages 31–2, and particularly [7] appendix M, Section M.6). It is the one which is commonly used in the 'small mass' or 'high energy' limit, since the (large) momentum term is then (block) diagonal. As usual, $\boldsymbol{\sigma} \equiv (\sigma_1, \sigma_2, \sigma_3)$ are the 2×2 Pauli matrices. Note that

$$\{\gamma_5, \beta\} = \{\gamma_5, \boldsymbol{\gamma}\} = 0. \tag{2.6}$$

We write

$$\Psi = \begin{pmatrix} \psi \\ \chi \end{pmatrix}. \tag{2.7}$$

The Dirac equation is then

$$(E - \boldsymbol{\sigma} \cdot \boldsymbol{p})\psi = m\chi \tag{2.8}$$

$$(E + \boldsymbol{\sigma} \cdot \boldsymbol{p})\chi = m\psi. \tag{2.9}$$

Notice that as $m \to 0$, (2.8) becomes $\boldsymbol{\sigma} \cdot \boldsymbol{p}\psi_0 = E\psi_0$, and $E \to |\boldsymbol{p}|$, so the zero mass limit of (2.8) is

$$(\boldsymbol{\sigma} \cdot \boldsymbol{p}/|\boldsymbol{p}|)\psi_0 = \psi_0, \tag{2.10}$$

which means that ψ_0 is an eigenstate of the helicity operator $\boldsymbol{\sigma} \cdot \boldsymbol{p}/|\boldsymbol{p}|$ with eigenvalue +1 ('positive helicity'). Similarly, the zero-mass limit of (2.9) shows that χ_0 has negative helicity.

For $m \neq 0$, ψ and χ of (2.8) and (2.9) are plainly not helicity eigenstates: indeed the mass term (in this representation) 'mixes' them. However, as we shall see shortly, it is these two-component objects, ψ and χ, that have well-defined (but different) Lorentz transformation properties. They are, indeed, examples of precisely the 2-component Weyl spinors we shall be dealing with.

Although not helicity eigenstates, ψ and χ are eigenstates of γ_5, in the sense that

$$\gamma_5 \begin{pmatrix} \psi \\ 0 \end{pmatrix} = \begin{pmatrix} \psi \\ 0 \end{pmatrix}, \quad \text{and} \quad \gamma_5 \begin{pmatrix} 0 \\ \chi \end{pmatrix} = -\begin{pmatrix} 0 \\ \chi \end{pmatrix}. \tag{2.11}$$

These two γ_5-eigenstates can be constructed from the original Ψ by using the projection operators P_{R} and P_{L} defined by

$$P_{\mathrm{R}} = \left(\frac{1+\gamma_5}{2}\right) = \begin{pmatrix} 1 & 0 \\ 0 & 0 \end{pmatrix} \tag{2.12}$$

and

$$P_{\mathrm{L}} = \left(\frac{1-\gamma_5}{2}\right) = \begin{pmatrix} 0 & 0 \\ 0 & 1 \end{pmatrix}. \tag{2.13}$$

Then

$$P_{\mathrm{R}}\Psi = \begin{pmatrix} \psi \\ 0 \end{pmatrix}, \quad P_{\mathrm{L}}\Psi = \begin{pmatrix} 0 \\ \chi \end{pmatrix}. \tag{2.14}$$

It is easy to check that $P_{\mathrm{R}}P_{\mathrm{L}} = 0$, $P_{\mathrm{R}}^2 = P_{\mathrm{R}}$, $P_{\mathrm{L}}^2 = P_{\mathrm{L}}$. The eigenvalue of γ_5 is called 'chirality'; ψ has chirality +1, and χ has chirality −1. In an unfortunate terminology, but one now too late to change, '+' chirality is denoted by 'R' (i.e right-handed) and '–' chirality by 'L' (i.e. left-handed), despite the fact that (as noted above) ψ and χ are *not* helicity eigenstates when $m \neq 0$. Anyway, a 'ψ' type 2-component spinor is often written as ψ_{R}, and a 'χ' type one as χ_{L}. For the moment, we shall not use these R and L subscripts, but shall understand that anything called ψ is an R-type spinor, and a χ is L-type.

Now, we said above that ψ and χ had well-defined Lorentz transformation character. Let's recall how this goes (see [7] Appendix M, Section M.6). There are basically two kinds of transformation: rotations and 'boosts' (i.e. pure velocity transformations). It is sufficient to consider *infinitesimal* transformations, which we can specify by their action on a 4-vector, for example the energy–momentum 4-vector $(E, \boldsymbol{p})$. Under an infinitesimal three-dimensional rotation,

$$E \to E' = E, \quad \boldsymbol{p} \to \boldsymbol{p}' = \boldsymbol{p} - \boldsymbol{\epsilon} \times \boldsymbol{p}, \tag{2.15}$$

where $\boldsymbol{\epsilon} = (\epsilon_1, \epsilon_2, \epsilon_3)$ are three infinitesimal parameters specifying the infinitesimal rotation; and under a velocity transformation

$$E \to E' = E - \boldsymbol{\eta} \cdot \boldsymbol{p}, \quad \boldsymbol{p} \to \boldsymbol{p}' = \boldsymbol{p} - \boldsymbol{\eta} E, \tag{2.16}$$

where $\boldsymbol{\eta} = (\eta_1, \eta_2, \eta_3)$ are three infinitesimal velocities. Under the Lorentz transformations thus defined, ψ and χ transform as follows (see equations (M.94) and (M.98) of [7], where however the top two components are called 'ϕ' rather than 'ψ'):

$$\psi \to \psi' = (1 + \mathrm{i}\boldsymbol{\epsilon} \cdot \boldsymbol{\sigma}/2 - \boldsymbol{\eta} \cdot \boldsymbol{\sigma}/2)\psi \tag{2.17}$$

and

$$\chi \to \chi' = (1 + \mathrm{i}\boldsymbol{\epsilon} \cdot \boldsymbol{\sigma}/2 + \boldsymbol{\eta} \cdot \boldsymbol{\sigma}/2)\chi. \tag{2.18}$$

Equations (2.17) and (2.18) are extremely important equations for us. They tell us how to construct the spinors ψ' and χ' for the rotated and boosted frame, in terms of the original spinors ψ and χ. That is to say, the ψ' and χ' specified by (2.17) and (2.18) satisfy the 'primed' analogues of (2.8) and (2.9), namely

$$(E' - \boldsymbol{\sigma} \cdot \boldsymbol{p}')\psi' = m\chi' \tag{2.19}$$

$$(E' + \boldsymbol{\sigma} \cdot \boldsymbol{p}')\chi' = m\psi'. \tag{2.20}$$

Let's pause to check this statement in a special case, that of a pure boost. Define $V_\eta = (1 - \boldsymbol{\eta} \cdot \boldsymbol{\sigma}/2)$. Then since $\boldsymbol{\eta}$ is infinitesimal, $V_\eta^{-1} = (1 + \boldsymbol{\eta} \cdot \boldsymbol{\sigma}/2)$. Now take (2.8), multiply from the left by V_η^{-1}, and insert the unit matrix $V_\eta^{-1} V_\eta$ as indicated:

$$\left[V_\eta^{-1}(E - \boldsymbol{\sigma} \cdot \boldsymbol{p})V_\eta^{-1}\right]V_\eta\psi = mV_\eta^{-1}\chi. \tag{2.21}$$

If (2.17) is right, we have $\psi' = V_\eta\psi$, and if (2.18) is right we have $\chi' = V_\eta^{-1}\chi$, in this pure boost case. So to establish the complete consistency between (2.17), (2.18) and (2.19), we need to show that

$$V_\eta^{-1}(E - \boldsymbol{\sigma} \cdot \boldsymbol{p})V_\eta^{-1} = (E' - \boldsymbol{\sigma} \cdot \boldsymbol{p}'), \tag{2.22}$$

that is,

$$(1 + \boldsymbol{\eta} \cdot \boldsymbol{\sigma}/2)(E - \boldsymbol{\sigma} \cdot \boldsymbol{p})(1 + \boldsymbol{\eta} \cdot \boldsymbol{\sigma}/2) = (E - \boldsymbol{\eta} \cdot \boldsymbol{p}) - \boldsymbol{\sigma} \cdot (\boldsymbol{p} - E\boldsymbol{\eta}) \tag{2.23}$$

to first order in $\boldsymbol{\eta}$, since the right-hand side of (2.23) is just $E' - \boldsymbol{\sigma} \cdot \boldsymbol{p}'$ from (2.16).

Exercise 2.1 Verify (2.23).

Returning now to equations (2.17) and (2.18), we note that ψ and χ actually behave the same under rotations (they have spin-1/2!), but differently under boosts. The interesting fact is that there are *two* kinds of 2-component spinors, distinguished by their different transformation character under boosts. Both are used in the Dirac 4-component spinor Ψ. In SUSY, however, the approach we shall mainly follow works with the 2-component Weyl spinors ψ and χ which (as we saw above) may also be labelled by 'R' and 'L' respectively. The alternative approach using 4-component Majorana spinors will be introduced in Section 2.5.

Before proceeding, we note another important feature of (2.17) and (2.18). Let V be the transformation matrix appearing in (2.17):

$$V = (1 + \mathrm{i}\boldsymbol{\epsilon}\cdot\boldsymbol{\sigma}/2 - \boldsymbol{\eta}\cdot\boldsymbol{\sigma}/2). \tag{2.24}$$

Then

$$V^{-1} = (1 - \mathrm{i}\boldsymbol{\epsilon}\cdot\boldsymbol{\sigma}/2 + \boldsymbol{\eta}\cdot\boldsymbol{\sigma}/2) \tag{2.25}$$

since we merely have to reverse the sense of the infinitesimal parameters, while

$$V^{\dagger} = (1 - \mathrm{i}\boldsymbol{\epsilon}\cdot\boldsymbol{\sigma}/2 - \boldsymbol{\eta}\cdot\boldsymbol{\sigma}/2) \tag{2.26}$$

using the hermiticity of the $\boldsymbol{\sigma}$'s. So

$$V^{\dagger^{-1}} = V^{-1^{\dagger}} = (1 + \mathrm{i}\boldsymbol{\epsilon}\cdot\boldsymbol{\sigma}/2 + \boldsymbol{\eta}\cdot\boldsymbol{\sigma}/2), \tag{2.27}$$

which is the matrix appearing in (2.18). Hence we may write, compactly,

$$\psi' = V\psi, \quad \chi' = V^{\dagger^{-1}}\chi = V^{-1^{\dagger}}\chi. \tag{2.28}$$

In summary, the R-type spinor ψ transforms by the matrix V, while the L-type spinor χ transforms by $V^{-1\dagger}$.

2.2 Constructing invariants and 4-vectors out of 2-component (Weyl) spinors

Let's start by recalling some things which should be familiar from a Dirac equation course. From the 4-component Dirac spinor Ψ we can form a Lorentz invariant

$$\bar{\Psi}\Psi = \Psi^{\dagger}\beta\Psi, \tag{2.29}$$

and a 4-vector

$$\bar{\Psi}\gamma^{\mu}\Psi = \Psi^{\dagger}\beta(\beta, \beta\boldsymbol{\alpha})\Psi = \Psi^{\dagger}(1, \boldsymbol{\alpha})\Psi. \tag{2.30}$$

In terms of our 2-component objects ψ and χ (2.29) becomes

$$\text{Lorentz invariant} \quad (\psi^\dagger \chi^\dagger)\begin{pmatrix} 0 & 1 \\ 1 & 0 \end{pmatrix}\begin{pmatrix} \psi \\ \chi \end{pmatrix} = \psi^\dagger \chi + \chi^\dagger \psi. \tag{2.31}$$

Using (2.28) it is easy to verify that the right-hand side of (2.31) is invariant. Indeed, perhaps more interestingly, each part of it is:

$$\psi^\dagger \chi \to \psi^{\dagger\prime} \chi' = \psi V^\dagger V^{\dagger^{-1}} \chi = \psi^\dagger \chi, \tag{2.32}$$

and similarly for $\chi^\dagger \psi$. Again, (2.30) becomes

$$\begin{aligned} \text{4-vector} \quad & (\psi^\dagger \chi^\dagger)\left[\begin{pmatrix} 1 & 0 \\ 0 & 1 \end{pmatrix}, \begin{pmatrix} \boldsymbol{\sigma} & 0 \\ 0 & -\boldsymbol{\sigma} \end{pmatrix}\right]\begin{pmatrix} \psi \\ \chi \end{pmatrix} \\ & = (\psi^\dagger \psi + \chi^\dagger \chi, \psi^\dagger \boldsymbol{\sigma} \psi - \chi^\dagger \boldsymbol{\sigma} \chi) \\ & \equiv \psi^\dagger \sigma^\mu \psi + \chi^\dagger \bar{\sigma}^\mu \chi, \end{aligned} \tag{2.33}$$

where we have introduced the important quantities

$$\sigma^\mu \equiv (1, \boldsymbol{\sigma}), \quad \bar{\sigma}^\mu = (1, -\boldsymbol{\sigma}), \tag{2.34}$$

in terms of which

$$\gamma^\mu = \begin{pmatrix} 0 & \bar{\sigma}^\mu \\ \sigma^\mu & 0 \end{pmatrix}. \tag{2.35}$$

As with the Lorentz invariant, it is actually the case that each of $\psi^\dagger \sigma^\mu \psi$ and $\chi^\dagger \bar{\sigma}^\mu \chi$ transforms, separately, as a 4-vector.

Exercise 2.2 Verify that last statement.

Indeed, since only the transformation character of the ψ's and χ's matters, quantities such as $\psi^{(1)\dagger} \sigma^\mu \psi^{(2)}$ and $\chi^{(1)\dagger} \bar{\sigma}^\mu \chi^{(2)}$ are also 4-vectors, just as $\bar{\Psi}^{(1)} \gamma^\mu \Psi^{(2)}$ is.

In this 'σ^μ, $\bar{\sigma}^\mu$' notation, the Dirac equation (2.8) and (2.9) becomes

$$\sigma^\mu p_\mu \psi = m\chi \tag{2.36}$$

$$\bar{\sigma}^\mu p_\mu \chi = m\psi. \tag{2.37}$$

So we can read off the useful news that '$\sigma^\mu p_\mu$' converts a ψ-type object to a χ-type one, and $\bar{\sigma}^\mu p_\mu$ converts a χ to a ψ – or, in slightly more proper language, the Lorentz transformation character of $\sigma^\mu p_\mu \psi$ is the same as that of χ, and the LT character of $\bar{\sigma}^\mu p_\mu \chi$ is the same as that of ψ.

Lastly in this re-play of Dirac formalism, the Dirac Lagrangian can be written in terms of ψ and χ:

$$\bar{\Psi}(\mathrm{i}\gamma^\mu \partial_\mu - m)\Psi = \psi^\dagger \mathrm{i}\sigma^\mu \partial_\mu \psi + \chi^\dagger \mathrm{i}\bar{\sigma}^\mu \partial_\mu \chi - m(\psi^\dagger \chi + \chi^\dagger \psi). \tag{2.38}$$

Note how $\bar{\sigma}^\mu$ belongs with χ, and σ^μ with ψ.

An interesting point may have occurred to the reader here: it is possible to form 4-vectors using only ψ's or only χ's (see Exercise 2.2), but the invariants introduced so far ($\psi^\dagger\chi$ and $\chi^\dagger\psi$) make use of both. So we might ask: *can we make an invariant out of just χ-type spinors, for instance?* This is important precisely because we want (for reasons outlined in the previous section) to construct theories involving the number of degrees of freedom present in one two-component Weyl (but not four-component Dirac) spinor. It is at this point that we part company with what is usually contained in standard Dirac courses.

Another way of putting our question is this: is it possible to construct a spinor from the components of, say, χ, which has the transformation character of a ψ? (and of course vice versa). If it is, then we can use it, with χ-type spinors, in place of ψ-type spinors when making invariants. The answer is that it is possible. Consider how the complex conjugate of χ, denoted by χ^*, transforms under Lorentz transformations. We have

$$\chi' = (1 + \mathrm{i}\boldsymbol{\epsilon}\cdot\boldsymbol{\sigma}/2 + \boldsymbol{\eta}\cdot\boldsymbol{\sigma}/2)\chi. \tag{2.39}$$

Taking the complex conjugate gives

$$\chi^{*\prime} = (1 - \mathrm{i}\boldsymbol{\epsilon}\cdot\boldsymbol{\sigma}^*/2 + \boldsymbol{\eta}\cdot\boldsymbol{\sigma}^*/2)\chi^*. \tag{2.40}$$

Now observe that $\sigma_1^* = \sigma_1$, $\sigma_2^* = -\sigma_2$, $\sigma_3^* = \sigma_3$, and that $\sigma_2\sigma_3 = -\sigma_3\sigma_2$ and $\sigma_1\sigma_2 = -\sigma_2\sigma_1$. It follows that

$$\sigma_2\chi^{*\prime} = \sigma_2(1 - \mathrm{i}\boldsymbol{\epsilon}\cdot(\sigma_1, -\sigma_2, \sigma_3)/2 + \boldsymbol{\eta}\cdot(\sigma_1, -\sigma_2, \sigma_3)/2)\chi^* \tag{2.41}$$

$$= (1 + \mathrm{i}\boldsymbol{\epsilon}\cdot\boldsymbol{\sigma}/2 - \boldsymbol{\eta}\cdot\boldsymbol{\sigma}/2)\sigma_2\chi^* \tag{2.42}$$

$$= V\sigma_2\chi^*, \tag{2.43}$$

referring to (2.24) for the definition of V, which is precisely the matrix by which ψ transforms.

We have therefore established the important result that

$$\sigma_2\chi^* \text{ transforms like a } \psi. \tag{2.44}$$

So let's at once introduce 'the ψ-like thing constructed from χ' via the definition

$$\psi_\chi \equiv \mathrm{i}\sigma_2\chi^*, \tag{2.45}$$

where the i has been put in for convenience (remember σ_2 involves i's). Then we are guaranteed by (2.32) that the quantity

$$\psi^\dagger_{\chi^{(1)}}\chi^{(2)} = \left(\mathrm{i}\sigma_2\chi^{(1)*}\right)^{*\mathrm{T}}\chi^{(2)} = \left(\mathrm{i}\sigma_2\chi^{(1)}\right)^{\mathrm{T}}\chi^{(2)} = \chi^{(1)\mathrm{T}}(-\mathrm{i}\sigma_2)\chi^{(2)} \tag{2.46}$$

where ^{T} denotes transpose, is Lorentz invariant, for any two χ-like things $\chi^{(1)}$, $\chi^{(2)}$, just as $\psi^\dagger\chi$ was. (Equally, so is $\chi^{(2)\dagger}\psi_{\chi^{(1)}}$.) Equation (2.46) is important, because it tells us *how to form the Lorentz invariant scalar product of two* χ's. This is the kind of product that we will need in SUSY transformations of the form (2.2).

Before proceeding, we note that the quantity $\chi^{(1)\mathrm{T}}(-\mathrm{i}\sigma_2)\chi^{(2)}$ is in a sense very familar. Let us write

$$\chi^{(1)} = \begin{pmatrix} \chi_{\uparrow}^{(1)} \\ \chi_{\downarrow}^{(1)} \end{pmatrix}, \quad \chi^{(2)} = \begin{pmatrix} \chi_{\uparrow}^{(2)} \\ \chi_{\downarrow}^{(2)} \end{pmatrix}, \tag{2.47}$$

adopting the 'spin-up', 'spin-down' notation used in elementary quantum mechanics. Then since

$$-\mathrm{i}\sigma_2 = \begin{pmatrix} 0 & -1 \\ 1 & 0 \end{pmatrix}, \tag{2.48}$$

we easily find

$$\chi^{(1)\mathrm{T}}(-\mathrm{i}\sigma_2)\chi^{(2)} = -\chi_{\uparrow}^{(1)}\chi_{\downarrow}^{(2)} + \chi_{\downarrow}^{(1)}\chi_{\uparrow}^{(2)}. \tag{2.49}$$

The right-hand side of (2.49) is recognized as (proportional to) the usual spin-0 combination of two spin-1/2 states. This means that it is invariant under spatial rotations. What the previous development shows is that it is actually also invariant under Lorentz transformations (i.e. boosts).

In particular, $\psi_{\chi}^{\dagger}\chi$ is Lorentz invariant, where the χ's are the same. This quantity is

$$(\mathrm{i}\sigma_2\chi^*)^{*\mathrm{T}}\chi = (\mathrm{i}\sigma_2\chi)^{\mathrm{T}}\chi = \chi^{\mathrm{T}}(-\mathrm{i}\sigma_2)\chi. \tag{2.50}$$

Let's write it out in detail. We have

$$\mathrm{i}\sigma_2 = \begin{pmatrix} 0 & 1 \\ -1 & 0 \end{pmatrix}, \quad \text{and} \quad \chi = \begin{pmatrix} \chi_1 \\ \chi_2 \end{pmatrix}, \tag{2.51}$$

so that

$$\mathrm{i}\sigma_2\chi = \begin{pmatrix} \chi_2 \\ -\chi_1 \end{pmatrix}, \quad \text{and} \quad (\mathrm{i}\sigma_2\chi)^{\mathrm{T}}\chi = \chi_2\chi_1 - \chi_1\chi_2. \tag{2.52}$$

The right-hand side of (2.52) vanishes if χ_1 and χ_2 are ordinary functions, but not if they are *anticommuting* quantities, as appear in (quantized) fermionic fields. So certainly this is a satisfactory invariant in terms of two-component quantized fields, or in terms of Grassmann numbers (see Appendix O of [7]). From now on, we shall assume that spinors are quantum fields. Strictly speaking, then, although the symbol ' * ' is perfectly suitable for the complex conjugates of the wavefunction parts in the free-field expansions, it should be understood as ' $\dagger\mathrm{T}$ ', which includes hermitian conjugation of quantum field operators and complex conjugation of wavefunctions.

We shall spell this out in more detail in Section 3.1, but – with the indicated understanding – we shall continue for the moment to use just the simple ' * '.

It is natural to ask: what about ψ^*? Performing manipulations analogous to those in (2.40–2.43), you can verify that

$$\sigma_2\psi^* \quad \text{transforms like} \quad \chi. \tag{2.53}$$

This licenses us to introduce a χ-type object constructed from a ψ, which we define by

$$\chi_\psi \equiv -\mathrm{i}\sigma_2\psi^*. \tag{2.54}$$

Then for any two ψ's $\psi^{(1)}$, $\psi^{(2)}$ say, we know that

$$\left(-\mathrm{i}\sigma_2\psi^{(1)*}\right)^{*\mathrm{T}}\psi^{(2)} = \left(-\mathrm{i}\sigma_2\psi^{(1)}\right)^{\mathrm{T}}\psi^{(2)} = \psi^{(1)\mathrm{T}}\mathrm{i}\sigma_2\psi^{(2)} \tag{2.55}$$

is an invariant. In particular, for the same ψ, the quantity

$$(-\mathrm{i}\sigma_2\psi)^{\mathrm{T}}\psi \tag{2.56}$$

is an invariant.

2.3 A more streamlined notation for Weyl spinors

It looks as if it is going to get pretty tedious keeping track of which two-component spinor is a χ-type one and which is ψ-type one, by writing things like $\chi^{(1)}, \chi^{(2)}, \ldots, \psi^{(1)}, \psi^{(2)}, \ldots$, all the time, and (even worse) things like $\psi^{\dagger}_{\chi^{(1)}}\chi^{(2)}$. A first step in the direction of a more powerful notation is to agree that the components of χ-type spinors have *lower indices*, as in (2.51). That is, anything written with lower indices is a χ-type spinor. So then we are free to name them how we please: $\chi_a, \xi_a, \ldots$, even ψ_a.

We can also streamline the cumbersome notation '$\psi_{\chi^{(1)}}{}^{\dagger}\chi^{(2)}$'. The point here is that this notation was – at this stage – introduced in order to construct invariants out of just χ-type things. But (2.46) tells us how to do this, in terms of the two χ's involved: you take one of them, say $\chi^{(1)}$, and form $\mathrm{i}\sigma_2\chi^{(1)}$. Then you take the matrix dot product (in the sense of '$u^{\mathrm{T}}v$') of this quantity and the second χ-type spinor. So, starting from a χ with lower indices, χ_a, let's define a χ with *upper indices* via (see equation (2.52))

$$\begin{pmatrix}\chi^1\\ \chi^2\end{pmatrix} \equiv \mathrm{i}\sigma_2\chi = \begin{pmatrix}\chi_2\\ -\chi_1\end{pmatrix}, \tag{2.57}$$

that is,

$$\chi^1 \equiv \chi_2, \quad \chi^2 \equiv -\chi_1. \tag{2.58}$$

Suppose now that ξ is a second χ-type spinor, and

$$\xi = \begin{pmatrix} \xi_1 \\ \xi_2 \end{pmatrix}. \tag{2.59}$$

Then we know that $(\mathrm{i}\sigma_2\chi)^{\mathrm{T}}\xi$ is a Lorentz invariant, and this is just

$$(\chi^1 \chi^2)\begin{pmatrix} \xi_1 \\ \xi_2 \end{pmatrix} = \chi^1\xi_1 + \chi^2\xi_2 = \chi^a\xi_a, \tag{2.60}$$

where a runs over the values 1 and 2. Equation (2.60) is a compact notation for this scalar product: it is a 'spinor dot product', analogous to the 'upstairs–downstairs' dot-products of special relativity, like $A^\mu B_\mu$. We can shorten the notation even further, indeed, to $\chi \cdot \xi$, or even to $\chi\xi$ if we know what we are doing. Note that if the components of χ and ξ commute, then it does not matter whether we write this invariant as $\chi \cdot \xi = \chi^1\xi_1 + \chi^2\xi_2$ or as $\xi_1\chi^1 + \xi_2\chi^2$. However, if they are anticommuting these will differ by a sign, and we need a convention as to which we take to be the 'positive' dot product. It is as in (2.60), which is remembered as '*summed-over χ-type (undotted) indices appear diagonally downwards, top left to bottom right*'.

The four-dimensional Lorentz-invariant dot product $A^\mu B_\mu$ of special relativity can also be written as $g^{\mu\nu}A_\nu B_\mu$, where $g^{\mu\nu}$ is the *metric tensor* of special relativity with components (in one common convention!) $g^{00} = +1$, $g^{11} = g^{22} = g^{33} = -1$, all others vanishing (see Appendix D of [7]). In a similar way we can introduce a metric tensor ϵ^{ab} for forming the Lorentz-invariant spinor dot product of two 2-component L-type spinors, by writing

$$\chi^a = \epsilon^{ab}\chi_b \tag{2.61}$$

(always summing on repeated indices, of course), so that

$$\chi^a\xi_a = \epsilon^{ab}\chi_b\xi_a. \tag{2.62}$$

For (2.61) to be consistent with (2.58), we require

$$\epsilon^{12} = +1, \epsilon^{21} = -1, \epsilon^{11} = \epsilon^{22} = 0. \tag{2.63}$$

Clearly ϵ^{ab}, regarded as a 2×2 matrix, is nothing but the matrix $\mathrm{i}\sigma_2$ of (2.51). We shall, however, continue to use the explicit '$\mathrm{i}\sigma_2$' notation for the most part, rather than the 'ϵ^{ab}' notation.

We can also introduce ϵ_{ab} via

$$\chi_a = \epsilon_{ab}\chi^b, \tag{2.64}$$

which is consistent with (2.58) if

$$\epsilon_{12} = -1, \epsilon_{21} = +1, \epsilon_{11} = \epsilon_{22} = 0. \tag{2.65}$$

Finally, you can verify that

$$\epsilon_{ab}\epsilon^{bc} = \delta_a^c, \tag{2.66}$$

as one would expect. It is important to note that these 'ϵ' metrics are *antisymmetric* under the interchange of the two indices a and b, whereas the SR metric $g^{\mu\nu}$ is symmetric under $\mu \leftrightarrow \nu$.

Exercise 2.3 (a) What is $\xi \cdot \chi$ in terms of $\chi \cdot \xi$ (assuming the components anti-commute)? (b) What is $\chi_a \xi^a$ in terms of $\chi^a \xi_a$? Do these both by brute force via components, and by using the ϵ dot product.

Given that χ transforms by $V^{-1\dagger}$ of (2.27), it is interesting to ask: how does the 'raised-index' version, $\mathrm{i}\sigma_2\chi$, transform?

Exercise 2.4 Show that $\mathrm{i}\sigma_2\chi$ transforms by V^*.

We can use the result of Exercise 2.4 to verify once more the invariance of $(\mathrm{i}\sigma_2\chi)^{\mathrm{T}}\xi$: $(\mathrm{i}\sigma_2\chi)^{\mathrm{T}}\xi \to (\mathrm{i}\sigma_2\chi)'^{\mathrm{T}}\xi' = (\mathrm{i}\sigma_2\chi)^{\mathrm{T}}(V^*)^{\mathrm{T}}V^{-1\dagger}\xi$. But $(V^*)^{\mathrm{T}} = V^\dagger$, and so the invariance is established.

We can therefore summarize the state of play so far by saying that *a downstairs χ-type spinor transforms by $V^{-1\dagger}$, while an upstairs χ-type spinor transforms by V^*.*

Clearly we also want an 'index' notation for ψ-type spinors. The general convention is that they are given 'dotted indices' i.e. we write things like $\psi^{\dot{a}}$. By convention, also, we decide that our ψ-type thing has an *upstairs* index, just as it was a convention that our χ-type thing had a downstairs index. Equation (2.55) tells us how to form scalar products out of two ψ-like things, $\psi^{(1)}$ and $\psi^{(2)}$, and invites us to define downstairs-indexed quantities

$$\begin{pmatrix} \psi_{\dot{1}} \\ \psi_{\dot{2}} \end{pmatrix} \equiv -\mathrm{i}\sigma_2\psi = \begin{pmatrix} 0 & -1 \\ 1 & 0 \end{pmatrix} \begin{pmatrix} \psi^{\dot{1}} \\ \psi^{\dot{2}} \end{pmatrix} \tag{2.67}$$

so that

$$\psi_{\dot{1}} \equiv -\psi^{\dot{2}}, \quad \psi_{\dot{2}} \equiv \psi^{\dot{1}}. \tag{2.68}$$

Then if ζ ('zeta') is a second ψ-type spinor, and

$$\zeta = \begin{pmatrix} \zeta^{\dot{1}} \\ \zeta^{\dot{2}} \end{pmatrix}, \tag{2.69}$$

we know that $(-\mathrm{i}\sigma_2\psi)^{\mathrm{T}}\zeta = \psi^{\mathrm{T}}\mathrm{i}\sigma_2\zeta$ is a Lorentz invariant, which is

$$(\psi_{\dot{1}}\psi_{\dot{2}})\begin{pmatrix}\zeta^{\dot{1}}\\ \zeta^{\dot{2}}\end{pmatrix} = \psi_{\dot{1}}\zeta^{\dot{1}} + \psi_{\dot{2}}\zeta^{\dot{2}} = \psi_{\dot{a}}\zeta^{\dot{a}}, \tag{2.70}$$

where $\dot{a}$ runs over the values 1, 2. Notice that with all these conventions, the 'positive' scalar product has been defined so that the *summed-over dotted indices appear diagonally upwards, bottom left to top right.*

As in Exercise 2.4, we can ask how (in terms of V) the downstairs dotted spinor $-\mathrm{i}\sigma_2\psi$ transforms.

Exercise 2.5 Show that $-\mathrm{i}\sigma_2\psi$ transforms by $V^{-1\mathrm{T}}$, and hence verify once again that $(-\mathrm{i}\sigma_2\psi)^{\mathrm{T}}\zeta$ is invariant.

We can introduce a metric tensor notation for the Lorentz-invariant scalar product of two 2-component R-type (dotted) spinors, too. We write

$$\psi_{\dot{a}} = \epsilon_{\dot{a}\dot{b}}\psi^{\dot{b}} \tag{2.71}$$

where, to be consistent with (2.68), we need

$$\epsilon_{\dot{1}\dot{2}} = -1,\ \epsilon_{\dot{2}\dot{1}} = +1,\ \epsilon_{\dot{1}\dot{1}} = \epsilon_{\dot{2}\dot{2}} = 0. \tag{2.72}$$

Then

$$\psi_{\dot{a}}\zeta^{\dot{a}} = \epsilon_{\dot{a}\dot{b}}\psi^{\dot{b}}\zeta^{\dot{a}}. \tag{2.73}$$

We can also define

$$\epsilon^{\dot{1}\dot{2}} = +1,\ \epsilon^{\dot{2}\dot{1}} = -1,\ \epsilon^{\dot{1}\dot{1}} = \epsilon^{\dot{2}\dot{2}} = 0, \tag{2.74}$$

with

$$\epsilon_{\dot{a}\dot{b}}\epsilon^{\dot{b}\dot{c}} = \delta_{\dot{a}}^{\dot{c}}. \tag{2.75}$$

Again, the ϵ's with dotted indices are antisymmetric under interchange of their indices.

We could of course think of shortening (2.70) further to $\psi \cdot \zeta$ or $\psi\zeta$, but without the dotted indices to tell us, we would not in general know whether such expressions referred to what we have been calling ψ- or χ-type spinors. So a ' $^{-}$ ' notation for ψ-type spinors is commonly used. That is to say, we define

$$\begin{aligned}\psi^{\dot{1}} &\equiv \bar{\psi}^{\dot{1}}\\ \psi^{\dot{2}} &\equiv \bar{\psi}^{\dot{2}}.\end{aligned} \tag{2.76}$$

Then (2.70) would be just $\bar{\psi} \cdot \bar{\zeta}$. It is worth emphasizing at once *that this 'bar' notation for dotted spinors has nothing to do with the 'bar' used in 4-component*

Dirac theory, as in (2.29), nor with the 'bar' often used to denote an antiparticle name, or field.

Exercise 2.6 (a) What is $\bar{\psi} \cdot \bar{\zeta}$ in terms of $\bar{\zeta} \cdot \bar{\psi}$ (assuming the components anti-commute)? (b) What is $\bar{\psi}_{\dot{a}}\bar{\zeta}^{\dot{a}}$ in terms of $\bar{\psi}^{\dot{a}}\bar{\zeta}_{\dot{a}}$? Do these by components and by using ϵ symbols.

Altogether, then, we have arrived at four types of two-component Weyl spinor: χ^a and χ_a transforming by V^* and $V^{-1\dagger}$, respectively, and $\bar{\psi}^{\dot{a}}$ and $\bar{\psi}_{\dot{a}}$ transforming by V and $V^{-1\mathrm{T}}$, respectively. The essential point is that invariants are formed by taking the matrix dot product between one quantity transforming by M say, and another transforming by $M^{-1\mathrm{T}}$.

Consider now χ_a^*: since χ_a transforms by $V^{-1\dagger}$, it follows that χ_a^* transforms by the complex conjugate of this matrix, which is $V^{-1\mathrm{T}}$. But this is exactly how a '$\bar{\psi}_{\dot{a}}$ transforms! So it is consistent to *define*

$$\bar{\chi}_{\dot{a}} \equiv \chi_a^*. \tag{2.77}$$

We can then raise the dotted index with the matrix $\mathrm{i}\sigma_2$, using the inverse of (2.67) – remember, once we have dotted indices, or bars, to tell us what kind of spinor we are dealing with, we no longer care what letter we use. In a similar way, $\bar{\psi}^{\dot{a}*}$ transforms by V^*, the same as χ^a, so we may write

$$\psi^a \equiv \bar{\psi}^{\dot{a}*} \tag{2.78}$$

and lower the index a by $-\mathrm{i}\sigma_2$.

It must be admitted that (2.78) creates something of a problem for us, given that we want to be free to continue to use the 'old' notation of Sections 2.1 and 2.2, as well as, from time to time, the new streamlined one. In the old notation, 'ψ' stands for an R-type dotted spinor with components $\psi^{\dot{1}}$, $\psi^{\dot{2}}$; but in the new notation, according to (2.78), the unbarred symbol 'ψ' should stand for an L-type undotted spinor (the 'old' ψ becoming the R-type dotted spinor $\bar{\psi}$). A similar difficulty does not, of course, arise in the case of the χ spinors, which only get barred when complex conjugated (see (2.77)). This is fortunate, since we shall be using χ- or L-type spinors almost exclusively. As regards the dotted R-type spinors, our convention will be that when we write dot products and other bilinears in terms of ψ and $\psi^\dagger$ (or ψ^*) we are using the 'old' notation, but when they are written in terms of ψ and $\bar{\psi}$ we are using the new one.

Definitions (2.77) and (2.78) allow us to write the 4-vectors $\psi^\dagger\sigma^\mu\psi$ and $\chi^\dagger\bar{\sigma}^\mu\chi$ in 'bar' notation. For example,

$$\chi^\dagger\bar{\sigma}^\mu\chi = \chi_a^*\bar{\sigma}^{\mu ab}\chi_b \equiv \bar{\chi}_{\dot{a}}\bar{\sigma}^{\mu\dot{a}b}\chi_b \equiv \bar{\chi}\bar{\sigma}^\mu\chi, \tag{2.79}$$

with the convention that the indices of $\bar{\sigma}^\mu$ are an upstairs dotted followed by an upstairs undotted (note that this adheres to the convention about how to sum over repeated dotted and undotted indices). Similarly (but rather less obviously)

$$\psi^\dagger \sigma^\mu \psi \equiv \psi \sigma^\mu \bar{\psi}, \tag{2.80}$$

with the indices of σ^μ being a downstairs undotted followed by a downstairs dotted.

As an illustration of a somewhat more complicated manipulation, we now show how to obtain the sometimes useful result

$$\bar{\xi}\bar{\sigma}^\mu \chi = -\chi \sigma^\mu \bar{\xi}. \tag{2.81}$$

The quantity on the left-hand side is $\bar{\xi}_{\dot{a}}\bar{\sigma}^{\mu \dot{a}b}\chi_b$. On the right-hand side, however, σ^μ carries downstairs indices, so clearly we must raise the indices of both $\bar{\xi}$ and χ. We write

$$\begin{aligned} \bar{\xi}_{\dot{a}}\bar{\sigma}^{\mu \dot{a}b}\chi_b &= \epsilon_{\dot{a}\dot{c}}\bar{\xi}^{\dot{c}}\bar{\sigma}^{\mu \dot{a}b}\epsilon_{bd}\chi^d \\ &= -\bar{\xi}^{\dot{c}}\epsilon_{\dot{c}\dot{a}}\bar{\sigma}^{\mu \dot{a}b}\epsilon_{bd}\chi^d. \end{aligned} \tag{2.82}$$

At this point it is easier to change to matrix notation, using the fact that both $\epsilon_{\dot{c}\dot{a}}$ and ϵ_{bd} are the same as (the matrix elements of) the matrix $-\mathrm{i}\sigma_2$. The next step is a useful exercise.

Exercise 2.7 Verify that

$$\sigma_2 \bar{\sigma}^\mu \sigma_2 = \sigma^{\mu \mathrm{T}}. \tag{2.83}$$

Equation (2.82) can therefore be written as

$$\bar{\xi}\bar{\sigma}^\mu \chi = \bar{\xi}^{\dot{c}}\sigma^{\mu \mathrm{T}}_{\dot{c}d}\chi^d = -\chi^d \sigma^\mu_{d\dot{c}}\bar{\xi}^{\dot{c}}, \tag{2.84}$$

where in the last step the minus sign comes from interchanging the order of the fermionic quantities. The right-hand side of (2.84) is precisely $-\chi\sigma^\mu\bar{\xi}$.

In the new notation, then, the familiar Dirac 4-component spinor (2.7) would be written as

$$\Psi = \begin{pmatrix} \bar{\psi}^{\dot{a}} \\ \chi_a \end{pmatrix}. \tag{2.85}$$

The conventions of different authors typically do not agree here. The notation we use is the same as that of Shifman [41] (see his equation (68) on page 335). Many other authors, for example Bailin and Love [42], use a choice for the Dirac matrices which is different from equations (2.4) and (2.5) herein, and which has the effect of interchanging the position, in Ψ, of the L (undotted) and R (dotted) parts – which, furthermore, they call 'ψ' and 'χ' respectively, the opposite way round from here – so that for them

$$\Psi_{\mathrm{BL}} = \begin{pmatrix} \psi_a \\ \bar{\chi}^{\dot{a}} \end{pmatrix}. \tag{2.86}$$

Furthermore, Bailin and Love's ϵ symbols, and hence their spinor scalar products, have the opposite sign from that used herein.

2.4 Dirac spinors using χ- (or L-) type spinors only

In the 'Weyl spinor' approach to SUSY, the simplest SUSY theory (which we shall meet in the next chapter) involves a complex scalar field and a 2-component spinor field. This is in fact the archetype of SUSY models leading to the MSSM. By convention, one uses χ-type spinors, i.e. undotted L-type spinors, no doubt because the V–A structure of the electroweak sector of the SM distinguishes the L parts of the fields, and one might as well give them a privileged status, although of course there are the R parts (dotted ψ-type spinors) as well. In a SUSY context, it is very convenient to be able to use only one kind of spinors, which in the MSSM is (for the reason just outlined) going to be L-type ones – but in that case how are we going to deal with the R parts of the SM fields?

Consider for example the electron field which we write in the unstreamlined notation as

$$\Psi^{(\mathrm{e})} = \begin{pmatrix} \psi_{\mathrm{e}} \\ \chi_{\mathrm{e}} \end{pmatrix}. \tag{2.87}$$

Instead of using the R-type electron field in the top two components, we can just as well use the *charge conjugate of the L-type positron* field, which is in fact of R-type, as we shall see. For a 4-component Dirac spinor, charge conjugation is defined by

$$\Psi_{\mathrm{C}} = C\bar{\psi}^{\mathrm{T}} = C_0\Psi^* \tag{2.88}$$

where[1]

$$C = -\mathrm{i}\gamma^2\beta = \begin{pmatrix} \mathrm{i}\sigma_2 & 0 \\ 0 & -\mathrm{i}\sigma_2 \end{pmatrix}, \quad C_0 = -\mathrm{i}\gamma^2 = \begin{pmatrix} 0 & \mathrm{i}\sigma_2 \\ -\mathrm{i}\sigma_2 & 0 \end{pmatrix}. \tag{2.89}$$

Thus if we write generally

$$\Psi = \begin{pmatrix} \psi \\ \chi \end{pmatrix} \tag{2.90}$$

[1] This choice of C_0 has the opposite sign from the one in equation (20.63) of [7] page 290; the present choice is more in conformity with SUSY conventions. We are sticking to the convention that the indices of the γ-matrices as defined in (2.5) appear upstairs; no significance should be attached to the position of the indices of the σ-matrices – it is common to write them downstairs.

then

$$\Psi_{\mathrm{C}} = \begin{pmatrix} \mathrm{i}\sigma_2 \chi^* \\ -\mathrm{i}\sigma_2 \psi^* \end{pmatrix}. \tag{2.91}$$

Note that the upper two components here are precisely ψ_χ of (2.45), and the lower two are χ_ψ of (2.54).

We can therefore define charge conjugation at the 2-component level by

$$\chi^{\mathrm{c}} \equiv \mathrm{i}\sigma_2 \chi^*, \quad \psi^{\mathrm{c}} \equiv -\mathrm{i}\sigma_2 \psi^*. \tag{2.92}$$

Now we recall (and will show explicitly in Section 2.5.2) that particle and antiparticle operators in Ψ are replaced by antiparticle and particle operators, respectively, in Ψ_{C}. In just the same way, χ and χ^{c} carry opposite values of conserved charges. Thus instead of (2.87) we may choose to write

$$\Psi^{(\mathrm{e})} = \begin{pmatrix} \chi^{\mathrm{c}}_{\bar{\mathrm{e}}} \\ \chi_{\mathrm{e}} \end{pmatrix}, \tag{2.93}$$

where

$$\chi^{\mathrm{c}}_{\bar{\mathrm{e}}} \equiv \mathrm{i}\sigma_2 \chi^*_{\bar{\mathrm{e}}} \tag{2.94}$$

and '$\bar{\mathrm{e}}$' stands for the antiparticle of e. Our previous work (cf. (2.45)) guarantees, of course, that the Lorentz transformation character of (2.93) is correct: that is, $\mathrm{i}\sigma_2 \chi^*_{\bar{\mathrm{e}}}$ is indeed a '$\psi^{\dot{a}}$'-type (i.e. R-type) object.

Particle fields are sometimes denoted simply by the particle label so that e_{L} is used in place of χ_{e} (the 'L' character must now be shown), and $\bar{e}^{\mathrm{c}}_{\mathrm{L}}$ in place of $\chi^{\mathrm{c}}_{\bar{\mathrm{e}}}$, but note that $\bar{e}^{\mathrm{c}}_{\mathrm{L}}$ is R-type!

In terms of the choice (2.93), a mass term for a Dirac fermion is (omitting now the 'L' subscripts from the χ's)

$$\begin{aligned} m\bar{\Psi}^{(\mathrm{e})}\Psi^{(\mathrm{e})} &= m\Psi^{(\mathrm{e})\dagger} \begin{pmatrix} 0 & 1 \\ 1 & 0 \end{pmatrix} \Psi^{(\mathrm{e})} = m((\mathrm{i}\sigma_2\chi_{\bar{\mathrm{e}}})^{\mathrm{T}} \chi^\dagger_{\mathrm{e}}) \begin{pmatrix} \chi_{\mathrm{e}} \\ \mathrm{i}\sigma_2 \chi^*_{\bar{\mathrm{e}}} \end{pmatrix} \\ &= m[\chi_{\bar{\mathrm{e}}} \cdot \chi_{\mathrm{e}} + \chi^\dagger_{\mathrm{e}} \mathrm{i}\sigma_2 \chi^*_{\bar{\mathrm{e}}}]. \end{aligned} \tag{2.95}$$

In the first term on the right-hand side of (2.95) we have used the quick 'dot' notation for two χ-type spinors introduced in Section 2.3. The second term can be similarly re-written in the bar notation:

$$\chi^{*\mathrm{T}}_{\mathrm{e}} \mathrm{i}\sigma_2 \chi^*_{\bar{\mathrm{e}}} = \bar{\chi}_{\mathrm{e}\dot{a}} \bar{\chi}^{\dot{a}}_{\bar{\mathrm{e}}} = \bar{\chi}_{\mathrm{e}} \cdot \bar{\chi}_{\bar{\mathrm{e}}}. \tag{2.96}$$

So the 'Dirac' mass has here been re-written wholly in terms of two L-type spinors, one associated with the e mode, the other with the $\bar{\mathrm{e}}$ mode.

2.5 Majorana spinors

2.5.1 Definition and simple bilinear equivalents

We stated in (2.54) that $\chi_\psi \equiv -\mathrm{i}\sigma_2\psi^*$ transforms like a χ-type object. It follows that it should be perfectly consistent with Dirac theory to assemble ψ and χ_ψ into a 4-component object:

$$\Psi_{\mathrm{M}}^{\psi} = \begin{pmatrix} \psi \\ -\mathrm{i}\sigma_2\psi^* \end{pmatrix}. \tag{2.97}$$

This must behave under Lorentz transformations just like an 'ordinary' Dirac 4-component object Ψ built from a ψ and a χ. However, Ψ_{M}^{ψ} of (2.97) has *fewer degrees of freedom* than an ordinary Dirac 4-component spinor Ψ, since it is fully determined by the 2-component object ψ. In a Dirac spinor Ψ involving a ψ and a χ, as in (2.7), there are two 2-component spinors, each of which is specified by four real quantities (each has two complex components), making eight in all. In Ψ_{M}^{ψ}, by contrast, there are only four real quantities, contained in the single (dotted) spinor ψ; explicitly,

$$\Psi_{\mathrm{M}}^{\psi} = \begin{pmatrix} \psi^{\dot{1}} \\ \psi^{\dot{2}} \\ -\psi^{\dot{2}*} \\ \psi^{\dot{1}*} \end{pmatrix}. \tag{2.98}$$

What this means physically becomes clearer when we consider the operation of *charge conjugation*, defined as in (2.88). For example,

$$\Psi_{\mathrm{M,C}}^{\psi} = \begin{pmatrix} 0 & \mathrm{i}\sigma_2 \\ -\mathrm{i}\sigma_2 & 0 \end{pmatrix}\begin{pmatrix} \psi^* \\ -\mathrm{i}\sigma_2\psi \end{pmatrix} = \begin{pmatrix} \psi \\ -\mathrm{i}\sigma_2\psi^* \end{pmatrix} = \Psi_{\mathrm{M}}^{\psi}. \tag{2.99}$$

So Ψ_{M}^{ψ} describes a spin-1/2 particle which is even under charge conjugation, that is, it is its own antiparticle. Such a particle is called a Majorana fermion, and (2.97) is a Majorana spinor field.

This charge-self-conjugate property is clearly the physical reason for the difference in the number of degrees of freedom in Ψ_{M}^{ψ} as compared with Ψ of (2.3). There are four physically distinguishable modes in a Dirac field, for example $\mathrm{e}_{\mathrm{L}}^-, \mathrm{e}_{\mathrm{R}}^-, \mathrm{e}_{\mathrm{L}}^+, \mathrm{e}_{\mathrm{R}}^+$. However, in a Majorana field there are only two, the antiparticle being the same as the particle; for example $\nu_{\mathrm{L}}, \nu_{\mathrm{R}}$, supposing, as is possible (see [7], Section 20.6), that neutrinos are Majorana particles.

We could also construct

$$\Psi_{\mathrm{M}}^{\chi} = \begin{pmatrix} \mathrm{i}\sigma_2\chi^* \\ \chi \end{pmatrix}, \tag{2.100}$$

which satisfies

$$\Psi^{\chi}_{\mathrm{M,C}} = \Psi^{\chi}_{\mathrm{M}}. \tag{2.101}$$

In this case,

$$\Psi^{\chi}_{\mathrm{M}} = \begin{pmatrix} \chi_2^* \\ -\chi_1^* \\ \chi_1 \\ \chi_2 \end{pmatrix}. \tag{2.102}$$

A formalism using χ's only must be equivalent to one using Ψ^{χ}_{M}'s only, and one using ψ's is equivalent to one using Ψ^{ψ}_{M}'s. The invariant '$\bar{\Psi}\Psi$' constructed from Ψ^{χ}_{M}, for instance, is

$$\bar{\Psi}^{\chi}_{\mathrm{M}}\Psi^{\chi}_{\mathrm{M}} = ((\mathrm{i}\sigma_2\chi^*)^{\dagger}\chi^{\dagger}) \begin{pmatrix} 0 & 1 \\ 1 & 0 \end{pmatrix} \begin{pmatrix} \mathrm{i}\sigma_2\chi^* \\ \chi \end{pmatrix} = \chi^{\mathrm{T}}(-\mathrm{i}\sigma_2)\chi + \chi^{\dagger}(\mathrm{i}\sigma_2)\chi^*. \tag{2.103}$$

The first term on the right-hand side of the last equality in (2.103) is just $\chi \cdot \chi$; the second is similar to the one in (2.96) and is $\bar{\chi} \cdot \bar{\chi}$. We have therefore established an equivalence (the first of several) between bilinears involving Weyl spinors and a Majorana bilinear:

$$\bar{\Psi}^{\chi}_{\mathrm{M}}\Psi^{\chi}_{\mathrm{M}} = \chi \cdot \chi + \bar{\chi} \cdot \bar{\chi}. \tag{2.104}$$

The expression $m\bar{\Psi}^{\chi}_{\mathrm{M}}\Psi^{\chi}_{\mathrm{M}}$ represents a possible mass term in a Lagrangian, for a Majorana fermion. More generally, when ξ is a χ-type spinor,

$$\bar{\Psi}^{\xi}_{\mathrm{M}}\Psi^{\chi}_{\mathrm{M}} = \xi \cdot \chi + \bar{\xi} \cdot \bar{\chi}. \tag{2.105}$$

Similarly, the invariant made from Ψ^{ψ}_{M} would be

$$\begin{aligned} \bar{\Psi}^{\psi}_{\mathrm{M}}\Psi^{\psi}_{\mathrm{M}} &= (\psi^{\dagger}(-\mathrm{i}\sigma_2\psi^*)^{\dagger}) \begin{pmatrix} 0 & 1 \\ 1 & 0 \end{pmatrix} \begin{pmatrix} \psi \\ -\mathrm{i}\sigma_2\psi^* \end{pmatrix} = \psi^{\dagger}(-\mathrm{i}\sigma_2)\psi^* + \psi^{\mathrm{T}}(\mathrm{i}\sigma_2)\psi \\ &= \psi \cdot \psi + \bar{\psi} \cdot \bar{\psi}, \end{aligned} \tag{2.106}$$

which can also serve as a mass term, and if η is a ψ-type (dotted) spinor then

$$\bar{\Psi}^{\eta}_{\mathrm{M}}\Psi^{\psi}_{\mathrm{M}} = \eta \cdot \psi + \bar{\eta} \cdot \bar{\psi}. \tag{2.107}$$

Note that all the terms in (2.104) and (2.106) would vanish if the field components did not anticommute.

The 4-component version of the Lorentz-invariant product of two Majorana spinors $\Psi_{1\mathrm{M}}$ and $\Psi_{2\mathrm{M}}$ has an interesting form. Consider

$$\bar{\Psi}_{1\mathrm{M}}\Psi_{2\mathrm{M}} = \Psi^{\dagger}_{1\mathrm{M}}\beta\Psi_{2\mathrm{M}}. \tag{2.108}$$

Equations (2.88) and (2.99) tell us that

$$\Psi_{1M} = -i\gamma^2\Psi_{1M}^*, \tag{2.109}$$

and hence

$$\Psi_{1M}^\dagger = \Psi_{1M}^T(-i\gamma^2) \tag{2.110}$$

using $\gamma^{2\dagger} = -\gamma^2$. It follows that

$$\Psi_{1M}^\dagger\beta\Psi_{2M} = \Psi_{1M}^T(-i\gamma^2\beta)\Psi_{2M} = \Psi_{1M}^T C\Psi_{2M}. \tag{2.111}$$

The matrix C therefore acts as a metric in forming the dot product of the two Ψ_M's. It is easy to check that (2.111) is the same as (2.103) when $\Psi_{1M} = \Psi_{2M} = \Psi_M^\chi$, and the same as (2.106) when $\Psi_{1M} = \Psi_{2M} = \Psi_M^\psi$.

We have seen how the Majorana invariant $\bar{\Psi}_M^\xi\Psi_M^\chi$ is expressible in terms of the 2-component spinors ξ and χ as $\xi\cdot\chi + \bar{\xi}\cdot\bar{\chi}$. We leave the following further equivalences as an important exercise.

Exercise 2.8 Verify

$$\bar{\Psi}_M^\xi\gamma_5\Psi_M^\chi = -\xi\cdot\chi + \bar{\xi}\cdot\bar{\chi} \tag{2.112}$$

$$\bar{\Psi}_M^\xi\gamma^\mu\Psi_M^\chi = \xi^\dagger\bar{\sigma}^\mu\chi - \chi^\dagger\bar{\sigma}^\mu\xi = \bar{\xi}\bar{\sigma}^\mu\chi - \bar{\chi}\bar{\sigma}^\mu\xi \tag{2.113}$$

$$\bar{\Psi}_M^\xi\gamma_5\gamma^\mu\Psi_M^\chi = \xi^\dagger\bar{\sigma}^\mu\chi + \chi^\dagger\bar{\sigma}^\mu\xi = \bar{\xi}\bar{\sigma}^\mu\chi + \bar{\chi}\bar{\sigma}^\mu\xi. \tag{2.114}$$

[Hint: use $\sigma_2\boldsymbol{\sigma}\sigma_2 = -\boldsymbol{\sigma}^T$, and the fact that a quantity such as $\xi^T\chi^*$, being itself a single-component object, is equal to its transpose, apart from a minus sign from changing the order of fermionic fields.]

Obtain analogous results for products built from ψ-type Majorana spinors.

From (2.105) and (2.112)–(2.114) we easily find

$$\xi\cdot\chi = \bar{\Psi}_M^\xi P_L\Psi_M^\chi \tag{2.115}$$

$$\bar{\xi}\cdot\bar{\chi} = \bar{\Psi}_M^\xi P_R\Psi_M^\chi \tag{2.116}$$

$$\bar{\xi}\bar{\sigma}^\mu\chi = \bar{\Psi}_M^\xi\gamma^\mu P_L\Psi_M^\chi \tag{2.117}$$

$$\bar{\chi}\bar{\sigma}^\mu\xi = -\bar{\Psi}_M^\xi\gamma^\mu P_R\Psi_M^\chi. \tag{2.118}$$

The last relation may be re-written, using (2.81), as

$$\xi\sigma^\mu\bar{\chi} = \bar{\Psi}_M^\xi\gamma^\mu P_R\Psi_M^\chi. \tag{2.119}$$

Relation (2.113) allows us to relate kinetic energy terms in the Weyl and Majorana formalisms. Beginning with the Majorana form (modelled on the Dirac one) we have

$$\int d^4x\, \bar{\Psi}^{\chi}_{\mathrm{M}}\gamma^{\mu}\partial_{\mu}\Psi^{\chi}_{\mathrm{M}} = \int d^4x\, (\chi^{\dagger}\bar{\sigma}^{\mu}\partial_{\mu}\chi - (\partial_{\mu}\chi)^{\dagger}\bar{\sigma}^{\mu}\chi)$$
$$= 2\int d^4x\, \chi^{\dagger}\bar{\sigma}^{\mu}\partial_{\mu}\chi, \tag{2.120}$$

where we have done a partial integration in the last step, throwing away the surface term. Hence, in a Lagrangian, a Weyl kinetic energy term $\chi^{\dagger}\mathrm{i}\bar{\sigma}^{\mu}\partial_{\mu}\chi$, which is simply the relevant bit of (2.38), is equivalent to the Majorana form $1/2\bar{\Psi}^{\chi}_{\mathrm{M}}\mathrm{i}\gamma^{\mu}\partial_{\mu}\Psi^{\chi}_{\mathrm{M}}$.

We have now discussed Lagrangian mass and kinetic terms for Majorana fields. How are these related to the corresponding terms for Dirac fields? Referring back to (2.38) we see that in the Dirac case a mass term *couples* the ψ (R-type) and χ (L-type) fields, as noted after equation (2.10). It cannot be constructed from either ψ or χ alone. This means that we cannot represent it in terms of just one Majorana field: we shall need two, one for the χ degrees of freedom, and one for the ψ. For that matter, neither can the kinetic terms in the Dirac Lagrangian (2.38). This must of course be so physically, because of the (oft-repeated) difference in the numbers of degrees of freedom involved. To take a particular case, then, if we want to represent the mass and kinetic terms of a (Dirac) electron field in terms of Majorana fields we shall need two of the latter:

$$\Psi^{\psi_{\mathrm{e}}}_{\mathrm{M}} = \begin{pmatrix} \psi_{\mathrm{e}} \\ -\mathrm{i}\sigma_2\psi^{*}_{\mathrm{e}} = \psi^{\mathrm{c}}_{\mathrm{e}} \end{pmatrix} \tag{2.121}$$

as in (2.97), and

$$\Psi^{\chi_{\mathrm{e}}}_{\mathrm{M}} = \begin{pmatrix} \mathrm{i}\sigma_2\chi^{*}_{\mathrm{e}} = \chi^{\mathrm{c}}_{\mathrm{e}} \\ \chi_{\mathrm{e}} \end{pmatrix} \tag{2.122}$$

as in (2.100), where ‘ c ’ is defined in (2.92). Note that the L-part of $\Psi^{\psi_{\mathrm{e}}}_{\mathrm{M}}$ consists of the charge-conjugate of the R-field $\psi^{\mathrm{c}}_{\mathrm{e}}$. Clearly

$$\Psi^{(\mathrm{e})} = P_{\mathrm{R}}\Psi^{\psi_{\mathrm{e}}}_{\mathrm{M}} + P_{\mathrm{L}}\Psi^{\chi_{\mathrm{e}}}_{\mathrm{M}}. \tag{2.123}$$

Exercise 2.9 then shows how to write Dirac ‘mass’ and ‘kinetic’ Lagrangian bilinears in terms of the indicated Majorana and Weyl quantities (as usual, necessary partial integrations are understood).

Exercise 2.9 Verify

(i)

$$\bar{\Psi}^{(\mathrm{e})}\Psi^{(\mathrm{e})} = \frac{1}{2}[\bar{\Psi}^{\psi_{\mathrm{e}}}_{\mathrm{M}}\Psi^{\chi_{\mathrm{e}}}_{\mathrm{M}} + \bar{\Psi}^{\chi_{\mathrm{e}}}_{\mathrm{M}}\Psi^{\psi_{\mathrm{e}}}_{\mathrm{M}}] \tag{2.124}$$

(ii)

$$\bar{\Psi}^{(e)}\not\partial\Psi^{(e)} = \frac{1}{2}\left[\bar{\Psi}_{\rm M}^{\psi_e}\not\partial\Psi_{\rm M}^{\psi_e} + \bar{\Psi}_{\rm M}^{\chi_e}\not\partial\Psi_{\rm M}^{\chi_e}\right]. \tag{2.125}$$

All the bilinear equivalences we have discussed will be useful when we come to locate the SM interactions inside the MSSM (Section 8.2) and when we perform elementary calculations involving MSSM superparticle interactions (Chapter 12).

2.5.2 *Quantization*

Up to now we have not needed to look more closely into how a Majorana field is quantized (other than that it is obviously fermionic in nature), but when we come to consider some SUSY calculations in Chapter 12 we shall need to understand (for instance) the forms of propagators for free Majorana particles. This is the main purpose of the present section.

Let us first recall how the usual 4-component Dirac field $\Psi_\alpha(\boldsymbol{x}, t)$ is quantized (here α is the spinor index). We shall follow the notational conventions used in Section 7.2 of [15]. The following equal time commutation relations are assumed:

$$\{\Psi_\alpha(\boldsymbol{x}, t), \Psi_\beta^\dagger(\boldsymbol{y}, t)\} = \delta(\boldsymbol{x} - \boldsymbol{y})\delta_{\alpha\beta}, \tag{2.126}$$

and

$$\{\Psi_\alpha(\boldsymbol{x}, t), \Psi_\beta(\boldsymbol{y}, t)\} = \{\Psi_\alpha^\dagger(\boldsymbol{x}, t), \Psi_\beta^\dagger(\boldsymbol{y}, t)\} = 0. \tag{2.127}$$

Ψ may be expanded in terms of creation and annihilation operators via

$$\Psi(\boldsymbol{x}, t) = \int \frac{\mathrm{d}^3\boldsymbol{k}}{(2\pi)^3\sqrt{2E_k}} \sum_{\lambda=1,2} [c_\lambda(k)u(k, \lambda)\mathrm{e}^{-\mathrm{i}k\cdot x} + d_\lambda^\dagger(k)v(k, \lambda)\mathrm{e}^{\mathrm{i}k\cdot x}], \tag{2.128}$$

where λ is a spin (or helicity) label, $E_k = (m^2 + \boldsymbol{k}^2)^{1/2}$, $c_\lambda(k)$ destroys a particle of 4-momentum k and spin λ, while $d_\lambda^\dagger(k)$ creates the corresponding antiparticle. Relations (2.126) and (2.127) are satisfied if the $c_\lambda(k)$'s obey the anticommutation relations

$$\left\{c_{\lambda_1}(k_1), c_{\lambda_2}^\dagger(k_2)\right\} = (2\pi)^3\delta(\boldsymbol{k}_1 - \boldsymbol{k}_2)\delta_{\lambda_1\lambda_2} \tag{2.129}$$

and

$$\left\{c_{\lambda_1}(k_1), c_{\lambda_2}(k_2)\right\} = \left\{c_{\lambda_1}^\dagger(k_1), c_{\lambda_2}^\dagger(k_2)\right\} = 0 \tag{2.130}$$

and similarly for the d's and $d^\dagger$'s. The charge conjugate field (cf. (2.88)) is

$$\begin{aligned}\Psi_{\rm C}(\boldsymbol{x},t) &\equiv -{\rm i}\gamma_2\Psi^{\dagger{\rm T}} \\ &= \int\frac{{\rm d}^3\boldsymbol{k}}{(2\pi)^3\sqrt{2E_k}}\sum_{\lambda=1,2}[c_\lambda^\dagger(k)(-{\rm i}\gamma_2 u^*(k,\lambda)){\rm e}^{{\rm i}k\cdot x} \\ &\quad + d_\lambda(k)(-{\rm i}\gamma_2 v^*(k,\lambda)){\rm e}^{-{\rm i}k\cdot x}]. \end{aligned} \tag{2.131}$$

It is straightforward to verify the results (see Section 20.5 of [7] – the change of sign in C_0 is immaterial)

$$-{\rm i}\gamma_2 v^*(k,\lambda) = u(k,\lambda), \quad -{\rm i}\gamma_2 u^*(k,\lambda) = v(k,\lambda), \tag{2.132}$$

which may also be written as

$$C\bar{v}^{\rm T} = u, \quad C\bar{u}^{\rm T} = v. \tag{2.133}$$

It follows that

$$\Psi_{\rm C}(\boldsymbol{x},t) = \int\frac{{\rm d}^3\boldsymbol{k}}{(2\pi)^3\sqrt{2E_k}}\sum_{\lambda=1,2}[d_\lambda(k)u(k,\lambda){\rm e}^{-{\rm i}k\cdot x} + c_\lambda^\dagger(k)v(k,\lambda){\rm e}^{{\rm i}k\cdot x}]. \tag{2.134}$$

Clearly (as stated earlier) the field $\Psi_{\rm C}$ is the same as Ψ but with particle and antiparticle operators interchanged.

As we have seen, a Majorana field is charge self-conjugate, which means that there is no distinction between particle and antiparticle, that is, $d_\lambda^\dagger(k) \equiv c_\lambda^\dagger(k)$ in (2.128), giving the equivalent expansion of a Majorana field $\Psi_{\rm M}(\boldsymbol{x},t)$:

$$\Psi_{\rm M}(\boldsymbol{x},t) = \int\frac{{\rm d}^3\boldsymbol{k}}{(2\pi)^3\sqrt{2E_k}}\sum_{\lambda=1,2}[c_\lambda(k)u(k,\lambda){\rm e}^{-{\rm i}k\cdot x} + c_\lambda^\dagger(k)v(k,\lambda){\rm e}^{{\rm i}k\cdot x}], \tag{2.135}$$

which of course satisfies

$$\Psi_{\rm M,C} = C\bar{\Psi}_{\rm M}^{\rm T} = \Psi_{\rm M}. \tag{2.136}$$

We now turn to the propagator question. We remind the reader that in quantum field theory all propagators are of the form 'vacuum expectation value of the time-ordered product of two fields'. Thus for a real scalar field $\phi(x)$ the propagator is $\langle 0|T(\phi(x_1)\phi(x_2))|0\rangle$, while for a Dirac field it is $\langle 0|T(\Psi_\alpha(x_1)\bar{\Psi}_\beta(x_2))|0\rangle$. As regards a Majorana field $\Psi_{\rm M}(x)$, it is in some way like a Dirac field (because of its spinorial character), but in another like the real scalar field (because in that case too there is no distinction between particle and antiparticle). The consequence of this is that for Majorana fields there are actually three non-vanishing propagator-type expressions: in addition to the Dirac-like propagator $\langle 0|T(\Psi_{{\rm M}\alpha}(x_1)\bar{\Psi}_{{\rm M}\beta}(x_2))|0\rangle$, there are also $\langle 0|T(\Psi_{{\rm M}\alpha}(x_1)\Psi_{{\rm M}\beta}(x_2))|0\rangle$ and $\langle 0|T(\bar{\Psi}_{{\rm M}\alpha}(x_1)\bar{\Psi}_{{\rm M}\beta}(x_2))|0\rangle$. Intuitively this must be the case, simply by virtue of the relation (2.136) between $\bar{\Psi}_M$ and Ψ_M. Indeed that relation can be used to obtain the second two propagators in terms of the first, as we shall now see.

It is plausible that the expression for the Dirac-like propagator is in fact the same as it would be for a Dirac field, namely

$$\langle 0|T(\Psi_{M\alpha}(x_1)\bar{\Psi}_{M\beta}(x_2))|0\rangle = S_{F\alpha\beta}(x_1 - x_2) \tag{2.137}$$

where $S_{F\alpha\beta}(x_1 - x_2)$ is the function whose Fourier transform (in the variable $x_1 - x_2$) is

$$\mathrm{i}(\not{k} - m + \mathrm{i}\epsilon)^{-1} = \frac{\mathrm{i}(\not{k} + m)}{k^2 - m^2 + \mathrm{i}\epsilon} \tag{2.138}$$

for a field of mass m (see, for example, Section 7.2 of [15]). The reader can check this (in a rather lengthy calculation) by inserting the expansion (2.135) in the left-hand side of (2.137), and using the anticommutation relations for the c and $c^\dagger$ operators, together with the vacuum conditions $c_\lambda(k)|0\rangle = \langle 0|c_\lambda^\dagger(k) = 0$. Consider now the quantity $\langle 0|\Psi_{M\alpha}(x_1)\Psi_{M\beta}(x_2)|0\rangle$. From (2.136) we have

$$\Psi_M(x_2) = C\beta\Psi_M^{\dagger\mathrm{T}}(x_2), \tag{2.139}$$

or in terms of components

$$\begin{aligned}\Psi_{M\beta}(x_2) &= C_{\beta\gamma}\beta_{\gamma\delta}\Psi^\dagger_{M\delta}(x_2)\\ &= \Psi^\dagger_{M\delta}(x_2)\beta_{\delta\gamma}C^{\mathrm{T}}_{\gamma\beta}\\ &= \bar{\Psi}_{M\gamma}(x_2)C^{\mathrm{T}}_{\gamma\beta}.\end{aligned} \tag{2.140}$$

Hence we obtain

$$\langle 0|T(\Psi_{M\alpha}(x_1)\Psi_{M\beta}(x_2))|0\rangle = \langle 0|T(\Psi_{M\alpha}(x_1)\bar{\Psi}_{M\gamma}(x_2)|0\rangle C^{\mathrm{T}}_{\gamma\beta} = S_{F\alpha\gamma}(x_1 - x_2)C^{\mathrm{T}}_{\gamma\beta}. \tag{2.141}$$

Exercise 2.10 Verify the relations

$$C^{\mathrm{T}} = -C = C^{-1}. \tag{2.142}$$

Hence show that (2.136) can also be written as

$$\bar{\Psi}_M = \Psi_M^{\mathrm{T}}C, \tag{2.143}$$

and use this result to show that

$$\langle 0|T(\bar{\Psi}_{\mathrm{M}\alpha}(x_1)\bar{\Psi}_{\mathrm{M}\beta}(x_2))|0\rangle = C^{\mathrm{T}}_{\alpha\gamma} S_{\mathrm{F}\gamma\beta}(x_1 - x_2). \tag{2.144}$$

When reducing matrix elements in covariant perturbation theory using Wick's theorem (see [15], Chapter 6), one has to remember to include these two non-Dirac-like contractions.

We are at last ready to take our first steps in SUSY.

3

Introduction to supersymmetry and the MSSM

3.1 Simple supersymmetry

In this section we will look at one of the simplest supersymmetric theories, one involving just two free fields: a complex spin-0 field ϕ and an L-type spinor field χ, both massless. The Lagrangian (density) for this system is

$$\mathcal{L} = \partial_\mu \phi^\dagger \partial^\mu \phi + \chi^\dagger \mathrm{i} \bar{\sigma}^\mu \partial_\mu \chi. \tag{3.1}$$

The ϕ part is familiar from introductory quantum field theory courses: the χ bit, as noted already, is just the appropriate part of the Dirac Lagrangian (2.38). The equation of motion for ϕ is of course $\Box\phi = 0$, while that for χ is $\mathrm{i}\bar{\sigma}^\mu \partial_\mu \chi = 0$ (compare (2.37)). We are going to try and find, by 'brute force', transformations in which the change in ϕ is proportional to χ (as in (2.2)), and the change in χ is proportional to ϕ, such that $\mathcal{L}$ is invariant – or, more precisely, such that the Action S is invariant, where

$$S = \int \mathcal{L}\, \mathrm{d}^4 x. \tag{3.2}$$

To ensure the invariance of S, it is sufficient that $\mathcal{L}$ changes by a total derivative, the integral of which is assumed to vanish at the boundaries of space–time.

As a preliminary, it is useful to get the *dimensions* of everything straight. The Action is the integral of the density $\mathcal{L}$ over all four-dimensional space, and is dimensionless in units $\hbar = c = 1$. In this system, there is only one independent dimension left, which we take to be that of mass (or energy), M (see Appendix B of [15]). Length has the same dimension as time (because $c = 1$), and both have the dimension of M^{-1} (because $\hbar = 1$). It follows that, for the Action to be dimensionless, $\mathcal{L}$ has dimension M^4. Since the gradients have dimension M, we can then read off the dimensions of ϕ and χ (denoted by $[\phi]$ and $[\chi]$):

$$[\phi] = \mathrm{M}, \quad [\chi] = \mathrm{M}^{3/2}. \tag{3.3}$$

Now, what are the SUSY transformations linking ϕ and χ? Several considerations can guide us to, if not the answer, then at least a good guess. Consider first the change in ϕ, $\delta_\xi \phi$, which has the form (already stated in (2.2))

$$\delta_\xi \phi = \text{parameter } \xi \times \text{other field } \chi, \tag{3.4}$$

where we shall take ξ to be independent of x.[1] On the left-hand side, we have a spin-0 field, which is invariant under Lorentz transformations. So we must construct a Lorentz invariant out of χ and the parameter ξ. One simple way to do this is to declare that ξ is also a χ- (or L-) type spinor, and use the invariant product (2.46). This gives

$$\delta_\xi \phi = \xi^{\mathrm{T}}(-\mathrm{i}\sigma_2)\chi, \tag{3.5}$$

or in the notation of Section 2.3

$$\delta_\xi \phi = \xi^a \chi_a = \xi \cdot \chi. \tag{3.6}$$

It is worth pausing to note some things about the parameter ξ. First, we repeat that it is a spinor. It doesn't depend on x, but it is not an invariant under Lorentz transformations: it transforms as a χ-type spinor, i.e. by $V^{-1\dagger}$. It has two components, of course, each of which is complex; hence four real numbers in all. These specify the transformation (3.5). Secondly, although ξ doesn't depend on x, and is not a field (operator) in that sense, we shall assume that its components *anticommute* with the components of spinor fields; that is, we assume they are Grassmann numbers (see [7] Appendix O). Lastly, since $[\phi] = \mathrm{M}$ and $[\chi] = \mathrm{M}^{3/2}$, to make the dimensions balance on both sides of (3.5) we need to assign the dimension

$$[\xi] = \mathrm{M}^{-1/2} \tag{3.7}$$

to ξ.

Now let us think what the corresponding $\delta_\xi \chi$ might be. This has to be something like

$$\delta_\xi \chi \sim \text{product of } \xi \text{ and } \phi. \tag{3.8}$$

On the left-hand side of (3.8) we have a quantity with dimensions $\mathrm{M}^{3/2}$, whereas on the right-hand side the algebraic product of ξ and ϕ has dimensions $\mathrm{M}^{-1/2+1} = \mathrm{M}^{1/2}$. Hence we need to introduce something with dimensions M^1 on the right-hand side. In this massless theory, there is only one possibility – the gradient operator ∂_μ, or more conveniently the momentum operator $\mathrm{i}\partial_\mu$. But now we have a 'loose' index

[1] That is to say, we shall be considering a *global* supersymmetry, as opposed to a *local* one, in which case ξ would depend on x. For the significance of the global/local distinction in gauge theories, see [15] and [7]. In the present case, making supersymmetry local leads to *supergravity*, which is beyond our scope.

μ on the right-hand side! The left-hand side is a spinor, and there is a spinor (ξ) also on the right-hand side, so we should probably get rid of the μ index altogether, by contracting it. We try

$$\delta_\xi \chi = (\mathrm{i}\sigma^\mu \partial_\mu \phi)\, \xi, \tag{3.9}$$

where σ^μ is given in (2.34). Note that the 2×2 matrices in σ^μ act on the 2-component column ξ to give, correctly, a 2-component column to match the left-hand side. But although both sides of (3.9) are 2-component column vectors, the right-hand side does not transform as a χ-type spinor. If we look back at (2.36) and (2.37), we see that the combination $\sigma^\mu \partial_\mu$ acting on a ψ transforms as a χ (and $\bar{\sigma}^\mu \partial_\mu$ on a χ transforms as a ψ). This suggests that we should let the $\sigma^\mu \partial_\mu \phi$ in (3.9) multiply a ψ-like thing, not the L-type ξ, in order to get something transforming as a χ. However, we know how to manufacture a ψ-like thing out of ξ! We just take (see (2.45)) $\mathrm{i}\sigma_2 \xi^*$. We therefore arrive at the guess

$$\delta_\xi \chi_a = A[\mathrm{i}\sigma^\mu(\mathrm{i}\sigma_2 \xi^*)]_a \partial_\mu \phi, \tag{3.10}$$

where A is some constant to be determined from the condition that $\mathcal{L}$ is invariant (up to a total derivative) under (3.5) and (3.10), and we have indicated the χ-type spinor index on both sides. Note that '$\partial_\mu \phi$' has no matrix structure and has been moved to the end.

Exercise 3.1 Check that in the bar notation of Section 2.3, (3.8) is (omitting the indices)

$$\delta_\xi \chi = A\mathrm{i}\sigma^\mu \bar{\xi} \partial_\mu \phi. \tag{3.11}$$

Equations (3.5) and (3.10) give the proposed SUSY transformations for ϕ and χ, but both are complex fields and we need to be clear what the corresponding transformations are for their hermitian conjugates $\phi^\dagger$ and $\chi^\dagger$. There are some notational concerns here that we should pause over. First, remember that ϕ and χ are quantum fields, even though we are not explicitly putting hats on them; on the other hand, ξ is not a field (it is x-independent). In the discussion of Lorentz transformations of spinors in Chapter 2, we used the symbol * to denote complex conjugation, it being tacitly understood that this really meant † when applied to creation and annihilation operators. Let us now spell this out in more detail. Consider the (quantum) field ϕ with a mode expansion

$$\phi = \int \frac{\mathrm{d}^3 \boldsymbol{k}}{(2\pi)^3 \sqrt{2\omega}} [a(k)\mathrm{e}^{-\mathrm{i}k\cdot x} + b^\dagger(k)\mathrm{e}^{\mathrm{i}k\cdot x}]. \tag{3.12}$$

Here the operator $a(k)$ destroys (say) a particle with 4-momentum k, and $b^\dagger(k)$ creates an antiparticle of 4-momentum k, while $\exp[\pm \mathrm{i}k \cdot x]$ are of course ordinary wavefunctions. For (3.12) the simple complex conjugation $*$ is not appropriate, since '$a^*(k)$' is not defined; instead, we want '$a^\dagger(k)$'. So instead of 'ϕ^*' we deal with $\phi^\dagger$, which is defined in terms of (3.12) by (a) taking the complex conjugate of the wavefunction parts and (b) taking the dagger of the mode operators. This gives

$$\phi^\dagger = \int \frac{d^3\boldsymbol{k}}{(2\pi)^3\sqrt{2\omega}}\,[a^\dagger(k)e^{ik\cdot x} + b(k)e^{-ik\cdot x}], \tag{3.13}$$

the conventional definition of the hermitian conjugate of (3.12).

For spinor fields like χ, on the other hand, the situation is slightly more complicated, since now in the analogue of (3.12) the scalar (spin-0) wavefunctions $\exp[\pm ik\cdot x]$ will be replaced by (free-particle) 2-component spinors. Thus, symbolically, the first (upper) component of the quantum field χ will have the form

$$\chi_1 \sim \text{mode operator} \times \text{first component of free-particle spinor of } \chi\text{-type}, \tag{3.14}$$

where we are of course using the 'downstairs, undotted' notation for the components of χ. In the same way as (3.13) we then define

$$\chi_1^\dagger \sim (\text{mode operator})^\dagger \times (\text{ first component of free-particle spinor})^*. \tag{3.15}$$

With this in hand, let us consider the hermitian conjugate of (3.5), that is $\delta_\xi\phi^\dagger$. Written out in terms of components (3.5) is

$$\delta_\xi\phi = (\xi_1\xi_2)\begin{pmatrix} 0 & -1 \\ 1 & 0 \end{pmatrix}\begin{pmatrix} \chi_1 \\ \chi_2 \end{pmatrix} = -\xi_1\chi_2 + \xi_2\chi_1. \tag{3.16}$$

We want to take the 'dagger' of this, but we are now faced with a decision about how to take the dagger of products of (anticommuting) spinor components, like $\xi_1\chi_2$. In the case of two matrices A and B, we know that $(AB)^\dagger = B^\dagger A^\dagger$. By analogy, we shall *define* the dagger to reverse the order of the spinors:

$$\delta_\xi\phi^\dagger = -\chi_2^\dagger\xi_1^* + \chi_1^\dagger\xi_2^*; \tag{3.17}$$

ξ isn't a quantum field and the '$*$' notation is suitable for it. Note that we here do *not* include a minus sign from reversing the order of the operators. Now (3.17) can be written in more compact form:

$$\begin{aligned}\delta_\xi\phi^\dagger &= \chi_1^\dagger\xi_2^* - \chi_2^\dagger\xi_1^* \\ &= (\chi_1^\dagger\chi_2^\dagger)\begin{pmatrix} 0 & 1 \\ -1 & 0 \end{pmatrix}\begin{pmatrix} \xi_1^* \\ \xi_2^* \end{pmatrix} \\ &= \chi^\dagger(i\sigma_2)\xi^*,\end{aligned} \tag{3.18}$$

where in the last line the † symbol, as applied to the 2-component spinor field χ, is understood in a matrix sense as well: that is

$$\chi^\dagger = \begin{pmatrix} \chi_1 \\ \chi_2 \end{pmatrix}^\dagger = (\chi_1^\dagger \chi_2^\dagger). \tag{3.19}$$

Equation (3.18) is a satisfactory outcome of these rather fiddly considerations because (a) we have seen exactly this spinor structure before, in (2.95), and we are assured its Lorentz transformation character is correct, and (b) it is nicely consistent with 'naively' taking the dagger of (3.5), treating it like a matrix product. In particular, the right-hand side of the last line of (3.18) can be written in the notation of Section 2.3 as $\bar{\chi} \cdot \bar{\xi}$ (making use of (2.68)) or equally as $\bar{\xi} \cdot \bar{\chi}$. Referring to (3.6) we therefore note the useful result

$$(\xi \cdot \chi)^\dagger = (\chi \cdot \xi)^\dagger = \bar{\xi} \cdot \bar{\chi} = \bar{\chi} \cdot \bar{\xi}. \tag{3.20}$$

Then (3.6) and (3.18) become

$$\delta_\xi \phi = \xi \cdot \chi = \chi \cdot \xi, \quad \delta_\xi \phi^\dagger = \bar{\xi} \cdot \bar{\chi} = \bar{\chi} \cdot \bar{\xi}. \tag{3.21}$$

In the same way, therefore, we can take the dagger of (3.10) to obtain

$$\delta_\xi \chi^\dagger = A \partial_\mu \phi^\dagger \xi^{\mathrm{T}} \mathrm{i}\sigma_2 \mathrm{i}\sigma^\mu, \tag{3.22}$$

where for later convenience we have here moved the $\partial_\mu \phi^\dagger$ to the front, and we have taken A to be real (which will be sufficient, as we shall see).

Exercise 3.2 Check that (3.22) is equivalent to

$$\delta_\xi \bar{\chi} = A \partial_\mu \phi^\dagger \bar{\sigma}^\mu \xi. \tag{3.23}$$

We are now ready to see if we can choose A so as to make $\mathcal{L}$ invariant under (3.5), (3.10), (3.18) and (3.22).

We have

$$\begin{aligned} \delta_\xi \mathcal{L} &= \partial_\mu(\delta_\xi \phi^\dagger)\partial^\mu \phi + \partial_\mu \phi^\dagger \partial^\mu(\delta_\xi \phi) + (\delta_\xi \chi^\dagger)\mathrm{i}\bar{\sigma}^\mu \partial_\mu \chi + \chi^\dagger \mathrm{i}\bar{\sigma}^\mu \partial_\mu(\delta_\xi \chi) \\ &= \partial_\mu(\chi^\dagger \mathrm{i}\sigma_2 \xi^*)\partial^\mu \phi + \partial_\mu \phi^\dagger \partial^\mu(\xi^{\mathrm{T}}(-\mathrm{i}\sigma_2)\chi) \\ &\quad + A(\partial_\mu \phi^\dagger \xi^{\mathrm{T}} \mathrm{i}\sigma_2 \mathrm{i}\sigma^\mu)\mathrm{i}\bar{\sigma}^\nu \partial_\nu \chi + A\chi^\dagger \mathrm{i}\bar{\sigma}^\nu \partial_\nu(\mathrm{i}\sigma^\mu \mathrm{i}\sigma_2 \xi^*)\partial_\mu \phi. \end{aligned} \tag{3.24}$$

Inspection of (3.24) shows that there are two types of term, one involving the parameters ξ^* and the other the parameters ξ^{T}. Consider the term involving $A\xi^*$. In it there appears the combination (pulling ∂_μ through the constant ξ^*)

$$\bar{\sigma}^\nu \partial_\nu \sigma^\mu \partial_\mu = (\partial_0 - \boldsymbol{\sigma} \cdot \boldsymbol{\nabla})(\partial_0 + \boldsymbol{\sigma} \cdot \boldsymbol{\nabla}) = \partial_0^2 - \boldsymbol{\nabla}^2 = \partial_\mu \partial^\mu. \tag{3.25}$$

We can therefore combine this and the other term in ξ^* from (3.24) to give

$$\delta_\xi \mathcal{L}|_{\xi^*} = \partial_\mu \chi^\dagger \mathrm{i}\sigma_2 \xi^* \partial^\mu \phi - \mathrm{i}A\chi^\dagger \partial_\mu \partial^\mu \sigma_2 \xi^* \phi. \tag{3.26}$$

This represents a change in $\mathcal{L}$ under our transformations, so it seems we have not succeeded in finding an invariance (or symmetry), since we cannot hope to cancel this change against the term involving ξ^{T}, which involves quite independent parameters. However, we must remember (see (3.2)) that the Action is the space–time integral of $\mathcal{L}$, and this will be invariant if we can arrange for the change in $\mathcal{L}$ to be a *total derivative*. Since ξ does not depend on x, we can indeed write (3.26) as a total derivative

$$\delta_\xi \mathcal{L}|_{\xi^*} = \partial_\mu (\chi^\dagger \mathrm{i}\sigma_2 \xi^* \partial^\mu \phi) \tag{3.27}$$

provided that

$$A = -1. \tag{3.28}$$

Similarly, if $A = -1$ the terms in ξ^{T} combine to give

$$\delta_\xi \mathcal{L}|_{\xi^{\mathrm{T}}} = \partial_\mu \phi^\dagger \partial^\mu (\xi^{\mathrm{T}}(-\mathrm{i}\sigma_2)\chi) + \partial_\mu \phi^\dagger \xi^{\mathrm{T}} \mathrm{i}\sigma_2 \sigma^\mu \bar{\sigma}^\nu \partial_\nu \chi. \tag{3.29}$$

The second term in (3.29) we can write as

$$\partial_\mu (\phi^\dagger \xi^{\mathrm{T}} \mathrm{i}\sigma_2 \sigma^\mu \bar{\sigma}^\nu \partial_\nu \chi) + \phi^\dagger \xi^{\mathrm{T}} (-\mathrm{i}\sigma_2) \sigma^\mu \bar{\sigma}^\nu \partial_\mu \partial_\nu \chi \tag{3.30}$$

$$= \partial_\mu (\phi^\dagger \xi^{\mathrm{T}} \mathrm{i}\sigma_2 \sigma^\mu \bar{\sigma}^\nu \partial_\nu \chi) + \phi^\dagger \xi^{\mathrm{T}} (-\mathrm{i}\sigma_2) \partial_\mu \partial^\mu \chi. \tag{3.31}$$

The second term of (3.31) and the first term of (3.29) now combine to give the total derivative

$$\partial_\mu (\phi^\dagger \xi^{\mathrm{T}} (-\mathrm{i}\sigma_2) \partial^\mu \chi), \tag{3.32}$$

so that finally

$$\delta_\xi \mathcal{L}|_{\xi^{\mathrm{T}}} = \partial_\mu (\phi^\dagger \xi^{\mathrm{T}} (-\mathrm{i}\sigma_2) \partial^\mu \chi) + \partial_\mu (\phi^\dagger \xi^{\mathrm{T}} \mathrm{i}\sigma_2 \sigma^\mu \bar{\sigma}^\nu \partial_\nu \chi), \tag{3.33}$$

which is also a total derivative. In summary, we have shown that under (3.5), (3.10), (3.18) and (3.22), with $A = -1$, $\mathcal{L}$ changes by a total derivative:

$$\delta_\xi \mathcal{L} = \partial_\mu (\chi^\dagger \mathrm{i}\sigma_2 \xi^* \partial^\mu \phi + \phi^\dagger \xi^{\mathrm{T}} (-\mathrm{i}\sigma_2) \partial^\mu \chi + \phi^\dagger \xi^{\mathrm{T}} \mathrm{i}\sigma_2 \sigma^\mu \bar{\sigma}^\nu \partial_\nu \chi) \tag{3.34}$$

and the Action is therefore invariant: we have a SUSY theory, in this sense. As we shall see in Chapter 4, the pair (ϕ, spin-0) and (χ, L-type spin-1/2) constitute a *left chiral supermultiplet* in SUSY.

Exercise 3.3 Show that (3.34) can also be written as

$$\delta_\xi \mathcal{L} = \partial_\mu(\chi^\dagger \mathrm{i}\sigma_2 \xi^* \partial^\mu \phi + \xi^{\mathrm{T}} \mathrm{i}\sigma_2 \sigma^\nu \bar{\sigma}^\mu \chi \partial_\nu \phi^\dagger + \xi^{\mathrm{T}}(-\mathrm{i}\sigma_2)\chi \partial^\mu \phi^\dagger). \tag{3.35}$$

The reader may well feel that it has been pretty heavy going, considering especially the simplicity, triviality almost, of the Lagrangian (3.1). A more professional notation would have been more efficient, of course, but there is a lot to be said for doing it the most explicit and straightforward way, first time through. As we proceed, we shall speed up the notation. In fact, interactions don't constitute an order of magnitude increase in labour, and the manipulations gone through in this simple example are quite representative.

3.2 A first glance at the MSSM

Before continuing with more formal work, we would like to whet the reader's appetite by indicating how the SUSY idea is applied to particle physics in the MSSM. The only type of SUSY theory we have discussed so far, of course, contains just one massless complex scalar field and one massless Weyl fermion field (which could be either L or R – we chose L). Such fields form a SUSY *supermultiplet*, called a *chiral* supermultiplet. Thus far, interactions have not been included: that will be done in Chapter 5. Other types of supermultiplet are also possible, as we shall learn in the next chapter (Section 4.4). For example, one can have a *vector* (or *gauge*) supermultiplet, in which a massless spin-1 field, which has two on-shell degrees of freedom, is partnered with a massless Weyl fermion field. The allowed (renormalizable) interactions for massless spin-1 fields are gauge interactions, and the theory can be made supersymmetric when Weyl fermion fields are included, as will be explained in Chapter 7. In fact, only these two types of supermultiplet are used in the MSSM. So we now need to consider how the fields of the SM, which comprise spin-0 Higgs fields, spin-$\frac{1}{2}$ quark and lepton fields, and spin-1 gauge fields, might be assigned to chiral and gauge supermultiplets. (Masses will eventually be generated by Higgs interactions, and by SUSY-breaking soft masses, as described in Section 9.2.)

A crucial point here is that SUSY transformations do not change $\mathrm{SU(3)_c}$, $\mathrm{SU(2)_L}$ or U(1) quantum numbers: that is to say, each SM field and its partner in a SUSY supermultiplet must have the same $\mathrm{SU(3)_c} \times \mathrm{SU(2)_L} \times \mathrm{U(1)}$ quantum numbers. Consider then the gluons, for example, which are the SM gauge bosons associated with local $\mathrm{SU(3)_c}$ symmetry. They belong (necessarily) to the eight-dimensional 'adjoint' representation of SU(3) (see [7] chapter 13, for example), and are flavour singlets. None of the SM fermions have these quantum numbers, so – to create a supersymmetric version of QCD – we are obliged to introduce a new SU(3) octet of Weyl fermions, called 'gluinos', which are the superpartners of the gluons. Similar considerations lead to the introduction of an $SU(2)_L$ triplet of Weyl fermions, called 'winos' ($\widetilde{W}^{\pm}$, $\widetilde{W}^0$), and a $U(1)_y$ 'bino' ($\widetilde{B}$).

Turning now to the SM fermions, consider first the left-handed lepton fields, for example the $SU(2)_L$ doublet

$$\begin{pmatrix} \nu_{eL} \\ e_L \end{pmatrix}. \tag{3.36}$$

These cannot be partnered by new spin-1 fields, since the latter would (in the interacting case) have to be gauge bosons, belonging to the three-dimensional adjoint representation of $SU(2)_L$, not the doublet representation. Instead, in a SUSY theory, we must partner the doublet (3.36) with a doublet of spin-0 bosons having the same SM quantum numbers. In the MSSM, this is done by introducing a new doublet of scalar fields to go with the lepton doublet (3.36), forming chiral supermultiplets:

$$\begin{pmatrix} \nu_{eL} \\ e_L \end{pmatrix} \text{ partnered by } \begin{pmatrix} \tilde{\nu}_{eL} \\ \tilde{e}_L \end{pmatrix} \tag{3.37}$$

where '$\tilde{\nu}$' is a scalar partner for the neutrino ('sneutrino'), and '$\tilde{e}$' is a scalar partner for the electron ('selectron'). Similarly, we will have smuons and staus, and their sneutrinos. These are all in chiral supermultiplets, and $SU(2)_L$ doublets, and (though bosons) they all carry the same lepton numbers as their SM partners.

What about quarks? They are a triplet of the $SU(3)_c$ colour gauge group, and no other SM particles are colour triplets. They cannot be partnered by new gauge fields, which must be in the octet representation of SU(3), not the triplet. So we will need new spin-0 partners for the quarks too, called squarks, which are colour triplets with the same baryon number as the quarks, belonging with them in chiral supermultiplets; they must also have the same electroweak quantum numbers as the quarks.

The electroweak interactions of both leptons and quarks are 'chiral', which means that the 'L' parts of the fields interact differently from the 'R' parts. The L parts belong to $SU(2)_L$ doublets, as above, while the R parts are $SU(2)_L$ singlets. So we need to arrange for scalar partners for the L and R parts separately: for example $(e_R, \tilde{e}_R)$, $(u_R, \tilde{u}_R)$, $(d_R, \tilde{d}_R)$, etc.; and

$$\begin{pmatrix} u_L \\ d_L \end{pmatrix}, \begin{pmatrix} \tilde{u}_L \\ \tilde{d}_L \end{pmatrix} \tag{3.38}$$

and so on.[2]

Finally, the Higgs sector: the scalar Higgs fields will need their own 'higgsinos', i.e. fermionic partners forming chiral supermultiplets. In fact, a crucial consequence

[2] As noted in Section 2.4, the 'particle R parts' will actually be represented by the charge conjugates of the 'antiparticle L parts'.

of making the SM supersymmetric, in the MSSM, is that, as we shall see in Chapter 8, *two separate* Higgs doublets are required. In the SM, Yukawa interactions involving the field

$$\phi = \begin{pmatrix} \phi^+ \\ \phi^0 \end{pmatrix}, \tag{3.39}$$

give masses to the $t_3 = -1/2$ components of the fermion doublets when ϕ^0 acquires a vev, while corresponding interactions with the charge-conjugate field

$$\phi_C \equiv \mathrm{i}\tau_2 \phi^{\dagger \mathrm{T}} = \begin{pmatrix} \bar{\phi}^0 \\ -\phi^- \end{pmatrix} \tag{3.40}$$

give masses to the $t_3 = +1/2$ components (see, for example, Section 22.6 of [7]). But in the supersymmetric version, the Yukawa interactions cannot involve both a complex scalar field ϕ and its hermitian conjugate $\phi^\dagger$ (see Section 5.1). Hence use of the charge-conjugate ϕ_C is forbidden by SUSY, and we need two independent Higgs chiral supermultiplets:

$$H_\mathrm{u} : \quad \begin{pmatrix} H_\mathrm{u}^+ \\ H_\mathrm{u}^0 \end{pmatrix}, \quad \begin{pmatrix} \tilde{H}_\mathrm{u}^+ \\ \tilde{H}_\mathrm{u}^0 \end{pmatrix} \tag{3.41}$$

and

$$H_\mathrm{d} : \quad \begin{pmatrix} H_\mathrm{d}^0 \\ H_\mathrm{d}^- \end{pmatrix}, \quad \begin{pmatrix} \tilde{H}_\mathrm{d}^0 \\ \tilde{H}_\mathrm{d}^- \end{pmatrix}. \tag{3.42}$$

This time, of course, the fields with tildes have spin 1/2. (The apparently 'wrong' labelling of H_u and H_d will be explained in Section 8.1.)

The chiral and gauge supermultiplets introduced here constitute the field content of the MSSM. The full theory includes supersymmetric interactions (Chapters 5, 7 and 8) and soft SUSY-breaking terms (see Section 9.2). It has been around for over 25 years: early reviews are given in [43], [44] and [45]; a more recent and very helpful 'supersymmetry primer' was provided by Martin [46], to which we shall make quite frequent reference in what follows. A comprehensive review may be found in [47]. Finally, there are two substantial monographs, by Drees *et al.* [48] and by Baer and Tata [49].

We will return to the MSSM in Chapter 8. For the moment, we should simply note that (a) none of the 'superpartners' has yet been seen experimentally, in particular they certainly cannot have the same mass as their SM partner states (as would normally be expected for a symmetry multiplet), so that (b) SUSY, as applied in the MSSM, must be broken somehow. We will include a brief discussion of SUSY breaking in Chapter 9, but a more detailed treatment is beyond the scope of this book.

4

The supersymmetry algebra and supermultiplets

A fundamental aspect of any symmetry (other than a U(1) symmetry) is the *algebra* associated with the *symmetry generators* (see for example Appendix M of [7]). For example, the generators T_i of SU(2) satisfy the commutation relations

$$[T_i, T_j] = \mathrm{i}\epsilon_{ijk} T_k, \tag{4.1}$$

where i, j and k run over the values 1, 2 and 3, and where the repeated index k is summed over; ϵ_{ijk} is the totally antisymmetric symbol such that $\epsilon_{123} = +1$, $\epsilon_{213} = -1$, etc. The commutation relations summarized in (4.1) constitute the 'SU(2) algebra', and it is of course exactly that of the angular momentum operators in quantum mechanics, in units $\hbar = 1$. Readers will be familiar with the way in which the whole theory of angular momentum in quantum mechanics – in particular, the SU(2) multiplet structure – can be developed just from these commutation relations. In the same way, in order to proceed in a reasonably systematic way with SUSY, and especially to understand what kind of 'supermultiplets' occur, we must know what the SUSY algebra is. In Section 1.3, we introduced the idea of generators of SUSY transformations, Q_a, and their associated algebra, which now involves anticommutation relations, was roughly indicated in (1.34). The main work of this chapter is to find the actual SUSY algebra, by a 'brute force' method once again, making use of what we have learned in Chapter 3.

4.1 One way of obtaining the SU(2) algebra

In Chapter 3, we arrived at recipes for SUSY transformations of spin-0 fields ϕ and $\phi^\dagger$, and spin-1/2 fields χ and $\chi^\dagger$. From these transformations, the algebra of the SUSY generators can be deduced. To understand the method, it is helpful to see it in action in a more familiar context, namely that of SU(2), as we now discuss.

Consider an SU(2) doublet of fields

$$q = \begin{pmatrix} u \\ d \end{pmatrix} \tag{4.2}$$

where u and d have equal mass, and identical interactions, so that the Lagrangian is invariant under (infinitesimal) transformations of the components of q of the form (see, for example, equation (12.95) of [7])

$$q \to q' = (1 - \mathrm{i}\boldsymbol{\epsilon} \cdot \boldsymbol{\tau}/2)q \equiv q + \delta_{\boldsymbol{\epsilon}} q, \tag{4.3}$$

where

$$\delta_{\boldsymbol{\epsilon}} q = -\mathrm{i}\boldsymbol{\epsilon} \cdot \boldsymbol{\tau}/2\, q. \tag{4.4}$$

Here, as usual, the three matrices $\boldsymbol{\tau} = (\tau_1, \tau_2, \tau_3)$ are the same as the Pauli $\boldsymbol{\sigma}$ matrices, and $\boldsymbol{\epsilon} = (\epsilon_1, \epsilon_2, \epsilon_3)$ are three real infinitesimal parameters specifying the transformation. For example, for $\boldsymbol{\epsilon} = (0, \epsilon_2, 0)$, $\delta_{\epsilon_2} q_1 = -(\epsilon_2/2)q_2$. The transformed fields q' satisfy the same anticommutation relations as the original fields q, so that q' and q are related by a unitary transformation

$$q' = UqU^{\dagger}. \tag{4.5}$$

For infinitesimal transformations, U has the general form

$$U_{\text{infl}} = (1 + \mathrm{i}\boldsymbol{\epsilon} \cdot \boldsymbol{T}) \tag{4.6}$$

where

$$\boldsymbol{T} = (T_1, T_2, T_3) \tag{4.7}$$

are the *generators* of infinitesimal SU(2) transformations; the unitarity of U implies that the $\boldsymbol{T}$'s are Hermitian. For infinitesimal transformations, therefore, we have (from (4.5) and (4.6))

$$\begin{aligned} q' &= (1 + \mathrm{i}\boldsymbol{\epsilon} \cdot \boldsymbol{T})q(1 - \mathrm{i}\boldsymbol{\epsilon} \cdot \boldsymbol{T}) \\ &= q + \mathrm{i}\boldsymbol{\epsilon} \cdot \boldsymbol{T}q - \mathrm{i}\boldsymbol{\epsilon} \cdot q\boldsymbol{T} \quad \text{to first order in } \boldsymbol{\epsilon} \\ &= q + \mathrm{i}\boldsymbol{\epsilon} \cdot [\boldsymbol{T}, q]. \end{aligned} \tag{4.8}$$

Hence from (4.3) and (4.4) we deduce (see equation (12.100) of [7])

$$\delta_{\boldsymbol{\epsilon}} q = \mathrm{i}\boldsymbol{\epsilon} \cdot [\boldsymbol{T}, q] = -\mathrm{i}\boldsymbol{\epsilon} \cdot \boldsymbol{\tau}/2\, q. \tag{4.9}$$

It is important to realize that the $\boldsymbol{T}$'s are themselves quantum field operators, constructed from the fields of the Lagrangian; for example in this simple case they would be

$$\boldsymbol{T} = \int q^{\dagger}(\boldsymbol{\tau}/2)q \, \mathrm{d}^3\boldsymbol{x} \tag{4.10}$$

as explained for example in Section 12.3 of [7], and re-derived in Section 4.3 below, equation (4.68).

Given an explicit formula for the generators, such as (4.10), we can proceed to calculate the commutation relations of the $\boldsymbol{T}$'s, knowing how the q's anticommute. The answer is that the $\boldsymbol{T}$'s obey the relations (4.1). However, there is another way to get these commutation relations, just by considering the small changes in the fields, as given by (4.9). Consider two such transformations

$$\delta_{\epsilon_1} q = \mathrm{i}\epsilon_1 [T_1, q] = -\mathrm{i}\epsilon_1 (\tau_1/2) q \tag{4.11}$$

and

$$\delta_{\epsilon_2} q = \mathrm{i}\epsilon_2 [T_2, q] = -\mathrm{i}\epsilon_2 (\tau_2/2) q. \tag{4.12}$$

We shall calculate the *difference* $(\delta_{\epsilon_1}\delta_{\epsilon_2} - \delta_{\epsilon_2}\delta_{\epsilon_1})q$ in two different ways: first via the second equality in (4.11) and (4.12), and then via the first equalities. Equating the two results will lead us to the algebra (4.1).

First, then, we use the second equality of (4.11) and (4.12) to obtain

$$\begin{aligned} \delta_{\epsilon_1}\delta_{\epsilon_2} q &= \delta_{\epsilon_1}\{-\mathrm{i}\epsilon_2(\tau_2/2\} q \\ &= -\mathrm{i}\epsilon_2(\tau_2/2)\delta_{\epsilon_1} q \\ &= -\mathrm{i}\epsilon_2(\tau_2/2). - \mathrm{i}\epsilon_1(\tau_1/2) q \\ &= -(1/4)\epsilon_1\epsilon_2\tau_2\tau_1 q. \end{aligned} \tag{4.13}$$

Note that in the last line we have changed the order of the ϵ parameters as we are free to do since they are ordinary numbers, but we cannot alter the order of the τ's since they are matrices which do not commute. Similarly,

$$\begin{aligned} \delta_{\epsilon_2}\delta_{\epsilon_1} q &= \delta_{\epsilon_2}\{-\mathrm{i}\epsilon_1(\tau_1/2\} q \\ &= -\mathrm{i}\epsilon_1(\tau_1/2)\delta_{\epsilon_2} q \\ &= -(1/4)\epsilon_1\epsilon_2\tau_1\tau_2 q. \end{aligned} \tag{4.14}$$

Hence

$$\begin{aligned} (\delta_{\epsilon_1}\delta_{\epsilon_2} - \delta_{\epsilon_2}\delta_{\epsilon_1}) q &= \epsilon_1\epsilon_2[\tau_1/2, \tau_2/2] q \\ &= \epsilon_1\epsilon_2 \mathrm{i}(\tau_3/2) q \\ &= -\mathrm{i}\epsilon_1\epsilon_2[T_3, q], \end{aligned} \tag{4.15}$$

where the second line follows from the fact that the quantities $\frac{1}{2}\tau_i$, as matrices, satisfy the algebra (4.1), and the third line results from the '$_3$' analogue of (4.11) and (4.12).

Now we calculate $(\delta_{\epsilon_1}\delta_{\epsilon_2} - \delta_{\epsilon_2}\delta_{\epsilon_1})q$ using the first equality of (4.11) and (4.12). We have

$$\begin{aligned}\delta_{\epsilon_1}\delta_{\epsilon_2}q &= \delta_{\epsilon_1}\{\mathrm{i}\epsilon_2[T_2, q]\}\\ &= \mathrm{i}\epsilon_2\delta_{\epsilon_1}\{[T_2, q]\}\\ &= \mathrm{i}\epsilon_1\mathrm{i}\epsilon_2[T_1, [T_2, q]].\end{aligned} \tag{4.16}$$

Similarly,

$$\delta_{\epsilon_2}\delta_{\epsilon_1}q = \mathrm{i}\epsilon_1\mathrm{i}\epsilon_2[T_2, [T_1, q]]. \tag{4.17}$$

Hence

$$(\delta_{\epsilon_1}\delta_{\epsilon_2} - \delta_{\epsilon_2}\delta_{\epsilon_1})q = -\epsilon_1\epsilon_2\{[T_1, [T_2, q]] - [T_2, [T_1, q]]\}. \tag{4.18}$$

Now we can rearrange the right-hand side of this equation by using the identity (which is easily checked by multiplying it all out)

$$[A, [B, C]] + [B, [C, A]] + [C, [A, B]] = 0. \tag{4.19}$$

We first write

$$[T_2, [T_1, q]] = -[T_2, [q, T_1]] \tag{4.20}$$

so that the two double commutators in (4.18) become

$$[T_1, [T_2, q]] - [T_2, [T_1, q]] = [T_1, [T_2, q]] + [T_2, [q, T_1]] = -[q, [T_1, T_2]], \tag{4.21}$$

where the last step follows by use of (4.19). Finally, then, (4.18) can be written as

$$(\delta_{\epsilon_1}\delta_{\epsilon_2} - \delta_{\epsilon_2}\delta_{\epsilon_1})q = -\epsilon_1\epsilon_2[[T_1, T_2], q], \tag{4.22}$$

which can be compared with (4.15). We deduce

$$[T_1, T_2] = \mathrm{i}T_3 \tag{4.23}$$

exactly as stated in (4.1).

This is the method we shall use to find the SUSY algebra, at least as far as it concerns the transformations for scalar and spinor fields found in Chapter 3.

4.2 Supersymmetry generators ('charges') and their algebra

In order to apply the preceding method, we need the SUSY analogue of (4.9). Equations (3.5) and (3.10) (with $A = -1$) provide us with the analogue of the second equality in (4.9), for $\delta_\xi\phi$ and for $\delta_\xi\chi$; what about the first? We want to write something like

$$\delta_\xi \phi \sim \mathrm{i}[\xi Q, \phi] = \xi^{\mathrm{T}}(-\mathrm{i}\sigma_2)\chi, \tag{4.24}$$

where Q is a SUSY generator. In the first (tentative) equality in (4.24), we must remember that ξ is a χ-type spinor quantity, and so it is clear that Q must be a spinor quantity also, or else one side of the equality would be bosonic and the other fermionic. In fact, since ϕ is a Lorentz scalar, we must combine ξ and Q into a Lorentz invariant. Let us suppose that Q transforms as a χ-type spinor also: then we know that $\xi^{\mathrm{T}}(-\mathrm{i}\sigma_2)Q$ is Lorentz invariant. So we shall write

$$\delta_\xi \phi = \mathrm{i}[\xi^{\mathrm{T}}(-\mathrm{i}\sigma_2)Q, \phi] = \xi^{\mathrm{T}}(-\mathrm{i}\sigma_2)\chi \tag{4.25}$$

or in the faster notation of Section 2.3

$$\delta_\xi \phi = \mathrm{i}[\xi \cdot Q, \phi] = \xi \cdot \chi . \tag{4.26}$$

We are going to calculate $(\delta_\eta \delta_\xi - \delta_\xi \delta_\eta)\phi$, so (since $\delta\phi \sim \chi$) we shall need (3.10) as well. This involves ξ^*, so to get the complete analogue of 'i$\epsilon \cdot \boldsymbol{T}$' we shall need to extend 'i$\xi \cdot Q$' to

$$\mathrm{i}(\xi^{\mathrm{T}}(-\mathrm{i}\sigma_2)Q + \xi^\dagger(\mathrm{i}\sigma_2)Q^*) = \mathrm{i}(\xi \cdot Q + \bar{\xi} \cdot \bar{Q}) . \tag{4.27}$$

We first calculate $(\delta_\eta \delta_\xi - \delta_\xi \delta_\eta)\phi$ using (3.5) and (3.10) (with $A = -1$):

$$\begin{aligned}(\delta_\eta \delta_\xi - \delta_\xi \delta_\eta)\phi &= \delta_\eta(\xi^{\mathrm{T}}(-\mathrm{i}\sigma_2\chi) - (\eta \leftrightarrow \xi)\\ &= \xi^{\mathrm{T}}(-\mathrm{i}\sigma_2)\mathrm{i}\sigma^\mu(-\mathrm{i}\sigma_2)\eta^*\partial_\mu\phi - (\eta \leftrightarrow \xi)\\ &= (\xi^{\mathrm{T}}c\sigma^\mu c\eta^* - \eta^{\mathrm{T}}c\sigma^\mu c\xi^*)\mathrm{i}\partial_\mu\phi,\end{aligned} \tag{4.28}$$

where we have introduced the notation

$$c \equiv -\mathrm{i}\sigma_2 = \begin{pmatrix} 0 & -1 \\ 1 & 0 \end{pmatrix}. \tag{4.29}$$

(4.28) can be written more compactly by using (see equation (2.83))

$$c\sigma^\mu c = -\bar{\sigma}^{\mu\mathrm{T}}. \tag{4.30}$$

Note that $\xi^{\mathrm{T}}\bar{\sigma}^{\mu\mathrm{T}}\eta^*$ is a single quantity (row vector times matrix times column vector) so it must equal its formal transpose, apart from a minus sign due to interchanging the order of anticommuting variables.[1] Hence

$$(\delta_\eta \delta_\xi - \delta_\xi \delta_\eta)\phi = (\eta^\dagger\bar{\sigma}^\mu\xi - \xi^\dagger\bar{\sigma}^\mu\eta)\mathrm{i}\partial_\mu\phi. \tag{4.31}$$

[1] Check this statement by looking at $(\eta^{\mathrm{T}}(-\mathrm{i}\sigma_2)\xi)^{\mathrm{T}}$, for instance.

Just to keep our 'Majorana' hand in, we note that (4.31) is simply

$$(\delta_\eta \delta_\xi - \delta_\xi \delta_\eta)\phi = \bar{\Psi}^\eta_{\rm M} \gamma^\mu \Psi^\xi_{\rm M} {\rm i}\partial_\mu \phi, \tag{4.32}$$

using (2.113).

On the other hand, we also have

$$\delta_\xi \phi = {\rm i}[\xi \cdot Q + \bar{\xi} \cdot \bar{Q}, \phi] \tag{4.33}$$

and so

$$(\delta_\eta \delta_\xi - \delta_\xi \delta_\eta)\phi = -\{[\eta \cdot Q + \bar{\eta} \cdot \bar{Q}, [\xi \cdot Q + \bar{\xi} \cdot \bar{Q}, \phi]] - [\xi \cdot Q + \bar{\xi} \cdot \bar{Q}, [\eta \cdot Q + \bar{\eta} \cdot \bar{Q}, \phi]]\}. \tag{4.34}$$

Just as in (4.21), the right-hand side of (4.34) can be rearranged using (4.19) and we obtain

$$\begin{aligned}[][\eta \cdot Q + \bar{\eta} \cdot \bar{Q}, \xi \cdot Q + \bar{\xi} \cdot \bar{Q}], \phi] &= (\eta^{\rm T} c \sigma^\mu c \xi^* - \xi^{\rm T} c \sigma^\mu c \eta^*){\rm i}\partial_\mu \phi \\ &= -(\eta^{\rm T} c \sigma^\mu c \xi^* - \xi^{\rm T} c \sigma^\mu c \eta^*)[P_\mu, \phi]\end{aligned} \tag{4.35}$$

where in the last step we have introduced the 4-momentum operator P_μ, which is also the generator of translations, such that

$$[P_\mu, \phi] = -{\rm i}\partial_\mu \phi \tag{4.36}$$

(we shall recall the proof of this equation in Chapter 6, see (6.9)).

It is *tempting* now to conclude that, just as in going from (4.15) and (4.22) to (4.23), we can infer from (4.35) the result

$$[\eta \cdot Q + \bar{\eta} \cdot \bar{Q}, \xi \cdot Q + \bar{\xi} \cdot \bar{Q}] = -(\eta^{\rm T} c \sigma^\mu c \xi^* - \xi^{\rm T} c \sigma^\mu c \eta^*) P_\mu. \tag{4.37}$$

However, for (4.37) to be true as an operator relation, it must hold when applied to all fields in the supermultiplet. But we have, so far, only established the right-hand side of (4.35) by considering the difference $\delta_\eta \delta_\xi - \delta_\xi \delta_\eta$ acting on ϕ (see (4.28)). Is it also true that

$$(\delta_\eta \delta_\xi - \delta_\xi \delta_\eta)\chi = (\xi^{\rm T} c \sigma^\mu c \eta^* - \eta^{\rm T} c \sigma^\mu c \xi^*){\rm i}\partial_\mu \chi \ ? \tag{4.38}$$

Unfortunately, the answer to this is *no*, as we shall see in Section 4.5, where we shall also learn how to repair the situation. For the moment, we proceed on the basis of (4.37).

In order to obtain, finally, the (anti)commutation relations of the Q's from (4.37), we need to get rid of the parameters η and ξ on both sides. First of all, we note that since the right-hand side of (4.37) has no terms in $\eta \ldots \xi$ or $\eta^* \ldots \xi^*$ we can deduce

$$[\eta \cdot Q, \xi \cdot Q] = [\bar{\eta} \cdot \bar{Q}, \bar{\xi} \cdot \bar{Q}] = 0. \tag{4.39}$$

The first commutator is

$$\begin{aligned} 0 &= (\eta^1 Q_1 + \eta^2 Q_2)(\xi^1 Q_1 + \xi^2 Q_2) - (\xi^1 Q_1 + \xi^2 Q_2)(\eta^1 Q_1 + \eta^2 Q_2) \\ &= -\eta^1\xi^1(2Q_1Q_1) - \eta^1\xi^2(Q_1Q_2 + Q_2Q_1) \\ &\quad - \eta^2\xi^1(Q_2Q_1 + Q_1Q_2) - \eta^2\xi^2(2Q_2Q_2), \end{aligned} \tag{4.40}$$

remembering that all quantities anticommute. Since all these combinations of parameters are independent, we can deduce

$$\{Q_a, Q_b\} = 0, \tag{4.41}$$

and similarly

$$\{Q_a^*, Q_b^*\} = 0. \tag{4.42}$$

Notice how, when the anticommuting quantities ξ and η are 'stripped away' from the Q and $\bar{Q}$, the commutators in (4.39) become *anti*commutators in (4.41) and (4.42).

Now let's look at the $[\eta \cdot Q, \bar{\xi} \cdot \bar{Q}]$ term in (4.37). Writing everything out longhand, we have

$$\bar{\xi} \cdot \bar{Q} = \xi^\dagger \mathrm{i}\sigma_2 Q^* = \xi_1^* Q_2^* - \xi_2^* Q_1^* \tag{4.43}$$

and

$$\eta \cdot Q = -\eta_1 Q_2 + \eta_2 Q_1. \tag{4.44}$$

So

$$\begin{aligned} [\eta \cdot Q, \bar{\xi} \cdot \bar{Q}] = \eta_1\xi_1^*(Q_2Q_2^* + Q_2^*Q_2) - \eta_1\xi_2^*(Q_2Q_1^* + Q_1^*Q_2) \\ - \eta_2\xi_1^*(Q_1Q_2^* + Q_2^*Q_1) + \eta_2\xi_2^*(Q_1Q_1^* + Q_1^*Q_1). \end{aligned} \tag{4.45}$$

Meanwhile, the right-hand side of (4.37) is

$$\begin{aligned} &-(\eta_1\eta_2)\begin{pmatrix} 0 & -1 \\ 1 & 0 \end{pmatrix} \sigma^\mu \begin{pmatrix} 0 & -1 \\ 1 & 0 \end{pmatrix} \begin{pmatrix} \xi_1^* \\ \xi_2^* \end{pmatrix} P_\mu \\ &= -(\eta_2 - \eta_1)\sigma^\mu \begin{pmatrix} -\xi_2^* \\ \xi_1^* \end{pmatrix} P_\mu \\ &= [\eta_2\xi_2^*(\sigma^\mu)_{11} - \eta_2\xi_1^*(\sigma^\mu)_{12} - \eta_1\xi_2^*(\sigma^\mu)_{21} + \eta_1\xi_1^*(\sigma^\mu)_{22}]\, P_\mu, \end{aligned} \tag{4.46}$$

where the subscripts on the matrices σ^μ denote the particular element of the matrix, as usual. Comparing (4.45) and (4.46) we deduce

$$\{Q_a, Q_b^*\} = (\sigma^\mu)_{ab} P_\mu. \tag{4.47}$$

We have been writing Q^* throughout, like ξ^* and η^*, but the Q's are quantum field operators and so (in accord with the discussion in Section 3.1) we should more properly write (4.47) as

$$\{Q_a, Q_b^\dagger\} = (\sigma^\mu)_{ab} P_\mu. \tag{4.48}$$

Once again, the commutator in (4.45) has led to an anticommutator in (4.48).

Equation (4.48) is the main result of this section, and is a most important equation; it provides the 'proper' version of (1.34). Although we have derived it by our customary brute force methods as applied to a particular (and very simple) case, it must be emphasized that equation (4.48) is indeed the correct SUSY algebra (up to normalization conventions[2]). Equation (4.48) shows (to repeat what was said in Section 1.3) that the SUSY generators are directly connected to the energy-momentum operator, which is the generator of translations in space–time. So it is justified to regard SUSY as some kind of extension of space–time symmetry, the Q's generating 'supertranslations'. We shall see further aspects of this in Chapter 6.

The foregoing results can easily be written in the more sophisticated notation of Section 2.3. In parallel with equation (2.77) we can define

$$\bar{Q}_{\dot{a}} \equiv Q_a^\dagger. \tag{4.49}$$

Then (4.42) is just

$$\{\bar{Q}_{\dot{a}}, \bar{Q}_{\dot{b}}\} = 0, \tag{4.50}$$

while (4.48) becomes

$$\{Q_a, \bar{Q}_{\dot{b}}\} = (\sigma^\mu)_{a\dot{b}} P_\mu. \tag{4.51}$$

Note that the indices of σ^μ follow correctly the convention mentioned after equation (2.80).

The SUSY algebra can also be written in Majorana form. Just as we can construct a 4-component Majorana spinor from a χ (or of course a ψ), so we can make a 4-component Majorana spinor charge Q_{M} from our L-type spinor charge Q, by

[2] Many authors normalize the SUSY charges differently, so that they get a '2' on the right-hand side. For completeness, we take the opportunity of this footnote to mention that more general SUSY algebras also exist, in which the single generator Q_a is replaced by N generators $Q_a^A (A = 1, 2, \ldots, N)$. Equation (4.48) is then replaced by $\{Q_a^A, Q_b^{B\dagger}\} = \delta^{AB}(\sigma^\mu)_{ab} P_\mu$. The more significant change occurs in the anticommutator (4.41), which becomes $\{Q_a^A, Q_b^B\} = \epsilon_{ab} Z^{AB}$, where $\epsilon_{12} = -1, \epsilon_{21} = +1, \epsilon_{11} = \epsilon_{22} = 0$ and the 'central charge' Z^{AB} is antisymmetric under $A \leftrightarrow B$. The reason why only the $N = 1$ case seems to have any immediate physical relevance will be explained at the end of Section 4.4.

setting (c.f. (2.100))

$$Q_{\mathrm{M}} = \begin{pmatrix} \mathrm{i}\sigma_2 Q^{\dagger\mathrm{T}} \\ Q \end{pmatrix} = \begin{pmatrix} Q_2^\dagger \\ -Q_1^\dagger \\ Q_1 \\ Q_2 \end{pmatrix}. \tag{4.52}$$

Let us call the components of this $Q_{\mathrm{M}\alpha}$, so that $Q_{\mathrm{M}1} = Q_2^\dagger$, $Q_{\mathrm{M}2} = -Q_1^\dagger$, etc. It is not completely obvious what the anticommutation relations of the $Q_{\mathrm{M}\alpha}$'s ought to be (given those of the Q_a's and $Q_a^\dagger$'s), but the answer turns out to be

$$\{Q_{\mathrm{M}\alpha}, Q_{\mathrm{M}\beta}\} = (\gamma^\mu(\mathrm{i}\gamma^2\gamma^0))_{\alpha\beta} P_\mu, \tag{4.53}$$

as can be checked with the help of (4.41), (4.42), (4.48) and (4.52). Note that '$-\mathrm{i}\gamma^2\gamma^0$' is the 'metric' we met in Section 2.5. The anticommutator (4.53) can be re-written rather more suggestively as

$$\{Q_{\mathrm{M}\alpha}, \bar{Q}_{\mathrm{M}\beta}\} = (\gamma^\mu)_{\alpha\beta} P_\mu \tag{4.54}$$

where (compare (2.111))

$$\bar{Q}_{\mathrm{M}\beta} = \left(Q_{\mathrm{M}}^{\mathrm{T}}(-\mathrm{i}\gamma^2\gamma^0)\right)_\beta = (Q_{\mathrm{M}}^\dagger\gamma^0)_\beta. \tag{4.55}$$

We note finally that the commutator of two P's is zero (translations commute), and that the commutator of a Q and a P also vanishes, since the Q's are independent of x:

$$[Q_a, P_\mu] = [Q_a^\dagger, P_\mu] = 0. \tag{4.56}$$

So all the commutation or anticommutation relations between Q's, $Q^\dagger$'s, and P's are now defined, and they involve only these quantities; we say that 'the supertranslation algebra is closed'.

4.3 The supersymmetry current

In the case of ordinary symmetries, the invariance of a Lagrangian under a transformation of the fields (characterized by certain parameters) implies the existence of a 4-vector j^μ (the 'symmetry current'), which is conserved: $\partial_\mu j^\mu = 0$. The generator of the symmetry is the 'charge' associated with this current, namely the spatial integral of j^0. An expression for j^μ is easily found (see, for example, [7], Section 12.3.1). Suppose the Lagrangian $\mathcal{L}$ is invariant under the transformation

$$\phi_r \to \phi_r + \delta\phi_r \tag{4.57}$$

where 'ϕ_r' stands generically for any field in $\mathcal{L}$, having several components labelled by r. Then

$$0 = \delta\mathcal{L} = \frac{\partial\mathcal{L}}{\partial\phi_r}\delta\phi_r + \frac{\partial\mathcal{L}}{\partial(\partial^\mu\phi_r)}\partial^\mu(\delta\phi_r) + \text{hermitian conjugate.} \tag{4.58}$$

But the equation of motion for ϕ_r is

$$\frac{\partial\mathcal{L}}{\partial\phi_r} = \partial_\mu\left(\frac{\partial\mathcal{L}}{\partial(\partial_\mu\phi_r)}\right). \tag{4.59}$$

Using (4.59) in (4.58) yields

$$\partial_\mu j^\mu = 0 \tag{4.60}$$

where

$$j^\mu = \frac{\partial\mathcal{L}}{\partial(\partial_\mu\phi_r)}\delta\phi_r + \text{hermitian conjugate.} \tag{4.61}$$

For example, consider the Lagrangian

$$\mathcal{L} = \bar{q}(\mathrm{i}\not{\partial} - m)q \tag{4.62}$$

where

$$q = \begin{pmatrix} u \\ d \end{pmatrix}. \tag{4.63}$$

This is invariant under the SU(2) transformation (4.4), which is characterized by three independent infinitesimal parameters, so there are three independent symmetries, three currents, and three generators (or charges). Consider for instance a transformation involving ϵ_1 alone. Then

$$\delta q = -\mathrm{i}\epsilon_1(\tau_1/2)q, \tag{4.64}$$

while from (4.62) we have

$$\frac{\partial\mathcal{L}}{\partial(\partial_\mu q)} = \bar{q}\,\mathrm{i}\gamma^\mu. \tag{4.65}$$

Hence from (4.61) and (4.64) we obtain the corresponding current as

$$\epsilon_1\bar{q}\gamma^\mu(\tau_1/2)q. \tag{4.66}$$

Clearly the constant factor ϵ_1 is irrelevant and can be dropped. Repeating the same steps for transformations associated with ϵ_2 and ϵ_3 we deduce the existence of the *isospin currents*

$$\boldsymbol{j}^\mu = \bar{q}\gamma^\mu(\boldsymbol{\tau}/2)q \tag{4.67}$$

and charges (generators)

$$\boldsymbol{T} = \int q^{\dagger}(\boldsymbol{\tau}/2)q \, \mathrm{d}^3\boldsymbol{x} \tag{4.68}$$

just as stated in (4.10).

We can apply the same procedure to find the *supersymmetry current* associated with the supersymmetry exhibited by the simple model considered in Section 3.1. However, there is an important difference between this example and the SU(2) model just considered: in the latter, the Lagrangian is indeed invariant under the transformation (4.3), but in the SUSY case we were only able to ensure that the Action was invariant, the Lagrangian changing by a total derivative, as given in (3.34) or (3.35). In this case, the '0' on the left-hand side of (4.58) must be replaced by $\partial_\mu K^\mu$ say, where K^μ is the expression in brackets in (3.34) or (3.35).

Furthermore, since the SUSY charges are spinors Q_a, we anticipate that the associated currents carry a spinor index too, so we write them as J_a^μ, where a is a spinor index. These will be associated with transformations characterized by the usual spinor parameters ξ. Similarly, there will be the hermitian conjugate currents associated with the parameters ξ^*.

Altogether, then, we can write (forming Lorentz invariants in the now familiar way)

$$\begin{aligned}
\xi^{\mathrm{T}}(-\mathrm{i}\sigma_2)J^\mu + \xi^{\dagger}\mathrm{i}\sigma_2 J^{\mu *} &= \frac{\partial \mathcal{L}}{\partial(\partial_\mu \phi)}\delta\phi + \delta\phi^{\dagger}\frac{\partial \mathcal{L}}{\partial(\partial_\mu \phi^{\dagger})} + \frac{\partial \mathcal{L}}{\partial(\partial_\mu \chi)}\delta\chi - K^\mu \\
&= \partial^\mu\phi^{\dagger}\xi^{\mathrm{T}}(-\mathrm{i}\sigma_2)\chi + \chi^{\dagger}\mathrm{i}\sigma_2\xi^*\partial^\mu\phi + \chi^{\dagger}\mathrm{i}\bar{\sigma}^\mu(-\mathrm{i}\sigma^\nu)\mathrm{i}\sigma_2\xi^*\partial_\nu\phi \\
&\quad - (\chi^{\dagger}\mathrm{i}\sigma_2\xi^*\partial^\mu\phi + \xi^{\mathrm{T}}\mathrm{i}\sigma_2\sigma^\nu\bar{\sigma}^\mu\chi\partial_\nu\phi^{\dagger} + \xi^{\mathrm{T}}(-\mathrm{i}\sigma_2)\chi\partial^\mu\phi^{\dagger}) \\
&= \chi^{\dagger}\bar{\sigma}^\mu\sigma^\nu\mathrm{i}\sigma_2\xi^*\partial_\nu\phi + \xi^{\mathrm{T}}(-\mathrm{i}\sigma_2)\sigma^\nu\bar{\sigma}^\mu\chi\partial_\nu\phi^{\dagger},
\end{aligned} \tag{4.69}$$

whence we read off the SUSY current as

$$J^\mu = \sigma^\nu\bar{\sigma}^\mu\chi\partial_\nu\phi^{\dagger}. \tag{4.70}$$

As expected, this current has two spinorial components, and it contains an unpaired fermionic operator χ.

Exercise 4.1 The supersymmetry charges (generators) are given by the spatial integral of the $\mu = 0$ component of the supersymmetry current (4.70), so that

$$Q_a = \int (\sigma^\nu\chi(y))_a \, \partial_\nu\phi^{\dagger}(y)\mathrm{d}^3\boldsymbol{y}. \tag{4.71}$$

Verify that these charges do indeed generate the required transformations of the fields, namely

(a)

$$\mathrm{i}[\xi\cdot Q, \phi(x)] = \xi\cdot\chi(x) \tag{4.72}$$

(you will need to use the bosonic equal-time commutation relations

$$[\phi(\boldsymbol{x}, t), \dot{\phi}^{\dagger}(\boldsymbol{y}, t)] = \mathrm{i}\delta^3(\boldsymbol{x} - \boldsymbol{y})), \tag{4.73}$$

and

(b) $$\mathrm{i}[\xi \cdot Q + \bar{\xi} \cdot \bar{Q}, \chi(x)] = -\mathrm{i}\sigma^{\mu}(\mathrm{i}\sigma_2\xi^*)\partial_{\mu}\phi(x) \tag{4.74}$$

(you will need the fermionic anti-commutation relations

$$\{\chi_a(\boldsymbol{x}, t), \chi_b^{\dagger}(\boldsymbol{y}, t)\} = \delta_{ab}\delta^3(\boldsymbol{x} - \boldsymbol{y}).) \tag{4.75}$$

4.4 Supermultiplets

We proceed to extract the physical consequences of (4.41), (4.42), (4.48) and (4.56). First, note from (4.56) that the operator P^2 commutes with all the generators Q_a, so that states in a supermultiplet, which are connected to each other by the action of the generators, must all have the same mass (and, more generally, the same 4-momentum). However, since the Q_a's are spinor operators, the action of a Q_a or $Q_a^{\dagger}$ on one state of spin j will produce a state with a spin differing from j by $\frac{1}{2}$. In fact, we know that under rotations (compare equation (4.9) for the case of isospin rotations, and equations (6.8) and (6.10) below for spatial translations)

$$\delta Q = -(\mathrm{i}\boldsymbol{\epsilon} \cdot \boldsymbol{\sigma}/2)Q = \mathrm{i}\boldsymbol{\epsilon} \cdot [\boldsymbol{J}, Q], \tag{4.76}$$

where the J's are the generators of rotations (i.e. angular momentum operators). For example, for a rotation about the 3-axis,

$$-\frac{1}{2}\sigma_3 Q = [J_3, Q], \tag{4.77}$$

which implies that

$$[J_3, Q_1] = -\frac{1}{2}Q_1, \quad [J_3, Q_2] = \frac{1}{2}Q_2. \tag{4.78}$$

It follows that if $|jm\rangle$ is a spin-j state with $J_3 = m$, then

$$(J_3 Q_1 - Q_1 J_3)|jm\rangle = -\frac{1}{2}Q_1|jm\rangle, \tag{4.79}$$

whence

$$J_3(Q_1|jm\rangle) = \left(m - \frac{1}{2}\right) Q_1|jm\rangle, \tag{4.80}$$

showing that $Q_1|jm\rangle$ has $J_3 = m - \frac{1}{2}$ – that is, Q_1 lowers the m-value by $\frac{1}{2}$ (like an annihilation operator for a 'u'-state). Similarly, Q_2 raises it by $\frac{1}{2}$ (like an annihilation operator for a 'd'-state). Likewise, since

$$[J_3, Q_1^\dagger] = \frac{1}{2} Q_1^\dagger, \tag{4.81}$$

we find that $Q_1^\dagger$ raises the m-value by $\frac{1}{2}$; and by the same token $Q_2^\dagger$ lowers it by $\frac{1}{2}$.

We now want to find the nature of the states that are 'connected' to each other by the application of the operators Q_a and $Q_a^\dagger$, that is, the analogue of the $(2j + 1)$-fold multiplet structure familiar in angular momentum theory. Our states will be labelled as $|p, \lambda\rangle$, where we take the 4-momentum eigenvalue to be $p = (E, 0, 0, E)$ since the fields are massless, and where λ is a helicity label, equivalent here to the eigenvalue of J_3. Let $|p, -j\rangle$ be a normalized eigenstate of J_3 with eigenvalue $\lambda = -j$, the minimum possible value of λ for given j. Then we must have

$$Q_2^\dagger |p, -j\rangle = 0 = Q_1 |p, -j\rangle. \tag{4.82}$$

This leaves only two states connected to $|p, -j\rangle$ by the SUSY generators, namely $Q_1^\dagger |p, -j\rangle$ and $Q_2 |p, -j\rangle$. The first of these must vanish. This follows by considering the SUSY algebra (4.48) with $a = b = 1$:

$$Q_1^\dagger Q_1 + Q_1 Q_1^\dagger = (\sigma^\mu)_{11} P_\mu. \tag{4.83}$$

The only components of σ^μ which have a non-vanishing '11' entry are $(\sigma^0)_{11} = 1$ and $(\sigma^3)_{11} = 1$, so we have

$$Q_1^\dagger Q_1 + Q_1 Q_1^\dagger = P_0 + P_3 = P^0 - P^3. \tag{4.84}$$

Hence, taking the expectation value in the state $|p, -j\rangle$, we find

$$\langle p, -j | Q_1^\dagger Q_1 + Q_1 Q_1^\dagger | p, -j\rangle = 0 \tag{4.85}$$

since the eigenvalue of $P^0 - P^3$ vanishes in this state. But from (4.82) we have $\langle p, -j | Q_1^\dagger = 0$, and hence we deduce that

$$\langle p, -j | Q_1 Q_1^\dagger | p, -j\rangle = 0. \tag{4.86}$$

It follows that either the state $Q_1^\dagger |p, -j\rangle$ has zero norm (which is not an acceptable state), or that

$$Q_1^\dagger |p, -j\rangle = 0, \tag{4.87}$$

as claimed.

This leaves just one state connected to our starting state, namely

$$Q_2 |p, -j\rangle. \tag{4.88}$$

We know that Q_2 raises λ by $1/2$, and hence

$$Q_2|p,-j\rangle \propto \left|p,-j+\frac{1}{2}\right\rangle. \tag{4.89}$$

Consider now the action of the generators on this new state $|p,-j+\frac{1}{2}\rangle$, which is proportional to $Q_2|p,-j\rangle$. Obviously, the application of Q_2 to it gives zero, since $Q_2Q_2=0$ from (4.41). Next, note that

$$Q_1Q_2|p-j\rangle = -Q_2Q_1|p,-j\rangle = 0, \tag{4.90}$$

using (4.41) and (4.82). Now consider

$$Q_1^\dagger Q_2|p,-j\rangle. \tag{4.91}$$

Given the chosen momentum eigenvalue, the $a=1, b=2$ element of (4.48) gives $Q_1Q_2^\dagger = -Q_2^\dagger Q_1$, and hence

$$Q_1^\dagger Q_2|p,-j\rangle = -Q_2Q_1^\dagger|p,-j\rangle = 0, \tag{4.92}$$

using (4.87). We are left with only $Q_2^\dagger$ to apply to $|p,-j+\frac{1}{2}\rangle$. This in fact just takes us back to the state we started from:

$$Q_2^\dagger\left|p,-j+\frac{1}{2}\right\rangle \propto Q_2^\dagger Q_2|p,-j\rangle \propto (2E - Q_2Q_2^\dagger)|p,-j\rangle \propto |p,-j\rangle, \tag{4.93}$$

where we have used (4.48) with $a=b=2$. So there are just two states in a supermultiplet of massless particles, one with helicity $-j$ and the other with helicity $-j+\frac{1}{2}$. However, any local Lorentz invariant quantum field theory must be invariant under the combined operation of **TCP**: this implies that for every supermultiplet of massless particle states with helicities $-j$ and $-j+\frac{1}{2}$ there must be a corresponding supermultiplet of massless antiparticle states with helicities j and $j-\frac{1}{2}$.[3]

Consider the case $j=\frac{1}{2}$. Then we have one supermultiplet consisting of the two states $|p,\lambda=-\frac{1}{2}\rangle$ and $|p,\lambda=0\rangle$. The second of these states cannot be associated with spin 1, since there is no $\lambda=0$ state for a massless spin-1 particle. Hence it must be a spin-0 state. This is, in fact, the *left chiral supermultiplet*, containing a massless left-handed spin-$\frac{1}{2}$ state and a massless scalar state. The corresponding fields are an L-type Weyl fermion χ and a complex scalar ϕ, as in the simple model of Section 3.1.

As already indicated in Section 3.2, we may assign the fermions of the SM to chiral supermultiplets, partnered by suitable squarks and sleptons. Consider, for example, the electron and neutrino states. The left-handed components form an

[3] We could equally well have started with the state of maximum helicity for given j, namely $|p,\lambda=j\rangle$, ending up with the supermultiplet $|p,\lambda=j\rangle$, $|p,\lambda=j-\frac{1}{2}\rangle$, and their **TCP**-conjugates.

$SU(2)_L$ doublet, which is partnered by a corresponding doublet of scalars (denoted by the same symbol as the SM states, but with a tilde), in left chiral supermultiplets:

$$\begin{pmatrix} \nu_{e\,L} \\ e_L \end{pmatrix} \quad \text{and} \quad \begin{pmatrix} \tilde{\nu}_{e\,L} \\ \tilde{e}_L \end{pmatrix}. \tag{4.94}$$

TCP-invariance of the Lagrangian guarantees the inclusion of the antiparticle states

$$\begin{pmatrix} \bar{e}_R \\ \bar{\nu}_{e\,R} \end{pmatrix} \quad \text{and} \quad \begin{pmatrix} \tilde{\bar{e}}_R \\ \tilde{\bar{\nu}}_{e\,R} \end{pmatrix}. \tag{4.95}$$

Note that the 'R' or 'L' label on the sleptons doesn't refer to their chirality (they are spinless) but rather to that of their superpartners. The right-handed component e_R is an $SU(2)_L$ singlet, however, and so cannot be partnered by the doublet selectron state $\tilde{e}_L$ introduced above. Instead, it is partnered by a new selectron state $\tilde{e}_R$, forming a right chiral supermultiplet:

$$e_R \quad \text{and} \quad \tilde{e}_R. \tag{4.96}$$

The corresponding antiparticle states

$$\bar{e}_L \quad \text{and} \quad \tilde{\bar{e}}_L \tag{4.97}$$

form a left chiral supermultiplet. Similar assignments are made for the other SM fermions.

In constructing the MSSM Lagrangian (see Section 8.1) it is conventional to describe all the SM fermions by L-type Weyl spinor fields. Thus for the electron we shall use fields χ_e (which destroys e_L and creates $\bar{e}_R$) and $\chi_{\bar{e}}$ (which destroys $\bar{e}_L$ and creates e_R). As we saw in Section 2.4, the R-type field ψ_e which destroys e_R and creates $\bar{e}_L$ is given in terms of $\chi_{\bar{e}}$ by $\psi_e = \mathrm{i}\sigma_2 \chi_{\bar{e}}^*$. For the accompanying selectron fields, we use $\tilde{e}_L$ (which destroys $\tilde{e}_L$ and creates $\tilde{\bar{e}}_R$) and $\tilde{\bar{e}}_L$ (which destroys $\tilde{\bar{e}}_L$ and creates $\tilde{e}_R$). Bearing in mind that $\tilde{\bar{e}}_L$ is the super-partner of the antiparticle of e_R, the field $\tilde{\bar{e}}_L$ can equally well be denoted by $\tilde{e}_R^\dagger$ (which creates $\tilde{e}_R$ and destroys the superpartner of the antiparticle of e_R). The other slepton and squark fields are treated similarly.

In the case $j = 1$, the supermultiplet consists of two states $|p, \lambda = -1\rangle$ and $|p, \lambda = -\frac{1}{2}\rangle$, which pairs a massless spin-1 state with a massless left-handed spin-$\frac{1}{2}$ state. This is the *vector*, or *gauge supermultiplet*. In terms of fields, the supermultiplet contains a massless gauge field and a massless Weyl spinor. The gauge bosons of the SM are assigned to gauge supermultiplets. In this case, **TCP**-invariance guarantees the appearance of both helicities, while the antiparticle states are contained in the same (adjoint) representation of the gauge group as the particle states (for example, the antiparticle of the W^+ is the W^-). In the MSSM, the SM gauge bosons are partnered by massless Weyl spinor states, also in the adjoint representation of the gauge group. When interactions are included, we arrive at supersymmetrized versions of the SM gauge theories (see Chapter 7).

This is an appropriate point to explain why only $N = 1$ SUSY (see the preceding footnote) has been considered. The reason is that in $N = 2$ SUSY the corresponding chiral multiplet contains four states: $\lambda = +\frac{1}{2}$, $\lambda = -\frac{1}{2}$ and two states with $\lambda = 0$. The phenomenological problem with this is that the R ($\lambda = \frac{1}{2}$) and L ($\lambda = -\frac{1}{2}$) states must transform in the same way under any gauge symmetry (similar remarks hold for all $N \geq 1$ supermultiplets). But we know that the $\mathrm{SU}(2)_\mathrm{L}$ gauge symmetry of the SM treats the L and R components of quark and lepton fields differently. So if we want to make a SUSY extension of the SM, it can only be the simple $N = 1$ SUSY, where we are free to treat the left chiral supermultiplet ($\lambda = -\frac{1}{2}$, $\lambda = 0$) differently from the right chiral supermultiplet ($\lambda = 0$, $\lambda = +\frac{1}{2}$). Further details of the representations for $N \geq 1$ are given in [42] Section 1.6, for example.

One other case of possible physical interest is the *gravity supermultiplet*, containing a spin-2 graviton state with $\lambda = -2$ and a spin-$\frac{3}{2}$ gravitino state with $\lambda = -\frac{3}{2}$. The interacting theory here is *supergravity*, which however lies beyond our scope.

We must now take up an issue raised after (4.36).

4.5 A snag, and the need for a significant complication

In Section 4.2 we arrived at the SUSY algebra by calculating the difference $\delta_\eta \delta_\xi - \delta_\xi \delta_\eta$ two different ways. We explicitly evaluated this difference as applied to ϕ, but in deducing the operator relation (4.37), it is crucial that a consistent result be obtained when $\delta_\eta \delta_\xi - \delta_\xi \delta_\eta$ is applied to χ. In fact, as noted after (4.38), this is not the case, as we now show. This will necessitate a significant modification of the SUSY transformations given so far, in order to bring about this desired consistency.

Consider first $\delta_\eta \delta_\xi \chi_a$, where we are indicating the spinor component explicitly:

$$\begin{aligned}\delta_\eta \delta_\xi \chi_a &= \delta_\eta(-\mathrm{i}\sigma^\mu(\mathrm{i}\sigma_2 \xi^*))_a \partial_\mu \phi \\ &= (\mathrm{i}\sigma^\mu(-\mathrm{i}\sigma_2 \xi^*))_a \partial_\mu \delta_\eta \phi \\ &= (\mathrm{i}\sigma^\mu(-\mathrm{i}\sigma_2 \xi^*))_a (\eta^\mathrm{T}(-\mathrm{i}\sigma_2)\partial_\mu \chi). \end{aligned} \tag{4.98}$$

There is an important identity involving products of three spinors, which we can use to simplify (4.98). The identity reads, for any three spinors λ, ζ and ρ,

$$\lambda_a(\zeta^\mathrm{T}(-\mathrm{i}\sigma_2)\rho) + \zeta_a(\rho^\mathrm{T}(-\mathrm{i}\sigma_2)\lambda) + \rho_a(\lambda^\mathrm{T}(-\mathrm{i}\sigma_2)\zeta) = 0, \tag{4.99}$$

or in the faster notation

$$\lambda_a(\zeta \cdot \rho) + \zeta_a(\rho \cdot \lambda) + \rho_a(\lambda \cdot \zeta) = 0. \tag{4.100}$$

Exercise 4.2 Check the identity (4.99).

We take, in (4.99),

$$\lambda_a = (\sigma^\mu(-\mathrm{i}\sigma_2)\xi^*)_a, \quad \zeta_a = \eta_a, \quad \rho_a = \partial_\mu \chi_a. \tag{4.101}$$

The right-hand side of (4.98) is then equal to

$$-\mathrm{i}\{\eta_a \partial_\mu \chi^{\mathrm{T}}(-\mathrm{i}\sigma_2)\sigma^\mu(-\mathrm{i}\sigma_2)\xi^* + \partial_\mu \chi_a(\sigma^\mu(-\mathrm{i}\sigma_2\xi^*))^{\mathrm{T}}(-\mathrm{i}\sigma_2)\eta\}. \tag{4.102}$$

But we know from (4.30) that the first term in (4.102) can be written as

$$\mathrm{i}\eta_a(\partial_\mu \chi^{\mathrm{T}}\bar{\sigma}^{\mu\mathrm{T}}\xi^*) = -\mathrm{i}\eta_a(\xi^\dagger\bar{\sigma}^\mu\partial_\mu\chi), \tag{4.103}$$

where to reach the second equality in (4.103) we have taken the formal transpose of the quantity in brackets, remembering the sign change from re-ordering the spinors. As regards the second term in (4.102), we again take the transpose of the quantity multiplying $\partial_\mu \chi_a$, so that it becomes

$$-\mathrm{i}\partial_\mu \chi_a(-\eta^{\mathrm{T}}\mathrm{i}\sigma_2\sigma^\mu(-\mathrm{i}\sigma_2)\xi^*) = -\mathrm{i}\eta^{\mathrm{T}}c\sigma^\mu c\xi^*\partial_\mu\chi_a. \tag{4.104}$$

After these manipulations, then, we have arrived at

$$\delta_\eta\delta_\xi\chi_a = -\mathrm{i}\eta_a(\xi^\dagger\bar{\sigma}^\mu\partial_\mu\chi) - \mathrm{i}\eta^{\mathrm{T}}c\sigma^\mu c\xi^*\partial_\mu\chi_a, \tag{4.105}$$

and so

$$\begin{aligned}(\delta_\eta\delta_\xi - \delta_\xi\delta_\eta)\chi_a = {} & (\xi^{\mathrm{T}}c\sigma^\mu c\eta^* - \eta^{\mathrm{T}}c\sigma^\mu c\xi^*)\mathrm{i}\partial_\mu\chi_a \\ & + \mathrm{i}\xi_a(\eta^\dagger\bar{\sigma}^\mu\partial_\mu\chi) - \mathrm{i}\eta_a(\xi^\dagger\bar{\sigma}^\mu\partial_\mu\chi).\end{aligned} \tag{4.106}$$

We now see the difficulty: the first term on the right-hand side of (4.106) is indeed exactly the same as (4.28) with ϕ replaced by χ, as hoped for in (4.38), *but there are in addition two unwanted terms.*

The two unwanted terms vanish when the equation of motion $\bar{\sigma}^\mu\partial_\mu\chi = 0$ is satisfied (for a massless field), i.e. 'on-shell'. But this is not good enough – we want a symmetry that applies for the internal (off-shell) lines in Feynman graphs, as well as for the on-shell external lines. Actually, we should not be too surprised that our naive SUSY of Section 4.2 has failed off-shell, for a reason that has already been touched upon: the numbers of degrees of freedom in ϕ and χ do not match up properly, the former having two (one complex field) and the latter four (two complex components). This suggests that we need to introduce another two degrees of freedom to supplement the two in ϕ – say a second complex scalar field F. We do this in the 'cheapest' possible way (provided it works), which is simply to add a term $F^\dagger F$ to the Lagrangian (3.1), so that F has no kinetic term:

$$\mathcal{L}_F = \partial_\mu\phi^\dagger\partial^\mu\phi + \chi^\dagger\mathrm{i}\bar{\sigma}^\mu\partial_\mu\chi + F^\dagger F. \tag{4.107}$$

The strategy now is to invent a SUSY transformation for the *auxiliary field* F, and the existing fields ϕ and χ, such that (a) $\mathcal{L}_F$ is invariant, at least up to a total derivative, and (b) the unwanted terms in $(\delta_\eta\delta_\xi - \delta_\xi\delta_\eta)\chi$ are removed.

We note that F has dimension M^2, suggesting that $\delta_\xi F$ should probably be of the form

$$\delta_\xi F \sim \xi\partial_\mu\chi, \tag{4.108}$$

which is consistent dimensionally. But as usual we need to ensure Lorentz covariance, and in this case that means that the right-hand side of (4.108) must be a Lorentz invariant. We know that $\bar{\sigma}^\mu\partial_\mu\chi$ transforms as a 'ψ'-type spinor (see (2.37)), and we know that an object of the form '$\xi^\dagger\psi$ is Lorentz invariant (see (2.31)). So we try (with a little hindsight)

$$\delta_\xi F = -\mathrm{i}\xi^\dagger\bar{\sigma}^\mu\partial_\mu\chi \tag{4.109}$$

and correspondingly

$$\delta_\xi F^\dagger = \mathrm{i}\partial_\mu\chi^\dagger\bar{\sigma}^\mu\xi. \tag{4.110}$$

The fact that these changes vanish if the equation of motion for χ is imposed (the on-shell condition) suggests that they might be capable of cancelling the unwanted terms in (4.106). Note also that, since ξ is independent of x, the changes in F and $F^\dagger$ are total derivatives: this will be important later (see the end of Section 6.3).

We must first ensure that the enlarged Lagrangian (4.107), or at least the corresponding Action, remains SUSY-invariant. Under the changes (4.109) and (4.110), the $F^\dagger F$ term in (4.107) changes by

$$(\delta_\xi F^\dagger)F + F^\dagger(\delta_\xi F) = (\mathrm{i}\partial_\mu\chi^\dagger\bar{\sigma}^\mu\xi)F - F^\dagger(\mathrm{i}\xi^\dagger\bar{\sigma}^\mu\partial_\mu\chi). \tag{4.111}$$

These terms have a structure very similar to the change in the χ term in (4.107), which is

$$\delta_\xi(\chi^\dagger\mathrm{i}\bar{\sigma}^\mu\partial_\mu\chi) = (\delta_\xi\chi^\dagger)\mathrm{i}\bar{\sigma}^\mu\partial_\mu\chi + \chi^\dagger\mathrm{i}\bar{\sigma}^\mu\partial_\mu(\delta_\xi\chi). \tag{4.112}$$

We see that if we choose

$$\delta_\xi\chi^\dagger = \text{previous change in } \chi^\dagger + F^\dagger\xi^\dagger \tag{4.113}$$

then the $F^\dagger$ part of the first term in (4.112) cancels the second term in (4.111). As regards the second term in (4.112), we write it as

$$\chi^\dagger\mathrm{i}\bar{\sigma}^\mu\partial_\mu(\delta_\xi\chi) = \chi^\dagger\mathrm{i}\bar{\sigma}^\mu\partial_\mu(\text{previous change in } \chi + \xi F), \tag{4.114}$$

where we have used the dagger of (4.113), namely

$$\delta_\xi\chi = \text{previous change in } \chi + \xi F. \tag{4.115}$$

The new term on the right-hand side of (4.114) can be written as

$$\chi^\dagger \mathrm{i}\bar{\sigma}^\mu \partial_\mu \xi F = \partial_\mu(\chi^\dagger \mathrm{i}\bar{\sigma}^\mu \xi F) - (\partial_\mu \chi^\dagger)\mathrm{i}\bar{\sigma}^\mu \xi F. \tag{4.116}$$

The first term of (4.116) is a total derivative, leaving the Action invariant, while the second cancels the first term in (4.111). The net result is that the total change in the last two terms of (4.107) amount to a harmless total derivative, together with the change in $\chi^\dagger \mathrm{i}\bar{\sigma}^\mu \partial_\mu \chi$ due to the previous changes in χ and $\chi^\dagger$. Since the transformation of ϕ has not been altered, the work of Section 3.1 then ensures the invariance (up to a total derivative) of the full Lagrangian (4.107).

Let us now re-calculate $(\delta_\eta \delta_\xi - \delta_\xi \delta_\eta)\chi$, including the new terms involving the auxiliary field F. Since the transformation of ϕ is unaltered, $\delta_\eta \delta_\xi \chi$ will be the same as before, in (4.105), together with an extra term

$$\delta_\eta(\xi_a F) = -\mathrm{i}\xi_a(\eta^\dagger \bar{\sigma}^\mu \partial_\mu \chi). \tag{4.117}$$

So $(\delta_\eta \delta_\xi - \delta_\xi \delta_\eta)\chi$ will be as before, in (4.106), together with the extra terms

$$\mathrm{i}\eta_a(\xi^\dagger \bar{\sigma}^\mu \partial_\mu \chi) - \mathrm{i}\xi_a(\eta^\dagger \bar{\sigma}^\mu \partial_\mu \chi). \tag{4.118}$$

These extra terms precisely cancel the unwanted terms in (4.106), as required. Similar results hold for the action of $(\delta_\eta \delta_\xi - \delta_\xi \delta_\eta)$ on ϕ and on F, and so with this enlarged structure including F we can indeed claim that (4.37) holds as an operator relation, being true when acting on any field of the theory.

For convenience, we collect together the SUSY transformations for ϕ, χ and F which we have finally arrived at:

$$\delta_\xi \phi = \xi \cdot \chi, \quad \delta_\xi \phi^\dagger = \bar{\xi} \cdot \bar{\chi}; \tag{4.119}$$

$$\delta_\xi F = -\mathrm{i}\xi^\dagger \bar{\sigma}^\mu \partial_\mu \chi, \quad \delta_\xi F^\dagger = \mathrm{i}\partial_\mu \chi^\dagger \bar{\sigma}^\mu \xi; \tag{4.120}$$

$$\delta_\xi \chi = -\mathrm{i}\sigma^\mu(\mathrm{i}\sigma_2 \xi^*)\partial_\mu \phi + \xi F, \quad \delta_\xi \chi^\dagger = \mathrm{i}\partial_\mu \phi^\dagger \xi^{\mathrm{T}}(-\mathrm{i}\sigma_2)\sigma^\mu + F^\dagger \xi^\dagger. \tag{4.121}$$

Exercise 4.3 Show that the changes in χ and $\chi^\dagger$ may be written as

$$\delta_\xi \chi = -\mathrm{i}\sigma^\mu \bar{\xi} \partial_\mu \phi + \xi F, \quad \delta_\xi \bar{\chi} = -\mathrm{i}\partial_\mu \phi^\dagger \bar{\sigma}^\mu \xi + F^\dagger \bar{\xi}. \tag{4.122}$$

Exercise 4.4 Verify that the supercurrent for the Lagrangian of (4.107) is still (4.70).

We end this chapter by translating the Lagrangian $\mathcal{L}_F$, and the SUSY transformations (4.119)–(4.122) under which the Action is invariant, into Majorana language, using the results of Section 2.5. Let us write

$$\phi = \frac{1}{\sqrt{2}}(A - \mathrm{i}B), \tag{4.123}$$

where A and B are real scalar fields, and similarly

$$F \to F - \mathrm{i}G. \tag{4.124}$$

The Lagrangian (4.107) then becomes

$$\mathcal{L}_{F,\mathrm{M}} = \frac{1}{2}\partial_\mu A \partial^\mu A + \frac{1}{2}\partial_\mu B \partial^\mu B + \frac{1}{2}\bar{\Psi}^\chi_\mathrm{M} \mathrm{i}\gamma^\mu \partial_\mu \Psi^\chi_\mathrm{M} + F^2 + G^2, \tag{4.125}$$

with the conventional normalization for the scalar fields, while clearly

$$\delta_\xi A = \frac{1}{\sqrt{2}} \bar{\Psi}^\xi_\mathrm{M} \Psi^\chi_\mathrm{M}, \quad \delta_\xi B = -\frac{\mathrm{i}}{\sqrt{2}} \bar{\Psi}^\xi_\mathrm{M} \gamma_5 \Psi^\chi_\mathrm{M}, \tag{4.126}$$

and

$$\delta_\xi F = -\frac{\mathrm{i}}{2} \bar{\Psi}^\xi_\mathrm{M} \gamma^\mu \partial_\mu \Psi^\chi_\mathrm{M}, \quad \delta_\xi G = \frac{1}{2} \bar{\Psi}^\xi_\mathrm{M} \gamma_5 \gamma^\mu \partial_\mu \Psi^\chi_\mathrm{M}. \tag{4.127}$$

As regards the transformations of χ and $\chi^\dagger$, we first rewrite them as

$$\delta_\xi \Psi^\chi_\mathrm{M} \equiv \begin{pmatrix} \delta_\xi(\mathrm{i}\sigma_2 \chi^*) \\ \delta_\xi \chi \end{pmatrix} = \begin{pmatrix} F + \mathrm{i}G & -\mathrm{i}\bar{\sigma}^\mu \partial_\mu \phi^\dagger \\ -\mathrm{i}\sigma^\mu \partial_\mu \phi & F - \mathrm{i}G \end{pmatrix} \begin{pmatrix} \mathrm{i}\sigma_2 \xi^* \\ \xi \end{pmatrix}. \tag{4.128}$$

The rest follows as Exercise 4.5.

Exercise 4.5 Verify that the transformation of Ψ^χ_M is

$$\delta_\xi \Psi^\chi_\mathrm{M} = F \Psi^\xi_\mathrm{M} + \mathrm{i}G\gamma_5 \Psi^\xi_\mathrm{M} - \frac{\mathrm{i}}{\sqrt{2}} \gamma^\mu \partial_\mu A \Psi^\xi_\mathrm{M} + \frac{1}{\sqrt{2}} \gamma_5 \gamma^\mu \partial_\mu B \Psi^\xi_\mathrm{M}. \tag{4.129}$$

The reader is warned that while these transformations have the same general structure as those given in other sources, definitions and conventions differ at many points.

5

The Wess–Zumino model

5.1 Interactions and the superpotential

The Lagrangian (4.107) describes a free (left) chiral supermultiplet, with a massless complex spin-0 field ϕ, a massless L-type spinor field χ, and a non-propagating complex field F. As we saw in Section 3.2, the MSSM places the quarks, leptons and Higgs bosons of the SM, labelled by gauge and flavour degrees of freedom, into chiral supermultiplets, partnered by the appropriate 'sparticles'. So our first step towards the MSSM is to generalize (4.107) to

$$\mathcal{L}_{\text{free WZ}} = \partial_\mu \phi_i^\dagger \partial^\mu \phi_i + \chi_i^\dagger \mathrm{i}\bar{\sigma}^\mu \partial_\mu \chi_i + F_i^\dagger F_i, \tag{5.1}$$

where the summed-over index i runs over internal degrees of freedom (e.g. flavour, and eventually gauge; see Chapter 7), and is not to be confused (in the case of χ_i) with the spinor component index. The corresponding Action is invariant under the SUSY transformations

$$\delta_\xi \phi_i = \xi \cdot \chi_i, \quad \delta_\xi \chi_i = -\mathrm{i}\sigma^\mu \mathrm{i}\sigma_2 \xi^* \partial_\mu \phi_i + \xi F_i, \quad \delta_\xi F_i = -\mathrm{i}\xi^\dagger \bar{\sigma}^\mu \partial_\mu \chi_i, \tag{5.2}$$

together with their hermitian conjugates.

The obvious next step is to introduce interactions in such a way as to preserve SUSY, that is, invariance of the Lagrangian (or the Action) under the transformations (5.2). This was first done (for this type of theory, in four dimensions) by Wess and Zumino [19] in the model named after them, to which this chapter is devoted; it is a fundamental component of the MSSM. We shall largely follow the account given by [46], Section 3.2.

We shall impose the important condition that the interactions should be renormalizable. This means that the mass dimension of all interaction terms must not be greater than 4, or, equivalently, that the coupling constants in the interaction terms should be dimensionless, or have positive dimension (see [15] Section 11.8). The most general possible set of renormalizable interactions among the fields ϕ, χ and

F is, in fact, rather simple:

$$\mathcal{L}_{\text{int}} = W_i(\phi, \phi^\dagger)F_i - \frac{1}{2}W_{ij}(\phi, \phi^\dagger)\chi_i \cdot \chi_j + \text{hermitian conjugate} \qquad (5.3)$$

where there is a sum on i and on j. Here W_i and W_{ij} are, for the moment, arbitrary functions of the bosonic fields; we shall see that they are actually related, and have a simple form. There is no term in the ϕ_i's alone, because under the transformation (5.2) this would become some function of the ϕ_i's multiplied by $\delta_\xi \phi_i = \xi \cdot \chi_i$ or $\delta_\xi \phi_i^\dagger = \bar{\xi} \cdot \bar{\chi}$; but these terms do not include any derivatives ∂_μ, or F_i or $F_i^\dagger$ fields, and it is clear by inspection of (5.2) that they couldn't be cancelled by the transformation of any other term.

As regards W_i and W_{ij}, we first note that since F_i has dimension 2, W_i cannot depend on χ_i, which has dimension 3/2, nor on any power of F_i other than the first, which is already included in (5.1). Indeed, W_i can involve no higher powers of ϕ_i and $\phi_i^\dagger$ than the second. Similarly, since $\chi_i \cdot \chi_j$ has dimension 3, W_{ij} can only depend on ϕ_i and $\phi_i^\dagger$, and contain no powers higher than the first. Furthermore, since $\chi_i \cdot \chi_j = \chi_j \cdot \chi_i$ (see Exercise 2.3), W_{ij} must be symmetric in the indices i and j.

Since we know that the Action for the 'free' part (5.1) is invariant under (5.2), we consider now only the change in $\mathcal{L}_{\text{int}}$ under (5.2), namely $\delta_\xi \mathcal{L}_{\text{int}}$. First, consider the part involving four spinors, which is

$$-\frac{1}{2}\frac{\partial W_{ij}}{\partial \phi_k}(\xi \cdot \chi_k)(\chi_i \cdot \chi_j) - \frac{1}{2}\frac{\partial W_{ij}}{\partial \phi_k^\dagger}(\bar{\xi} \cdot \bar{\chi}_k)(\chi_i \cdot \chi_j) + \text{hermitian conjugate.} \qquad (5.4)$$

Neither of these terms can be cancelled by the variation of any other term. However, the first term will vanish provided that

$$\frac{\partial W_{ij}}{\partial \phi_k} \text{ is symmetric in } i,\ j \text{ and } k. \qquad (5.5)$$

The reason is that the identity (4.99) (with $\lambda \to \chi_k, \zeta \to \chi_i, \rho \to \chi_j$) implies

$$(\xi \cdot \chi_k)(\chi_i \cdot \chi_j) + (\xi \cdot \chi_i)(\chi_j \cdot \chi_k) + (\xi \cdot \chi_j)(\chi_k \cdot \chi_i) = 0, \qquad (5.6)$$

from which it follows that if (5.5) is true, then the first term in (5.4) will vanish identically. However, there is no corresponding identity for the 4-spinor product in the second term of (5.4). The only way to get rid of this second term, and thus preserve SUSY for such interactions, is to say that W_{ij} cannot depend on $\phi_k^\dagger$, only on ϕ_k.[1] Thus we now know that W_{ij} must have the form

$$W_{ij} = M_{ij} + y_{ijk}\phi_k \tag{5.7}$$

where the matrix M_{ij} (which has the dimensions and significance of a mass) is symmetric in i and j, and where the 'Yukawa couplings' y_{ijk} are symmetric in i, j and k. It is convenient to write (5.7) as

$$W_{ij} = \frac{\partial^2 W}{\partial\phi_i \partial\phi_j} \tag{5.8}$$

which is automatically symmetric in i and j, and where[2] (bearing in mind the symmetry properties of W_{ij})

$$W = \frac{1}{2}M_{ij}\phi_i\phi_j + \frac{1}{6}y_{ijk}\phi_i\phi_j\phi_k. \tag{5.9}$$

Exercise 5.1 Justify (5.9).

Next, consider those parts of $\delta_\xi \mathcal{L}_{\rm int}$ which contain one derivative ∂_μ. These are (recall $c = -{\rm i}\sigma_2$)

$$W_i(-{\rm i}\xi^\dagger\bar{\sigma}^\mu\partial_\mu\chi_i) - \frac{1}{2}W_{ij}\{\chi_i^{\rm T}c{\rm i}\sigma^\mu c\xi^*\}\partial_\mu\phi_j + \frac{1}{2}W_{ij}\xi^\dagger c{\rm i}\sigma^{{\rm T}\mu}\partial_\mu\phi_i c\chi_j + {\rm h.c.}, \tag{5.10}$$

where h.c. means hermitian conjugate. Consider the expression in curly brackets, $\{\chi_i^{\rm T}\ldots\xi^*\}$. Since this is a single quantity (after evaluating the matrix products), it is equal to its transpose, which is

$$-\xi^\dagger c{\rm i}\sigma^{\mu{\rm T}}c\chi_i = \xi^\dagger{\rm i}\bar{\sigma}^\mu\chi_i, \tag{5.11}$$

where the first minus sign comes from interchanging two fermionic quantities, and the second equality uses the result $c\sigma^{\mu{\rm T}}c = -\bar{\sigma}^\mu$ (cf. (4.30)). So the second term in (5.10) is

$$-\frac{1}{2}W_{ij}{\rm i}\xi^\dagger\bar{\sigma}^\mu\chi_i\partial_\mu\phi_j, \tag{5.12}$$

and the third term is

$$\frac{1}{2}W_{ij}\xi^\dagger c{\rm i}\sigma^{\mu{\rm T}}c\chi_j\partial_\mu\phi_i = -\frac{1}{2}W_{ij}{\rm i}\xi^\dagger\bar{\sigma}^\mu\chi_j\partial_\mu\phi_i. \tag{5.13}$$

[1] This is a point of great importance for the MSSM: as mentioned at the end of Section 3.2, the SM uses both the Higgs field ϕ and its charge conjugate, which is related to $\phi^\dagger$ by (3.40), but in the MSSM we shall need to have two separate ϕ's.

[2] A linear term of the form $A_l\phi_l$ could be added to (5.9), consistently with (5.8) and (5.7). This is relevant to one model of SUSY breaking, see Section 9.1.

These two terms add to give

$$-W_{ij}\mathrm{i}\xi^\dagger\bar{\sigma}^\mu\chi_i\partial_\mu\phi_j = -\mathrm{i}\xi^\dagger\bar{\sigma}^\mu\chi_i\partial_\mu\left(\frac{\partial W}{\partial\phi_i}\right), \tag{5.14}$$

where in the second equality we have used

$$\partial_\mu\left(\frac{\partial W}{\partial\phi_i}\right) = \frac{\partial^2 W}{\partial\phi_i\partial\phi_j}\partial_\mu\phi_j = W_{ij}\partial_\mu\phi_j. \tag{5.15}$$

Altogether, then, (5.10) has become

$$-\mathrm{i}W_i\xi^\dagger\bar{\sigma}^\mu\partial_\mu\chi_i - \mathrm{i}\xi^\dagger\bar{\sigma}^\mu\chi_i\partial_\mu\left(\frac{\partial W}{\partial\phi_i}\right). \tag{5.16}$$

This variation cannot be cancelled by anything else, and our only chance of saving SUSY is to have it equal a total derivative (giving an invariant Action, as usual). The condition for (5.16) to be a total derivative is that W_i should have the form

$$W_i = \frac{\partial W}{\partial\phi_i}, \tag{5.17}$$

in which case (5.16) becomes

$$\partial_\mu\left\{\frac{\partial W}{\partial\phi_i}(-\mathrm{i}\xi^\dagger\bar{\sigma}^\mu\chi_i)\right\}. \tag{5.18}$$

Referring to (5.9), we see that the condition (5.17) implies

$$W_i = M_{ij}\phi_j + \frac{1}{2}y_{ijk}\phi_j\phi_k \tag{5.19}$$

together with a possible constant term A_i (see the preceding footnote).

Exercise 5.2 Verify that the remaining terms in $\delta_\xi\mathcal{L}$ do cancel.

In summary, we have found conditions on W_i and W_{ij} (namely equations (5.17) and (5.8) with W given by (5.9)) such that the interactions (5.3) give an Action which is invariant under the SUSY transformations (5.2). The quantity W, from which both W_i and W_{ij} are derived, encodes all the allowed interactions, and is clearly a central part of the model; for reasons that will become clearer in the following chapter, it is called the 'superpotential'.

Exercise 5.3 Verify that the supersymmetry current for the Lagrangian of (5.1) together with (5.3) is

$$J^\mu = \sigma^\nu\bar{\sigma}^\mu\chi_i\partial_\nu\phi_i^\dagger - \mathrm{i}F_i\sigma^\mu\mathrm{i}\sigma_2\chi_i^{\dagger\mathrm{T}}. \tag{5.20}$$

Consider now the part of the complete Lagrangian (including (5.1)) containing F_i and $F_i^\dagger$, which is just $F_iF_i^\dagger + W_iF_i + W_i^\dagger F_i^\dagger$. Since this contains no gradients, the Euler–Lagrange (E–L) equations for F_i and $F_i^\dagger$ are simply

$$\frac{\partial \mathcal{L}}{\partial F_i} = 0, \text{ or } F_i^\dagger + W_i = 0. \tag{5.21}$$

Hence $F_i = -W_i^\dagger$, and similarly $F_i^\dagger = -W_i$. These relations, coming from the E–L equations, involve (again) no derivatives, and hence the canonical commutation relations will not be affected, and it is permissible to replace F_i and $F_i^\dagger$ in the Lagrangian by these values determined from the E–L equations. This results in the complete [Wess–Zumino (W–Z) [19]] Lagrangian now having the form

$$\mathcal{L}_{\mathrm{WZ}} = \mathcal{L}_{\mathrm{free\ WZ}} - |W_i|^2 - \frac{1}{2}\{W_{ij}\chi_i \cdot \chi_j + \text{h.c.}\}. \tag{5.22}$$

It is worth spending a little time looking in more detail at the model of (5.22). For simplicity we shall discuss just one supermultiplet, dropping the indices i and j. In that case, (5.9) becomes

$$W = \frac{1}{2}M\phi^2 + \frac{1}{6}y\phi^3, \tag{5.23}$$

and hence

$$W_i = \frac{\partial W}{\partial \phi} = M\phi + \frac{1}{2}y\phi^2 \tag{5.24}$$

and

$$W_{ij} = \frac{\partial^2 W}{\partial \phi^2} = M + y\phi. \tag{5.25}$$

First, consider the terms which are quadratic in the fields ϕ and χ, which correspond to kinetic and mass terms (rather than interactions proper). This will give us an opportunity to learn about mass terms for 2-component spinors. The quadratic terms for a single supermultiplet are

$$\begin{aligned}\mathcal{L}_{\mathrm{WZ,quad}} = {} & \partial_\mu\phi^\dagger\partial^\mu\phi + \chi^\dagger \mathrm{i}\bar{\sigma}^\mu\partial_\mu\chi - MM^*\phi^\dagger\phi \\ & -\frac{1}{2}M\chi^{\mathrm{T}}(-\mathrm{i}\sigma_2)\chi - \frac{1}{2}M^*\chi^\dagger(\mathrm{i}\sigma_2)\chi^{\dagger\mathrm{T}},\end{aligned} \tag{5.26}$$

where we have reverted to the explicit forms of the spinor products. In (5.26), $\chi^\dagger$ is as given in (3.19), while evidently

$$\chi^{\dagger\mathrm{T}} = \begin{pmatrix}\chi_1^\dagger \\ \chi_2^\dagger\end{pmatrix}, \tag{5.27}$$

where '1' and '2', of course, label the spinor components. The E–L equation for $\phi^\dagger$ is

$$\partial_\mu \left(\frac{\partial \mathcal{L}}{\partial(\partial_\mu \phi^\dagger)} \right) - \frac{\partial \mathcal{L}}{\partial \phi^\dagger} = 0, \tag{5.28}$$

which leads immediately to

$$\partial_\mu \partial^\mu \phi + |M|^2 \phi = 0, \tag{5.29}$$

which is just the standard free Klein–Gordon equation for a spinless field of mass $|M|$.

In considering the analogous E–L equation for (say) $\chi^\dagger$, we need to take care in evaluating (functional) derivatives of $\mathcal{L}$ with respect to fields such as χ or $\chi^\dagger$ which anticommute. Consider the term $-(1/2)M\chi \cdot \chi$ in (5.26), which is

$$-\frac{1}{2}M(\chi_1 \chi_2) \begin{pmatrix} 0 & -1 \\ 1 & 0 \end{pmatrix} \begin{pmatrix} \chi_1 \\ \chi_2 \end{pmatrix} = -\frac{1}{2}M(-\chi_1\chi_2 + \chi_2\chi_1) = -M\chi_2\chi_1 = +M\chi_1\chi_2. \tag{5.30}$$

We define

$$\frac{\partial}{\partial \chi_1}(\chi_1 \chi_2) = \chi_2, \tag{5.31}$$

and then necessarily

$$\frac{\partial}{\partial \chi_2}(\chi_1 \chi_2) = -\chi_1. \tag{5.32}$$

Hence

$$\frac{\partial}{\partial \chi_1}\left\{-\frac{1}{2}M\chi \cdot \chi\right\} = M\chi_2, \tag{5.33}$$

and

$$\frac{\partial}{\partial \chi_2}\left\{-\frac{1}{2}M\chi \cdot \chi\right\} = -M\chi_1. \tag{5.34}$$

Equations (5.33) and (5.34) can be combined as

$$\frac{\partial}{\partial \chi_a}\left\{-\frac{1}{2}M\chi \cdot \chi\right\} = M(\mathrm{i}\sigma_2 \chi)_a. \tag{5.35}$$

Exercise 5.4 Show similarly that

$$\frac{\partial}{\partial \chi_a^\dagger}\left\{-\frac{1}{2}M^* \chi^\dagger \mathrm{i}\sigma_2 \chi^{\dagger\mathrm{T}}\right\} = M^*(-\mathrm{i}\sigma_2 \chi^{\dagger\mathrm{T}})_a. \tag{5.36}$$

We are now ready to consider the E–L equation for $\chi^\dagger$, which is

$$\partial_\mu\left(\frac{\partial\mathcal{L}}{\partial(\partial_\mu\chi_{\dot a}^\dagger)}\right)-\frac{\partial\mathcal{L}}{\partial\chi_{\dot a}^\dagger}=0. \tag{5.37}$$

Using just the quadratic parts (5.26) this yields

$$\mathrm{i}\bar\sigma^\mu\partial_\mu\chi=M^*\mathrm{i}\sigma_2\chi^{\dagger\mathrm{T}}. \tag{5.38}$$

As a notational check, we know from Section 2.3 that χ transforms by $V^{-1\dagger}$, and hence $\chi^{\dagger\mathrm{T}}$ transforms by $V^{-1\mathrm{T}}$, which is the same as a 'lower dotted' spinor of type $\psi_{\dot a}$. The lower dotted index is raised by the matrix $\mathrm{i}\sigma_2$. Hence the right-hand side of (5.38) transforms like a $\psi^{\dot a}$ spinor, and this is consistent with the left-hand side, by (2.37).

Exercise 5.5 Similarly, show that

$$\mathrm{i}\sigma^\mu\partial_\mu(\mathrm{i}\sigma_2\chi^{\dagger\mathrm{T}})=M\chi. \tag{5.39}$$

It follows from (5.38) and (5.39) that

$$\begin{aligned}\mathrm{i}\sigma^\mu\partial_\mu(\mathrm{i}\bar\sigma^\nu\partial_\nu\chi)&=\mathrm{i}\sigma^\mu\partial_\mu(M^*\mathrm{i}\sigma_2\chi^{\dagger\mathrm{T}})\\&=|M|^2\chi.\end{aligned} \tag{5.40}$$

So, using (3.25) on the left-hand side we have simply

$$\partial_\mu\partial^\mu\chi+|M|^2\chi=0, \tag{5.41}$$

which shows that the χ field also has mass $|M|$. So we have verified that the quadratic parts (5.26) describe a free spin-0 and spin-1/2 field which are degenerate, both having mass $|M|$. It is perhaps worth pointing out that, although we started (for simplicity) with massless fields, we now see that it is perfectly possible to have massive supersymmetric theories, the bosonic and fermionic superpartners having (of course) the same mass.

Next, let us consider briefly the interaction terms in (5.22), again just for the case of one chiral superfield. These terms are

$$-\left|M\phi+\frac{1}{2}y\phi^2\right|^2-\frac{1}{2}\{(M+y\phi)\chi\cdot\chi+\text{h.c.}\}. \tag{5.42}$$

In addition to the quadratic parts $|M|^2\phi^\dagger\phi$ and $-(1/2)M\chi\cdot\chi+\text{h.c.}$ which we have just discussed, (5.42) contains three true interactions, namely

(i) a 'cubic' interaction among the ϕ fields,

$$-\frac{1}{2}(My^*\phi\phi^{\dagger 2}+M^*y\phi^2\phi^\dagger); \tag{5.43}$$

(ii) a 'quartic' interaction among the ϕ fields,

$$-\frac{1}{4}|y|^2\phi^2\phi^{\dagger 2}; \tag{5.44}$$

(iii) a Yukawa-type coupling between the ϕ and χ fields,

$$-\frac{1}{2}\{y\phi\chi\cdot\chi + \text{h.c.}\}. \tag{5.45}$$

It is noteworthy that the same coupling parameter y enters into the cubic and quartic bosonic interactions (5.43) and (5.44), as well as the Yukawa-like fermion–boson interaction (5.45). In particular, the quartic coupling constant appearing in (5.44) is equal to the square of the Yukawa coupling in (5.45). This is exactly the relationship noted in (1.21), as being required for the cancellation (between bosonic and fermionic contributions) of quadratic divergences in a bosonic self-energy.

We shall demonstrate such a cancellation explicitly in the next section, for the W–Z model. For this purpose, it is convenient to express the Lagrangian in Majorana form, with ϕ given by (4.123). We take the parameters M and y to be real. The quadratic parts (5.26) are then (cf. (4.125))

$$\frac{1}{2}\bar{\Psi}^{\chi}_{\mathrm{M}}(\mathrm{i}\gamma^\mu\partial_\mu - M)\Psi^{\chi}_{\mathrm{M}} + \frac{1}{2}\partial^\mu A\partial_\mu A - \frac{1}{2}M^2A^2 + \frac{1}{2}\partial^\mu B\partial_\mu B - \frac{1}{2}M^2B^2, \tag{5.46}$$

showing that the fermion and the two real scalars have the same mass M, while the interactions (5.43) and (5.44) become

$$\mathcal{L}_{\mathrm{c}} = -MgA(A^2 + B^2) \tag{5.47}$$

and

$$\mathcal{L}_{\mathrm{q}} = -\frac{1}{2}g^2(A^2 + B^2)^2, \tag{5.48}$$

where we have defined $g = y/2\sqrt{2}$. We leave the third interaction as Exercise 5.6.

Exercise 5.6 Verify that the interaction (5.45) becomes

$$\mathcal{L}_{\mathrm{y}} = -g\left[A\bar{\Psi}^{\chi}_{\mathrm{M}}\Psi^{\chi}_{\mathrm{M}} + \mathrm{i}B\bar{\Psi}^{\chi}_{\mathrm{M}}\gamma_5\Psi^{\chi}_{\mathrm{M}}\right]. \tag{5.49}$$

We note that the γ_5 coupling in the second term of (5.49) shows that B is a pseudoscalar field (see, for example, Section 20.3 of [7]); A is a scalar field.

5.2 Cancellation of quadratic divergences in the W–Z model

We shall consider the one-loop ($O(g^2)$) contributions to the perturbative expansion of A-particle propagator, defined as $\langle\Omega|T(A(x)A(y))|\Omega\rangle$, where $|\Omega\rangle$ is the ground state (vacuum) of the interacting theory, and T is the time-ordering operator. The general expression for the propagator is (see, for example, [50] Section 4.4)

$$\langle\Omega|T(A(x)A(y))|\Omega\rangle = \frac{\langle 0|T\{A(x)A(y)\exp[\mathrm{i}\int \mathrm{d}^4z\mathcal{L}'(z)]\}|0\rangle}{\langle 0|T\{\exp[\mathrm{i}\int \mathrm{d}^4z\mathcal{L}'(z)]\}|0\rangle} \tag{5.50}$$

where $\mathcal{L}'$ is the interaction Lagrangian density, and it is understood that all fields on the right-hand side of (5.50) are in the interaction picture. In the present case, $\mathcal{L}'$ is just the sum of the three terms (5.47), (5.48) and (5.49). Perturbation theory proceeds by expanding the exponentials in (5.50) in powers of the coupling constant g, and by reducing the resulting time-ordered products by Wick's theorem. We recall that the role of the denominator in (5.50) is to remove contributions to the numerator from all vacuum to vacuum processes that are disconnected from the points x and y; we therefore need only consider the connected contributions to the numerator.

To lowest order (g^0) the right-hand side of (5.50) is just the free A propagator $D_{\mathrm{A}}(x-y)$, which has the Fourier (momentum–space) expansion

$$D_{\mathrm{A}}(x-y) = \int \frac{\mathrm{d}^4k}{(2\pi)^4}\mathrm{e}^{-\mathrm{i}k\cdot(x-y)}\frac{\mathrm{i}}{k^2-M^2}, \tag{5.51}$$

where the addition of the infinitesimal quantity $\mathrm{i}\epsilon$ in the denominator is understood. The terms of order g from (5.47) and (5.49) both vanish. At order g^2, contributions arise from expanding the exponential of (5.48) to first order (it already contains a factor of g^2), and from expanding the exponential of the sum of of (5.47) and (5.49) to second order. The contribution from (5.48) is

$$-\mathrm{i}\frac{g^2}{2}\langle 0|T(A(x)A(y)\int \mathrm{d}^4z[A^4(z)+2A^2(z)B^2(z)+B^4(z)]|0\rangle. \tag{5.52}$$

In the Wick reduction of the B^4 term in (5.52), one $B(z)$ can only be paired with another, which leads to a disconnected contribution. In the A^2B^2 term, we may contract $A(x)$ with the first $A(z)$ and $A(y)$ with the second, or the other way around; these contributions are identical. The two B's must be contracted together. The A^2B^2 term in (5.52) therefore becomes

$$-2\mathrm{i}g^2\int \mathrm{d}^4z\, D_{\mathrm{A}}(x-z)D_{\mathrm{A}}(y-z)D_{\mathrm{B}}(z-z) \tag{5.53}$$

where D_{B} is the B propagator, also given by (5.51). Substituting the Fourier expansions of D_{A} and D_{B} into (5.53) we obtain

$$-2\mathrm{i}g^2\int \mathrm{d}^4z\int \frac{\mathrm{d}^4p}{(2\pi)^4}\frac{\mathrm{i}\,\mathrm{e}^{-\mathrm{i}p\cdot(x-z)}}{p^2-M^2}\int \frac{\mathrm{d}^4q}{(2\pi)^4}\frac{\mathrm{i}\,\mathrm{e}^{-\mathrm{i}q\cdot(y-z)}}{q^2-M^2}\int \frac{\mathrm{d}^4k}{(2\pi)^4}\frac{\mathrm{i}\,\mathrm{e}^{-\mathrm{i}k\cdot(z-z)}}{k^2-M^2}. \tag{5.54}$$

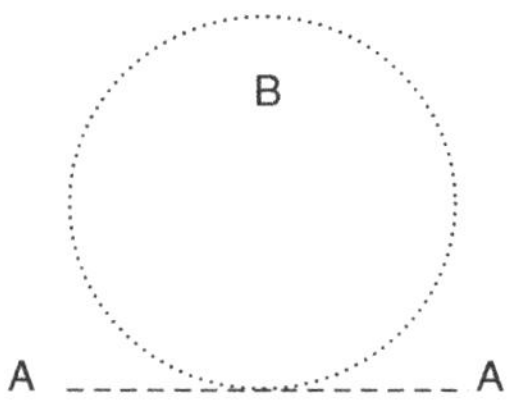

Figure 5.1 B-loop contribution to the A self-energy.

The integration over z yields $(2\pi)^4\delta^4(p+q)$, allowing the q-integration to be done. (5.54) then becomes

$$\int \frac{d^4p}{(2\pi)^4} e^{-ip\cdot(x-y)} \frac{i}{p^2-M^2}\left(-i\Pi_A^{(B)}\right)\frac{i}{p^2-M^2}, \tag{5.55}$$

where

$$-i\Pi_A^{(B)} = 2g^2 \int \frac{d^4k}{(2\pi)^4} \frac{1}{k^2-M^2} \tag{5.56}$$

is the B-loop contribution to the A self-energy (see, for example, [15] Section 10.1). This corresponds to the diagram of Figure 5.1, which is of the same type as Figure 1.1; as expected, the integral in (5.56) is essentially the same as in (1.9). Simple power-counting (four powers of k in the numerator, two in the denominator) suggests that the integral is quadratically divergent, but before proceeding with the remaining $O(g^2)$ contributions to the A propagator it will be useful to evaluate the integral in (5.56) explicitly.

The integral to be evaluated is

$$\int \frac{d^3k}{(2\pi)^4} \int dk^0 \frac{1}{(k^0)^2 - \boldsymbol{k}^2 - M^2 + i\epsilon}. \tag{5.57}$$

One way of proceeding, explained in Section 10.3 of [15], is to perform the k^0 integral by contour integration. Borrowing the result given in equation (10.48) of that reference, we find that (5.57) is equal to

$$-\pi i \int \frac{d^3\boldsymbol{k}}{(2\pi)^4} \frac{1}{(\boldsymbol{k}^2+M^2)^{1/2}}. \tag{5.58}$$

Changing to polar coordinates in $\boldsymbol{k}$-space, we may write this as

$$\frac{-i}{4\pi^2}\int_0^{\Lambda} \frac{u^2 du}{(u^2+M^2)^{1/2}}, \tag{5.59}$$

where $u = |\boldsymbol{k}|$, and we have now included the integration limits, with a cut-off Λ at the upper end. The integral in (5.59) may be evaluated by elementary means,

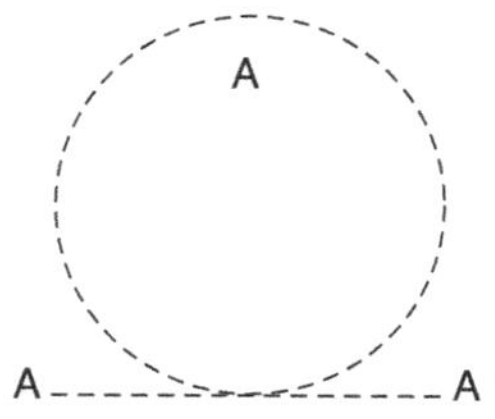

Figure 5.2 A-loop contribution to the A self-energy.

leading to the result

$$\Pi_{\mathrm{A}}^{(\mathrm{B})} = 2g^2 \frac{1}{8\pi^2} \left[\Lambda^2 (1 + M^2/\Lambda^2)^{1/2} - M^2 \ln \left(\frac{\Lambda + \Lambda(1 + M^2/\Lambda^2)^{1/2}}{M} \right) \right]. \tag{5.60}$$

For large values of Λ ($\gg M$) the square roots may be expanded in powers of M/Λ; (5.60) then reduces to

$$\Pi_{\mathrm{A}}^{(\mathrm{B})} = 2g^2 \frac{1}{8\pi^2} [\Lambda^2 - M^2 \ln(\Lambda/M) + \text{finite terms as } \Lambda \to \infty]. \tag{5.61}$$

We have confirmed that the leading divergence is quadratic (in powers of the cut-off), and that its coefficient is independent of the mass M appearing in the denominator of (5.56). This mass does however enter into the next-to-leading (logarithmic) divergence.

We return to the remaining term in (5.52), which is

$$-\mathrm{i}\frac{g^2}{2} \langle 0 | T\{A(x)A(y) \int A(z)A(z)A(z)A(z)\, \mathrm{d}^4 z\} | 0 \rangle. \tag{5.62}$$

The connected contributions arise through contracting $A(x)$ with any one of the four $A(z)$'s and $A(y)$ with any one of the remaining three $A(z)$'s, leaving one $A(z)A(z)$ contraction. These 12 contributions are identical, and (5.62) becomes

$$-6\mathrm{i}g^2 \int D_{\mathrm{A}}(x - z) D_{\mathrm{A}}(y - z) D_{\mathrm{A}}(z - z). \tag{5.63}$$

Following the same steps as in (5.53)–(5.56), we find that (5.63) has the same form as (5.55) but with $-\mathrm{i}\Pi_{\mathrm{A}}^{(\mathrm{B})}$ replaced by $-\mathrm{i}\Pi_{\mathrm{A}}^{(\mathrm{A})}$ where

$$-\mathrm{i}\Pi_{\mathrm{A}}^{(\mathrm{A})} = 6g^2 \int \frac{\mathrm{d}^4 k}{(2\pi)^4} \frac{1}{k^2 - M^2}, \tag{5.64}$$

which corresponds to the self-energy diagram of Figure 5.2. The contribution of the boson loops to the quadratically divergent part of the A self-energy is

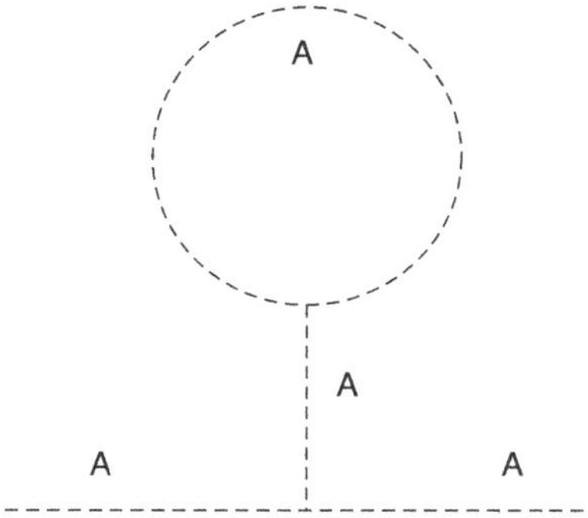

Figure 5.3 A-loop tadpole contribution to the A propagator.

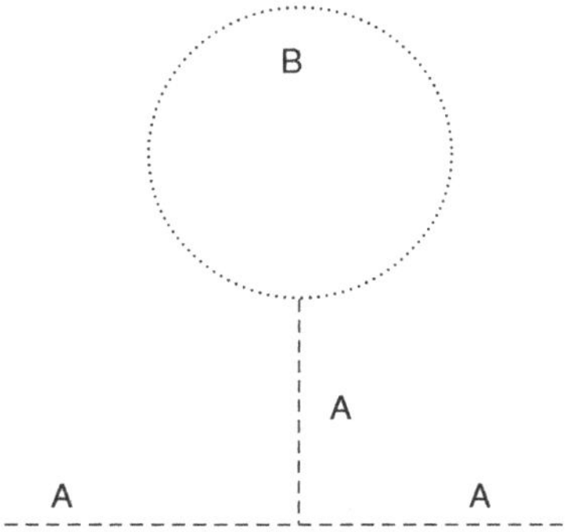

Figure 5.4 B-loop tadpole contribution to the A propagator.

therefore

$$\Pi_{\mathrm{A}}^{(\mathrm{b,\,quad})} = 8g^2 \frac{1}{8\pi^2}\Lambda^2. \tag{5.65}$$

We now consider the $O(g^2)$ terms arising from the second-order term in the expansion of the exponential of the sum of (5.47) and (5.49), that is from

$$-\frac{1}{2}\int \mathrm{d}^4z\,\mathrm{d}^4z'\,\langle 0|T\{A(x)A(y)(\mathcal{L}_{\mathrm{c}}(z)+\mathcal{L}_{\mathrm{y}}(z))(\mathcal{L}_{\mathrm{c}}(z')+\mathcal{L}_{\mathrm{y}}(z'))\}|0\rangle. \tag{5.66}$$

Since there are two terms in each of $\mathcal{L}_{\mathrm{c}}$ and $\mathcal{L}_{\mathrm{y}}$, there are 16 products of the form '$A(x)A(y)f(z)g(z')$' in (5.66), each with a large number of terms in their Wick expansion. It is helpful to think first in terms of (connected) diagrams, which can then be associated with terms in (5.66) to be evaluated. First of all, there are three 'tadpole' diagrams shown in Figures 5.3, 5.4 and 5.5. The first of these has the structure $D_{\mathrm{A}}(x-z)D_{\mathrm{A}}(y-z)D_{\mathrm{A}}(z-z')D_{\mathrm{A}}(z'-z')$, which arises in the Wick

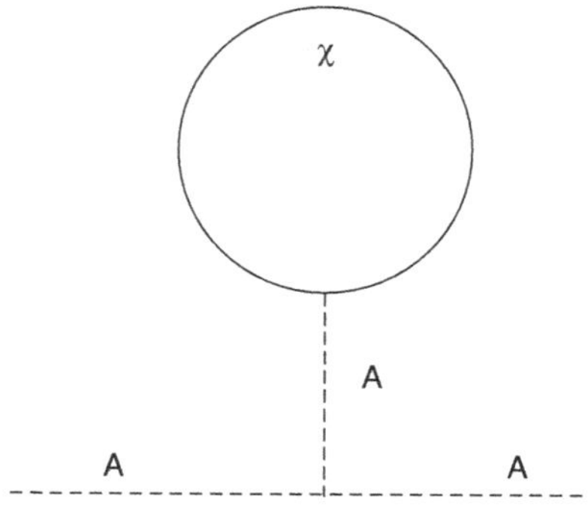

Figure 5.5 χ-loop tadpole contribution to the A propagator.

expansion of the term

$$-\frac{1}{2}M^2 g^2 \int d^4z\, d^4z'\, \langle 0|T\{A(x)A(y)A^3(z)A^3(z')\}|0\rangle. \tag{5.67}$$

This structure is obtained by contracting $A(x)$ with any one of the three $A(z)$'s, and $A(y)$ with either of the two remaining $A(z)$'s; then the last $A(z)$ can be contracted with any of the three $A(z')$'s, leaving one $A(z')A(z')$ pair. This gives 18 identical contributions, and there are a further 18 in which z and z' (which are integration variables) are interchanged. Thus Figure 5.3 corresponds to the amplitude

$$-18M^2 g^2 \int d^4z\, d^4z'\, D_A(x-z)D_A(y-z)D_A(z-z')D_A(z'-z'). \tag{5.68}$$

Exercise 5.7 By inserting the Fourier expansions for the D_A's and performing the integrals over z and z' show that (5.68) can be written in the form (5.55) with $-i\Pi_A^{(B)}$ replaced by

$$-i\Pi_A^{(t,\,A)} = -18g^2 \int \frac{d^4k}{(2\pi)^4} \frac{1}{k^2 - M^2}. \tag{5.69}$$

We note that (5.69) contains a quadratic divergence.

The second tadpole contribution is from Figure 5.4 which has the structure $D_A(x-z)D_A(y-z)D_A(z-z')D_B(z'-z')$ (it is clear that all three tadpoles share the first three factors). This arises from the Wick expansion of the term

$$-\frac{1}{2}M^2 g^2 2 \int d^4z\, d^4z'\, \langle 0|T\{A(x)A(y)A^3(z)A(z')B^2(z')\}|0\rangle. \tag{5.70}$$

We can contract $A(x)$ with any of the three $A(z)$'s, and then $A(y)$ with either of the two remaining ones; this leaves just one way for the last $A(z)$ to be contracted with

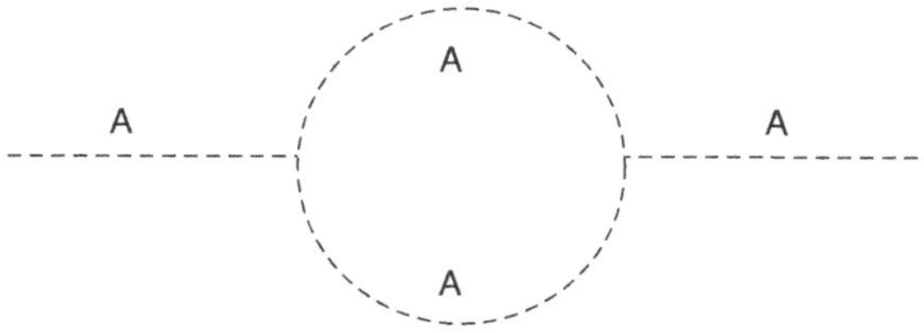

Figure 5.6 A-loop contribution to the A propagator.

$A(z')$, and $B(z')$ with $B(z')$. Hence (5.70) contributes

$$-6M^2g^2\int \mathrm{d}^4z\,\mathrm{d}^4z'\,D_\mathrm{A}(x-z)D_\mathrm{A}(y-z)D_\mathrm{A}(z-z')D_\mathrm{B}(z'-z'), \tag{5.71}$$

which leads to the self-energy contribution

$$-\mathrm{i}\Pi_\mathrm{A}^{(\mathrm{t,\,B})} = -6g^2\int\frac{\mathrm{d}^4k}{(2\pi)^4}\frac{1}{k^2-M^2}, \tag{5.72}$$

also containing a quadratic divergence.

The third tadpole arises from the reduction of the term

$$-\frac{1}{2}2Mg^2\int \mathrm{d}^4z\,\mathrm{d}^4z'\,\langle 0|T\{A(x)A(y)A^3(z)A(z')\bar{\Psi}_\mathrm{M}^\chi(z')\Psi_\mathrm{M}^\chi(z')\}|0\rangle. \tag{5.73}$$

Once again, there are six ways of getting the structure of Figure 5.5, and the associated self-energy contribution is

$$\begin{aligned}-\mathrm{i}\Pi_\mathrm{A}^{(\mathrm{t,}\,\chi)} &= 6Mg^2\frac{1}{M^2}\int\frac{\mathrm{d}^4k}{(2\pi)^4}\mathrm{Tr}\frac{1}{\not{k}-M}\\ &= 24g^2\int\frac{\mathrm{d}^4k}{(2\pi)^4}\frac{1}{k^2-M^2},\end{aligned} \tag{5.74}$$

the minus sign relative to (5.69) and (5.72) being characteristic of a fermionic loop, and arising from the re-ordering of the fermionic fields for the contraction (2.137). This contribution of the fermion-loop tadpole therefore exactly *cancels* the combined contributions (5.69) and (5.72) of the boson-loop tadpoles; in particular, their quadratic divergences are cancelled.

There remain the non-tadpole connected graphs from (5.66). There are two purely bosonic ones, shown in Figures 5.6 and 5.7. The corresponding self-energies are both proportional to (see, for example, Section 10.1.1 of [15])

$$\int\frac{\mathrm{d}^4k}{(2\pi)^4}\frac{1}{(k^2-M^2)}\frac{1}{((k-p)^2-M^2)}. \tag{5.75}$$

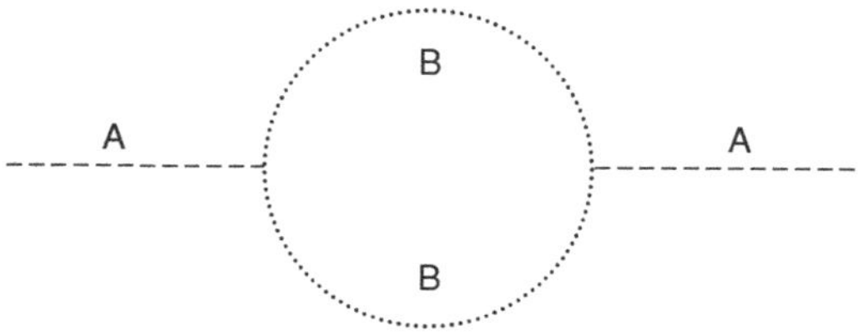

Figure 5.7 B-loop contribution to the A propagator.

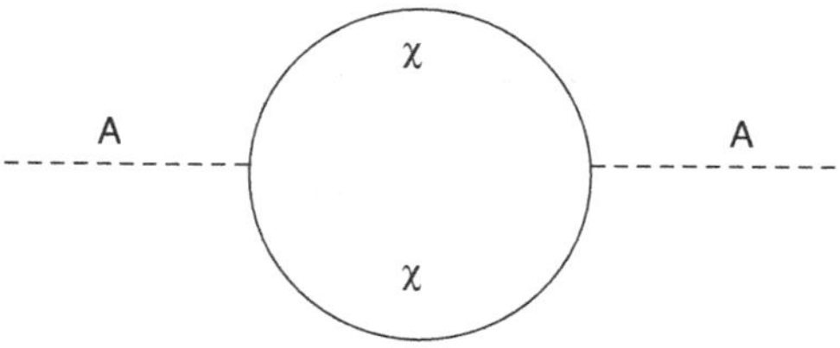

Figure 5.8 χ-loop contribution to the A propagator.

This integral is only logarithmically divergent (four powers of k in the numerator and in the denominator), and we do not need to consider these contributions any further.

We are left with Figure 5.8, which arises from the term

$$\frac{1}{2}\int d^4z\, d^4z' \langle 0|T\{A(x)A(y)\big(-\mathrm{i}gA(z)\bar{\Psi}^{\chi}_{\mathrm{M}}(z)\Psi^{\chi}_{\mathrm{M}}(z)\big)\big(-\mathrm{i}gA(z')\bar{\Psi}^{\chi}_{\mathrm{M}}(z')\Psi^{\chi}_{\mathrm{M}}(z')\big)\}|0\rangle, \tag{5.76}$$

since the term '$B^2(\bar{\Psi}\Psi)^2$' gives a disconnected piece, while the term '$A\bar{\Psi}\Psi B\bar{\Psi}\Psi$' contains an odd number of A or B fields and vanishes. In (5.76), $A(x)$ may be contracted with $A(z)$ and $A(y)$ with $A(z')$, or vice versa. These contributions are the same, so that (5.76) becomes

$$-g^2\int\int d^4z\, d^4z'\, D_{\mathrm{A}}(x-z)D_{\mathrm{A}}(y-z')\langle 0|T\big(\bar{\Psi}^{\chi}_{\mathrm{M}\alpha}(z)\Psi^{\chi}_{\mathrm{M}\alpha}(z)\bar{\Psi}^{\chi}_{\mathrm{M}\beta}(z')\Psi^{\chi}_{\mathrm{M}\beta}(z')\big)|0\rangle, \tag{5.77}$$

where we have indicated the spinor indices explicitly. We must now recall the discussion of Section 2.5.2 concerning propagators (contractions) for Majorana fields. The T-product in (5.77) yields two distinct contractions:

$$\begin{aligned}
&\langle 0|T\big(\bar{\Psi}^{\chi}_{\mathrm{M}\alpha}(z)\Psi^{\chi}_{\mathrm{M}\alpha}(z)\bar{\Psi}^{\chi}_{\mathrm{M}\beta}(z')\Psi^{\chi}_{\mathrm{M}\beta}(z')\big)|0\rangle \\
&\quad = -\langle 0|T\big(\Psi^{\chi}_{\mathrm{M}\alpha}(z)\bar{\Psi}^{\chi}_{\mathrm{M}\beta}(z')\big)|0\rangle\langle 0|T\big(\Psi^{\chi}_{\mathrm{M}\beta}(z')\bar{\Psi}^{\chi}_{\mathrm{M}\alpha}(z)\big)|0\rangle \\
&\quad\quad + \langle 0|T\big(\bar{\Psi}^{\chi}_{\mathrm{M}\alpha}(z)\bar{\Psi}^{\chi}_{\mathrm{M}\beta}(z')\big)|0\rangle\langle 0|T\big(\Psi^{\chi}_{\mathrm{M}\beta}(z')\Psi^{\chi}_{\mathrm{M}\alpha}(z)\big)|0\rangle,
\end{aligned} \tag{5.78}$$

where the signs of the terms on the right-hand side are determined by the number of corresponding interchanges of fermionic fields. The first term on the right-hand side of (5.78) is, from (2.137),

$$-S_{\mathrm{F}\alpha\beta}(z-z')S_{\mathrm{F}\beta\alpha}(z'-z) = -\mathrm{Tr}(S_{\mathrm{F}}(z-z')S_{\mathrm{F}}(z'-z)) \tag{5.79}$$

where 'Tr' means the sum over the diagonal elements of the matrix product $S_{\mathrm{F}}S_{\mathrm{F}}$. The second term in (5.78) is, from (2.141) and (2.144),

$$\begin{aligned} C^{\mathrm{T}}_{\alpha\gamma}S_{\mathrm{F}\gamma\beta}(z-z')S_{\mathrm{F}\beta\delta}(z'-z)C^{\mathrm{T}}_{\delta\alpha} &= \mathrm{Tr}(C^{\mathrm{T}}S_{\mathrm{F}}(z-z')S_{\mathrm{F}}(z'-z)C^{\mathrm{T}}) \\ &= \mathrm{Tr}(C^{\mathrm{T}}C^{\mathrm{T}}S_{\mathrm{F}}(z-z')S_{\mathrm{F}}(z'-z)) = -\mathrm{Tr}(S_{\mathrm{F}}(z-z')S_{\mathrm{F}}(z'-z)) \end{aligned} \tag{5.80}$$

using (2.142). The two terms in (5.78) are therefore the same, and the contribution of (5.77) is

$$+2g^2\int \mathrm{d}^4z\,\mathrm{d}^4z'\,D_{\mathrm{A}}(x-z)D_{\mathrm{A}}(y-z')\mathrm{Tr}(S_{\mathrm{F}}(z-z')S_{\mathrm{F}}(z'-z)). \tag{5.81}$$

The next step is left to Exercise 5.8.

Exercise 5.8 By inserting the Fourier expansions for D_{A} and S_{F} (see (2.138)), show that (5.81) can be written in the form (5.55) with $-\mathrm{i}\Pi^{(\mathrm{B})}_{\mathrm{A}}$ replaced by $-\mathrm{i}\Pi^{(\chi)}_{\mathrm{A}}$ where

$$-\mathrm{i}\Pi^{(\chi)}_{\mathrm{A}} = -2g^2\mathrm{Tr}\int \frac{\mathrm{d}^4k}{(2\pi)^4}\frac{1}{(\not{k}-M)}\frac{1}{(\not{k}-\not{p}-M)} \tag{5.82}$$

is the χ-loop contribution to the A self-energy.

It is clear that (5.82) contains a quadratic divergence (four powers of k in the numerator, two in the denominator). Following Appendix D of [45], we may isolate it as follows. We have

$$\begin{aligned} -\mathrm{i}\Pi^{(\chi)}_{\mathrm{A}} &= -2g^2\mathrm{Tr}\int \frac{\mathrm{d}^4k}{(2\pi)^4}\frac{\mathrm{Tr}[(\not{k}+M)(\not{k}-\not{p}+M)]}{(k^2-M^2)((k-p)^2-M^2)} \\ &= -8g^2\int \frac{\mathrm{d}^4k}{(2\pi)^4}\frac{k^2-k\cdot p+M^2}{(k^2-M^2)((k-p)^2-M^2)} \\ &= -4g^2\int \frac{\mathrm{d}^4k}{(2\pi)^4}\frac{[(k^2-M^2)+((k-p)^2-M^2)-p^2+4M^2]}{(k^2-M^2)((k-p)^2-M^2)} \\ &= -4g^2\int \frac{\mathrm{d}^4k}{(2\pi)^4}\frac{1}{k^2-M^2} - 4g^2\int \frac{\mathrm{d}^4k'}{(2\pi)^4}\frac{1}{k'^2-M^2} + \text{remainder}, \end{aligned} \tag{5.83}$$

where we have changed variable to $k' = k - p$ in the second term, and where the 'remainder' is at most logarithmically divergent. The quadratically divergent part of the χ-loop contribution to the A self-energy is therefore

$$\Pi^{(\chi,\,\mathrm{quad})}_{\mathrm{A}} = -8g^2\frac{1}{8\pi^2}\Lambda^2. \tag{5.84}$$

Quite remarkably, we see from (5.65) and (5.84) that the contribution from the fermion (χ) loop exactly cancels that from the boson loops. The dedicated reader may like to check that the quadratic divergences also cancel in the one-loop corrections to the B self-energy. Another example is provided by Exercise 5.9.

Exercise 5.9 Show that in the W–Z model the bosonic and fermionic contributions to the zero point energy exactly cancel each other. (This is a particular case of the general result that the vacuum energy of a SUSY-invariant theory vanishes; see Section 9.1.)

In their original paper, Wess and Zumino [19] remarked (with an acknowledgement to B. W. Lee) on the fact that their model turned out to have fewer divergences than a conventional renormalizable theory: the interactions were of standard renormalizable types, but there were special relations between the masses and coupling constants. They noted the cancellation of quadratic divergences in the A and B self-energies, and also pointed out that the logarithmic divergence of the vertex correction to the spinor–scalar and spinor–pseudoscalar interactions in (5.49) was also cancelled, leaving a finite vertex correction. They verified these statements in a one-loop approximation, using the theory with the auxiliary fields eliminated – the procedure we have followed in reproducing one of their results.

However, Wess and Zumino [19] then went on to explore (at one-loop level) the divergence structure of their model before the auxiliary fields (i.e. F and G of (4.125)) are eliminated. It then transpired that there were even more cancellations in this case, and that the only renormalization constant needed was a logarithmically divergent wavefunction renormalization, the same for all fields in the theory. For example, no mass corrections for the A or B particles were generated: the quadratic divergences in the self-energies cancelled as before, but also the remaining logarithmically divergent contribution was proportional to p^2, and hence associated with a wavefunction (or field-strength) renormalization (see, for example, Section 10.1.3 of [15]).

These one-loop results of Wess and Zumino [19] were extended to two loops by Iliopoulos and Zumino [51], who also gave a general proof, to all orders in perturbation theory, to show that the single, logarithmically divergent, wavefunction renormalization constant was sufficient to renormalize the theory, when analyzed without eliminating the auxiliary fields. What this means is that only the kinetic energy terms are renormalized, there being no renormalization of the other terms at all; that is to say, there is no renormalization of the superpotential W. This is one form of the 'SUSY non-renormalization theorem', which is now understood to hold generally (in perturbation theory) for any SUSY-invariant theory. This theorem was first established by 'supergraph' methods [52], which allow Feynman graphs involving all the fields in one supermultiplet, including auxiliary fields, to be calculated simultaneously. The first step towards this formalism is the introduction of 'superfields', which group together these supermultiplet components into one object. This will be the subject of the following chapter. However, the supergraph proof of the SUSY non-renormalization theorem is beyond our scope; we refer interested readers to Chapter 6 of [48].

6

Superfields

Thus far we have adopted (pretty much) a 'brute force', or 'do-it-yourself' approach, retreating quite often to explicit matrix expressions, and arriving at SUSY-invariant Lagrangians by direct construction. We might well wonder whether there is not a more general procedure which would somehow automatically generate SUSY-invariant interactions. Such a procedure is indeed available within the superfield approach, to which we now turn. This formalism has other advantages too. First, it gives us more insight into SUSY transformations, and their linkage with space–time translations; second, the appearance of the auxiliary field F is better motivated; and finally, and in practice rather importantly, the superfield notation is widely used in discussions of the MSSM.

6.1 SUSY transformations on fields

By way of a warm-up exercise, let's recall some things about space–time translations. A translation of coordinates takes the form

$$x'^{\mu} = x^{\mu} + a^{\mu} \tag{6.1}$$

where a^{μ} is a constant 4-vector. In the unprimed coordinate frame, observers use states $|\alpha\rangle$, $|\beta\rangle$, ..., and deal with amplitudes of the form $\langle\beta|\phi(x)|\alpha\rangle$, where $\phi(x)$ is scalar field. In the primed frame, observers evaluate ϕ at x', and use states $|\alpha\rangle' = U|\alpha\rangle$, ..., where U is unitary, in such a way that their matrix elements (and hence transition probabilities) are equal to those calculated in the unprimed frame:

$$\langle\beta|U^{-1}\phi(x')U|\alpha\rangle = \langle\beta|\phi(x)|\alpha\rangle. \tag{6.2}$$

Since this has to be true for all pairs of states, we can deduce

$$U^{-1}\phi(x')U = \phi(x) \tag{6.3}$$

or

$$U\phi(x)U^{-1} = \phi(x') = \phi(x+a). \tag{6.4}$$

For an infinitesimal translation, $x'^{\mu} = x^{\mu} + \epsilon^{\mu}$, we may write

$$U = 1 + \mathrm{i}\epsilon_{\mu}P^{\mu} \tag{6.5}$$

where the four operators P^{μ} are the *generators* of this transformation (cf. (4.6)); (6.4) then becomes

$$\begin{aligned}(1+\mathrm{i}\epsilon_{\mu}P^{\mu})\phi(x)(1-\mathrm{i}\epsilon_{\mu}P^{\mu}) &= \phi(x^{\mu}+\epsilon^{\mu}) \\ &= \phi(x^{\mu}) + \epsilon^{\mu}\frac{\partial\phi}{\partial x^{\mu}};\end{aligned} \tag{6.6}$$

that is,

$$\phi(x) + \delta\phi(x) = \phi(x) + \epsilon^{\mu}\partial_{\mu}\phi(x), \tag{6.7}$$

where (cf. (4.9))

$$\delta\phi(x) = \mathrm{i}\epsilon_{\mu}[P^{\mu}, \phi(x)] = \epsilon_{\mu}\partial^{\mu}\phi(x). \tag{6.8}$$

We therefore obtain the fundamental relation

$$\mathrm{i}[P^{\mu}, \phi(x)] = \partial^{\mu}\phi(x). \tag{6.9}$$

In (6.9) the P^{μ} are constructed from field operators – for example P^0 is the Hamiltonian, which is the spatial integral of the appropriate Hamiltonian density – and the canonical commutation relations of the fields must be consistent with (6.9). We used (6.9) in Section 4.2; see (4.36).

We can also look at (6.8) another way: we can say

$$\delta\phi = \epsilon_{\mu}\partial^{\mu}\phi = -\mathrm{i}\epsilon_{\mu}\hat{P}^{\mu}\phi, \tag{6.10}$$

where $\hat{P}^{\mu}$ is a *differential operator* acting on the *argument* of ϕ. Clearly $\hat{P}^{\mu} = \mathrm{i}\partial^{\mu}$ as usual.

We are now going to carry out analogous steps using SUSY transformations. This will entail enlarging the space of coordinates x^{μ} on which the fields can depend to include also *fermionic* degrees of freedom – specifically, spinor degrees of freedom θ and θ^*. Fields which depend on these spinorial degrees of freedom as well as on x are called *superfields*, and the extended space of x^{μ}, θ and θ^* is called *superspace*. Just as the operators P^{μ} generate (via the unitary operator U of (6.4)) a shift in the space–time argument of ϕ, so we expect to be able to construct analogous unitary operators from Q and $Q^{\dagger}$, which should similarly effect shifts in the spinorial arguments of the field. Actually, we shall see that the matter is rather more interesting than that, because a shift will also be induced in the space–time argument x; this is to be expected, given the link between the SUSY generators and the space–time translation generators P^μ embodied in the SUSY algebra (4.48). Having constructed these operators and seen what shifts they induce, we shall then look at the analogue of (6.10), and arrive at a differential operator representation of the SUSY generators, say $\hat{Q}$ and $\hat{Q}^\dagger$, the differentials in this case being with respect to the spinor degrees of freedom of superspace (i.e. θ and θ^*). We can close the circle by checking that the generators $\hat{Q}$ and $\hat{Q}^\dagger$ defined this way do indeed satisfy the SUSY algebra (4.48) (this step being analogous to checking that the angular momentum operators $\hat{\boldsymbol{L}} = -\mathrm{i}\boldsymbol{x} \times \boldsymbol{\nabla}$ obey the SU(2) algebra).

The basic idea is simple. We may write (6.4) as

$$\mathrm{e}^{\mathrm{i}x\cdot P}\phi(0)\mathrm{e}^{-\mathrm{i}x\cdot P} = \phi(x). \tag{6.11}$$

In analogy to this, let us consider a 'U' for a SUSY transformation which has the form

$$U(x, \theta, \theta^*) = \mathrm{e}^{\mathrm{i}x\cdot P}\mathrm{e}^{\mathrm{i}\theta\cdot Q}\mathrm{e}^{\mathrm{i}\bar{\theta}\cdot\bar{Q}}. \tag{6.12}$$

Here Q and Q^* (or $Q^{\dagger\mathrm{T}}$) are the (spinorial) SUSY generators met in Section 4.2, and θ and θ^* are spinor degrees of freedom associated with these SUSY 'translations'. Note that, as usual,

$$\theta\cdot Q \equiv \theta^{\mathrm{T}}(-\mathrm{i}\sigma_2)Q, \tag{6.13}$$

and

$$\bar{\theta}\cdot\bar{Q} \equiv \theta^\dagger(\mathrm{i}\sigma_2)Q^{\dagger\mathrm{T}}. \tag{6.14}$$

When the field $\phi(0)$ is transformed via '$U(x, \theta, \theta^*)\phi(0)U^{-1}(x, \theta, \theta^*)$', we expect to obtain a ϕ which is a function of x, but also now of the 'fermionic coordinates' θ and θ^*, so we shall write it as Φ, a superfield:

$$U(x, \theta, \theta^*)\Phi(0)U^{-1}(x, \theta, \theta^*) = \Phi(x, \theta, \theta^*). \tag{6.15}$$

Now consider the product of two ordinary spatial translation operators:

$$\mathrm{e}^{\mathrm{i}x\cdot P}\mathrm{e}^{\mathrm{i}a\cdot P} = \mathrm{e}^{\mathrm{i}(x+a)\cdot P}, \tag{6.16}$$

since all the components of P commute. We say that this product of translation operators 'induces the transformation $x \to x + a$ in parameter (coordinate) space'. We are going to generalize this by multiplying two U's of the form (6.12) together, and asking: *what transformations are induced in the space–time coordinates, and in the spinorial degrees of freedom*?

Such a product is

$$U(a, \xi, \xi^*)U(x, \theta, \theta^*) = \mathrm{e}^{\mathrm{i}a\cdot P}\mathrm{e}^{\mathrm{i}\xi\cdot Q}\mathrm{e}^{\mathrm{i}\bar{\xi}\cdot\bar{Q}}\mathrm{e}^{\mathrm{i}x\cdot P}\mathrm{e}^{\mathrm{i}\theta\cdot Q}\mathrm{e}^{\mathrm{i}\bar{\theta}\cdot\bar{Q}}. \tag{6.17}$$

Unlike in (6.16), it is *not* possible simply to combine all the exponents here, because the operators Q and $Q^\dagger$ do not commute – rather, they satisfy the algebra (4.48). However, as noted in Section 4.2, the components of P do commute with those of Q and $Q^\dagger$, so we can freely move the operator $\exp[\mathrm{i}x \cdot P]$ through the operators to the left of it, and combine it with $\exp[\mathrm{i}a \cdot P]$ to yield $\exp[\mathrm{i}(x + a) \cdot P]$, as in (6.16). The non-trivial part is

$$\mathrm{e}^{\mathrm{i}\xi\cdot Q}\mathrm{e}^{\mathrm{i}\bar{\xi}\cdot\bar{Q}}\mathrm{e}^{\mathrm{i}\theta\cdot Q}\mathrm{e}^{\mathrm{i}\bar{\theta}\cdot\bar{Q}}. \tag{6.18}$$

To simplify this, we use the Baker–Campbell–Hausdorff (B–C–H) identity:

$$\mathrm{e}^A\mathrm{e}^B = \mathrm{e}^{A + B + \frac{1}{2}[A,B] + \frac{1}{6}[[A,B],B] + \cdots}. \tag{6.19}$$

Let's apply (6.19) to the first two products in (6.18), taking $A = \mathrm{i}\xi \cdot Q$ and $B = \mathrm{i}\bar{\xi} \cdot \bar{Q}$. We get

$$\mathrm{e}^{\mathrm{i}\xi\cdot Q}\mathrm{e}^{\mathrm{i}\bar{\xi}\cdot\bar{Q}} = \mathrm{e}^{\mathrm{i}\xi\cdot Q + \mathrm{i}\bar{\xi}\cdot\bar{Q} - \frac{1}{2}[\xi\cdot Q, \bar{\xi}\cdot\bar{Q}] + \cdots} \tag{6.20}$$

Writing out the commutator in detail, we have

$$\begin{aligned}
[\xi \cdot Q, \bar{\xi} \cdot \bar{Q}] &= [\xi^1 Q_1 + \xi^2 Q_2, \xi_1^* Q_2^\dagger - \xi_2^* Q_1^\dagger] \\
&= [\xi^1 Q_1 + \xi^2 Q_2, -\xi^{2*} Q_2^\dagger - \xi^{1*} Q_1^\dagger] \\
&= [\xi^a Q_a, -\xi^{b*} Q_b^\dagger] \\
&= -\xi^a Q_a \xi^{b*} Q_b^\dagger + \xi^{b*} Q_b^\dagger \xi^a Q_a \\
&= \xi^a \xi^{b*} (Q_a Q_b^\dagger + Q_b^\dagger Q_a) \\
&= \xi^a \xi^{b*} (\sigma^\mu)_{ab} P_\mu
\end{aligned} \tag{6.21}$$

using (4.48). This means that life is not so bad after all: since P commutes with Q and $Q^\dagger$, there are no more terms in the B–C–H identity to calculate, and we have established the result

$$\mathrm{e}^{\mathrm{i}\xi\cdot Q}\mathrm{e}^{\mathrm{i}\bar{\xi}\cdot\bar{Q}} = \mathrm{e}^{\mathrm{i}A\cdot P}\mathrm{e}^{\mathrm{i}(\xi\cdot Q + \bar{\xi}\cdot\bar{Q})}, \tag{6.22}$$

where

$$A^\mu = \frac{1}{2}\mathrm{i}\xi^a(\sigma^\mu)_{ab}\xi^{b*}. \tag{6.23}$$

Note that we have moved the $\exp[\mathrm{i}A \cdot P]$ expression to the front, using the fact that P commutes with Q and $Q^\dagger$.

We pause in the development to comment immediately on (6.22): under this kind of transformation, the spacetime coordinate acquires an additional shift, namely A^{μ}, which is built out of the spinor parameters ξ and ξ^*.

Exercise 6.1 Explain why $\xi^a(\sigma^{\mu})_{ab}\xi^{b*}$ is a 4-vector.

[Hint: We know from Section 2.2 that the quantity $\xi^{\dagger}\bar{\sigma}^{\mu}\xi$ is a 4-vector, but the combination $\xi^a(\sigma^{\mu})_{ab}\xi^{b*}$ isn't quite the same, apparently. Actually it turns out to be the same, apart from a minus sign. First note that $\xi^a(\sigma^{\mu})_{ab}\xi^{b*} = -\bar{\xi}^{\dot{b}}(\sigma^{\mu})_{a\dot{b}}\xi^a$. Now lower both the indices on the ξ's using the ϵ symbols. You reach the expression $\bar{\xi}_{\dot{c}}\epsilon^{\dot{c}\dot{b}}(\sigma^{\mu\mathrm{T}})_{\dot{b}a}\epsilon^{ad}\xi_d$. Now use the relation between ϵ and $\mathrm{i}\sigma_2$ (i.e. the matrix $-c$ of Section 4.2), together with (4.30) to show that the expression is equal to $-\bar{\xi}\bar{\sigma}^{\mu}\xi$, or equivalently $-\xi^{\dagger}\bar{\sigma}^{\mu}\xi$.]

Continuing on with the reduction of (6.18), we consider

$$\mathrm{e}^{\mathrm{i}\xi\cdot Q}\mathrm{e}^{\mathrm{i}\bar{\xi}\cdot\bar{Q}}\mathrm{e}^{\mathrm{i}\theta\cdot Q}\mathrm{e}^{\mathrm{i}\bar{\theta}\cdot\bar{Q}} = \mathrm{e}^{\mathrm{i}A\cdot P}\mathrm{e}^{\mathrm{i}(\xi\cdot Q+\bar{\xi}\cdot\bar{Q})}\mathrm{e}^{\mathrm{i}\theta\cdot Q}\mathrm{e}^{\mathrm{i}\bar{\theta}\cdot\bar{Q}}, \tag{6.24}$$

and apply B–C–H to the second and third terms in the product on the right-hand side:

$$\begin{aligned}\mathrm{e}^{\mathrm{i}(\xi\cdot Q+\bar{\xi}\cdot\bar{Q})}\mathrm{e}^{\mathrm{i}\theta\cdot Q} &= \mathrm{e}^{\mathrm{i}(\xi\cdot Q+\bar{\xi}\cdot\bar{Q}+\theta\cdot Q)-\frac{1}{2}[\xi\cdot Q+\bar{\xi}\cdot\bar{Q},\theta\cdot Q]+\cdots}\\ &= \mathrm{e}^{\mathrm{i}(\xi\cdot Q+\bar{\xi}\cdot\bar{Q}+\theta\cdot Q)+\frac{1}{2}\theta^a(\sigma^{\mu})_{ab}\xi^{b*}P_{\mu}},\end{aligned} \tag{6.25}$$

using (6.21) and (4.39). The expression (6.18) is now

$$\mathrm{e}^{-\frac{1}{2}\xi^a(\sigma^{\mu})_{ab}\xi^{b*}P_{\mu}+\frac{1}{2}\theta^a(\sigma^{\mu})_{ab}\xi^{b*}P_{\mu}}\mathrm{e}^{\mathrm{i}(\xi\cdot Q+\bar{\xi}\cdot\bar{Q}+\theta\cdot Q)}\mathrm{e}^{\mathrm{i}\bar{\theta}\cdot\bar{Q}}. \tag{6.26}$$

We now apply B–C–H 'backwards' to the penultimate factor:

$$\mathrm{e}^{\mathrm{i}(\xi\cdot Q+\bar{\xi}\cdot\bar{Q}+\theta\cdot Q)} = \mathrm{e}^{\mathrm{i}(\xi+\theta)\cdot Q}\mathrm{e}^{\mathrm{i}\bar{\xi}\cdot\bar{Q}}\mathrm{e}^{\frac{1}{2}[(\xi+\theta)\cdot Q,\bar{\xi}\cdot\bar{Q}]}. \tag{6.27}$$

Evaluating the commutator as before leads to the final result

$$\mathrm{e}^{\mathrm{i}\xi\cdot Q}\mathrm{e}^{\mathrm{i}\bar{\xi}\cdot\bar{Q}}\mathrm{e}^{\mathrm{i}\theta\cdot Q}\mathrm{e}^{\mathrm{i}\bar{\theta}\cdot\bar{Q}} = \mathrm{e}^{\mathrm{i}[-\mathrm{i}\theta^a(\sigma^{\mu})_{ab}\xi^{b*}P_{\mu}]}\mathrm{e}^{\mathrm{i}(\xi+\theta)\cdot Q}\mathrm{e}^{\mathrm{i}(\bar{\xi}+\bar{\theta})\cdot\bar{Q}} \tag{6.28}$$

where in the final product we have again used (4.39) to add the exponents.

Exercise 6.2 Check (6.28).

Inspecting (6.28), we infer that the product $U(a,\xi,\xi^*)U(x,\theta,\theta^*)$ induces the transformations

$$\begin{aligned}&0\to\theta\to\theta+\xi\\ &0\to\theta^*\to\theta^*+\xi^*\\ &0\to x^{\mu}\to x^{\mu}+a^{\mu}-\mathrm{i}\theta^a(\sigma^{\mu})_{ab}\xi^{b*}.\end{aligned} \tag{6.29}$$

That is to say,

$$\begin{aligned} & U(a, \xi, \xi^*)U(x, \theta, \theta^*)\Phi(0)U^{-1}(x, \theta, \theta^*)U^{-1}(a, \xi, \xi^*) \\ & = U(a, \xi, \xi^*)\Phi(x, \theta, \theta^*)U^{-1}(a, \xi, \xi^*) \\ & = \Phi(x^\mu + a^\mu - \mathrm{i}\theta^a(\sigma^\mu)_{ab}\xi^{b*}, \theta + \xi, \theta^* + \xi^*). \end{aligned} \tag{6.30}$$

We now proceed with the second part of our SUSY extension of ordinary translations, namely the analogue of equation (6.10).

6.2 A differential operator representation of the SUSY generators

Equation (6.10) provided us with a differential operator representation of the generators of translations, by considering an infinitesimal displacement (the reader might care to recall similar steps for infinitesimal rotations, which lead to the usual representation of the angular momentum operators as $\hat{\boldsymbol{L}} = -\mathrm{i}\boldsymbol{x} \times \boldsymbol{\nabla}$). Analogous steps applied to (6.30) will lead to an explicit representation of the SUSY generators as certain differential operators. We will then check that they satisfy the anticommutation relations (4.48), just as the angular momentum operators satisfy the familiar SU(2) algebra.

We regard (6.30) as the result of applying the transformation parametrized by a, ξ, ξ^* to the field $\Phi(x, \theta, \theta^*)$. For an infinitesimal such transformation associated with ξ and ξ^*, the change in Φ is

$$\delta\Phi = -\mathrm{i}\theta^a(\sigma^\mu)_{ab}\xi^{b*}\partial_\mu\Phi + \xi^a\frac{\partial\Phi}{\partial\theta^a} + \xi^*_a\frac{\partial\Phi}{\partial\theta^*_a}. \tag{6.31}$$

Before proceeding, we check the notational consistency of (6.31). In Section 2.3 we stated the convention for summing over undotted labels, which was 'diagonally from top left to bottom right, as in $\xi^a\chi_a$'. For (6.31) to be consistent with this convention, it should be the case that the derivative $\partial/\partial\theta^a$ behaves as a 'χ_a'-type object. A quick way of seeing that this is likely to be correct is simply to calculate

$$\frac{\partial}{\partial\theta^a}(\theta^b\theta_b). \tag{6.32}$$

Consider $a = 1$. Now $\theta^b\theta_b = -2\theta^1\theta^2$ and so

$$\frac{\partial}{\partial\theta^1}(\theta^b\theta_b) = -2\theta^2 = 2\theta_1. \tag{6.33}$$

Similarly,

$$\frac{\partial}{\partial\theta^2}(\theta^b\theta_b) = 2\theta_2, \tag{6.34}$$

or generally

$$\frac{\partial}{\partial \theta^a}(\theta \cdot \theta) = 2\theta_a, \tag{6.35}$$

which at least checks the claim in this simple case. Similarly we stated the convention for products of dotted indices as $\psi_{\dot{a}}\zeta^{\dot{a}}$, and we related dotted-index quantities to complex conjugated quantities, via $\bar{\chi}_{\dot{a}} \equiv \chi_a^*$. Consider the last term in (6.31): since $\xi_a^* \equiv \bar{\xi}_{\dot{a}}$, it should be the case that $\partial/\partial\theta_a^*$ behaves as a '$\bar{\zeta}^{\dot{a}}$'-type (or equivalently as a 'ζ^{a*}') object.

Exercise 6.3 Check this by considering $\partial/\partial\theta_a^*(\bar{\theta} \cdot \bar{\theta})$.

In analogy with (6.10), we want to write (6.31) as

$$\delta\Phi = (-\mathrm{i}\xi \cdot \hat{Q} - \mathrm{i}\bar{\xi} \cdot \bar{\hat{Q}})\Phi = (-\mathrm{i}\xi^a \hat{Q}_a - \mathrm{i}\xi_a^* \hat{Q}^{\dagger a})\Phi. \tag{6.36}$$

Comparing (6.31) with (6.36), it is easy to identify $\hat{Q}_a$ as

$$\hat{Q}_a = \mathrm{i}\frac{\partial}{\partial \theta^a}. \tag{6.37}$$

There is a similar term in $\hat{Q}^{\dagger a}$, namely

$$\hat{Q}^{\dagger a} = \mathrm{i}\frac{\partial}{\partial \theta_a^*}, \tag{6.38}$$

and in addition another contribution given by

$$-\mathrm{i}\xi_a^* \hat{Q}^{\dagger a}\Phi = -\mathrm{i}\theta^a(\sigma^\mu)_{ab}\xi^{b*}\partial_\mu\Phi. \tag{6.39}$$

Our present objective is to verify that these $\hat{Q}$ operators satisfy the SUSY anticommutation relations (4.48). To do this, we need to deal with the lower-index operators $\hat{Q}_a^\dagger$ rather than $\hat{Q}^{\dagger a}$.

Exercise 6.4 Check that (6.38) can be converted to

$$\hat{Q}_a^\dagger = -\mathrm{i}\frac{\partial}{\partial \theta^{a*}}. \tag{6.40}$$

As regards (6.39), we use $\xi_a^* \hat{Q}^{\dagger a} = -\xi^{a*}\hat{Q}_a^\dagger$ (see Exercise 2.6 (b) after equation (2.76)), and $\theta^a \xi^{b*} = -\xi^{b*}\theta^a$, followed by an interchange of the indices a and b to give finally

$$\hat{Q}_a^\dagger = -\mathrm{i}\frac{\partial}{\partial \theta^{a*}} + \theta^b(\sigma^\mu)_{ba}\partial_\mu. \tag{6.41}$$

It is now a useful exercise to check that the explicit representations (6.37) and (6.41) do indeed result in the required relations

$$[\hat{Q}_a, \hat{Q}_b^\dagger] = \mathrm{i}(\sigma^\mu)_{ab}\partial_\mu = (\sigma^\mu)_{ab}\hat{P}_\mu, \tag{6.42}$$

as well as $[\hat{Q}_a, \hat{Q}_b] = [\hat{Q}_a^\dagger, \hat{Q}_b^\dagger] = 0$. We have therefore produced a representation of the SUSY generators in terms of fermionic parameters, and derivatives with respect to them, which satisfies the SUSY algebra (4.48).

6.3 Chiral superfields, and their (chiral) component fields

Suppose now that a superfield $\Phi(x, \theta, \theta^*)$ does not in fact depend on θ^*, only on x and θ: $\Phi(x, \theta)$.[1] Consider the expansion of such a Φ in powers of θ. Due to the fermionic nature of the variables θ, which implies that $(\theta_1)^2 = (\theta_2)^2 = 0$, there will only be three terms in the expansion, namely a term independent of θ, a term linear in θ and a term involving $\frac{1}{2}\theta \cdot \theta = -\theta_1\theta_2$:

$$\Phi(x, \theta) = \phi(x) + \theta \cdot \chi(x) + \frac{1}{2}\theta \cdot \theta F(x). \tag{6.43}$$

This is the most general form of such a superfield (which depends only on x and θ), and it depends on three *component fields*, ϕ, χ and F. We have of course deliberately given these component fields the same names as those in our previous chiral supermultiplet. We shall now verify that the transformation law (6.36) for the superfield Φ, with $\hat{Q}$ given by (6.37) and $\hat{Q}^\dagger$ by (6.41), implies precisely the previous transformations (5.2) for the component fields ϕ, χ and F, thus justifying this identification.

We have

$$\begin{aligned}\delta\Phi &= (-\mathrm{i}\xi^a\hat{Q}_a - \mathrm{i}\xi_a^*\hat{Q}^{\dagger a})\Phi = (-\mathrm{i}\xi^a\hat{Q}_a + \mathrm{i}\xi^{a*}\hat{Q}_a^\dagger)\Phi \\ &= \left(\xi^a\frac{\partial}{\partial\theta^a} + \xi^{a*}\frac{\partial}{\partial\theta^{a*}} + \mathrm{i}\xi^{a*}\theta^b(\sigma^\mu)_{ba}\partial_\mu\right)\left[\phi(x) + \theta^c\chi_c + \frac{1}{2}\theta\cdot\theta F\right] \\ &\equiv \delta_\xi\phi + \theta^a\delta_\xi\chi_a + \frac{1}{2}\theta\cdot\theta\delta_\xi F.\end{aligned} \tag{6.44}$$

We evaluate the derivatives in the second line as follows. First, we have

$$\frac{\partial}{\partial\theta^a}\left[\theta^c\chi_c + \frac{1}{2}\theta\cdot\theta\right] = \chi_a + \theta_a, \tag{6.45}$$

[1] Such a superfield is usually called a 'left-chiral superfield', because (see (6.43)) it contains only the L-type spinor χ, and not the R-type spinor ψ (or $\chi^\dagger$).

using (6.35), so that the $\xi^a \partial/\partial\theta^a$ term yields

$$\xi^a \chi_a + \theta^a \xi_a F. \tag{6.46}$$

Next, the term in $\partial/\partial\theta^{a*}$ vanishes since Φ doesn't depend on θ^*. The remaining term is

$$\mathrm{i}\xi^{a*}\theta^b(\sigma^\mu)_{ba}\partial_\mu\phi + \mathrm{i}\xi^{a*}\theta^b(\sigma^\mu)_{ba}\theta^c\partial_\mu\chi_c; \tag{6.47}$$

note that the fermionic nature of θ precludes any cubic term in θ. The first term in (6.47) can alternatively be written as

$$-\mathrm{i}\theta^b(\sigma^\mu)_{ba}\xi^{a*}\partial_\mu\phi. \tag{6.48}$$

Referring to (6.44) we can therefore identify the part independent of θ as

$$\delta_\xi\phi = \xi^a\chi_a, \tag{6.49}$$

and the part linear in θ as

$$\theta^a\delta_\xi\chi_a = \theta^a(\xi_a F - \mathrm{i}(\sigma^\mu)_{ab}\xi^{b*}\partial_\mu\phi). \tag{6.50}$$

Since (6.50) has to be true for all θ we can remove the θ^a throughout, and then (6.49) and (6.50) indeed reproduce (5.2) for the fields ϕ and χ (recall that $(\mathrm{i}\sigma_2\xi^*)_b = \xi^{b*}$).

We are left with the second term of (6.47), which is bilinear in θ, and which ought to yield $\delta_\xi F$. We manipulate this term as follows. First, we write the general product $\theta^a\theta^b$ in terms of the scalar product $\theta\cdot\theta$ by using the result of Exercise 6.5 which follows.

Exercise 6.5 Show that $\theta^a\theta^b = -\frac{1}{2}\epsilon^{ab}\theta\cdot\theta$, where $\epsilon^{12}=1, \epsilon^{21}=-1, \epsilon^{11}=\epsilon^{22}=0$; also that $\theta^{a*}\theta^{b*} = +\frac{1}{2}\epsilon^{ab}\bar{\theta}\cdot\bar{\theta}$.

The second term in (6.47) is then

$$-\mathrm{i}\xi^{a*}(\sigma^{\mu\mathrm{T}})_{ab}\epsilon^{bc}\partial_\mu\chi_c\frac{1}{2}\theta\cdot\theta. \tag{6.51}$$

Comparing this with (6.44) we deduce

$$\delta_\xi F = -\mathrm{i}\xi^{a*}(\sigma^{\mu\mathrm{T}})_{ab}\epsilon^{bc}\partial_\mu\chi_c. \tag{6.52}$$

Exercise 6.6 Verify that this is in fact the same as the $\delta_\xi F$ given in (5.2) (remember that '$\xi^\dagger$' means (ξ_1^*, ξ_2^*), not (ξ^{1*}, ξ^{2*})).

So the chiral superfield $\Phi(x,\theta)$ of (6.43) contains the component fields ϕ, χ and F transforming correctly under SUSY transformations; we say that the chiral superfield provides a linear representation of the SUSY algebra. Note that *three* component fields (ϕ, χ and F) are required for this result: here is a more 'deductive' justification for the introduction of the field F.

We close this rather formal section with a most important observation: *the change in the F field, (6.52), is actually a total derivative, since the parameters ξ are independent of x; it follows that, in general, the 'F-component' of a chiral superfield, in the sense of the expansion (6.43), will always transform by a total derivative, and will therefore automatically correspond to a SUSY-invariant Action.*

We now consider products of chiral superfields, and show how to exploit the italicized remark so as to obtain SUSY-invariant interactions; in particular, those of the W–Z model introduced in Chapter 5.

6.4 Products of chiral superfields

Let Φ_i be a left-chiral superfield where, as in Chapter 5, the suffix i labels the gauge and flavour degrees of freedom of the component fields. Φ_i has an expansion of the form (6.43):

$$\Phi_i(x,\theta) = \phi_i(x) + \theta\cdot\chi_i(x) + \frac{1}{2}\theta\cdot\theta F_i(x). \tag{6.53}$$

Consider now the product of two such superfields:

$$\Phi_i\Phi_j = \left(\phi_i + \theta\cdot\chi_i + \frac{1}{2}\theta\cdot\theta F_i\right)\left(\phi_j + \theta\cdot\chi_j + \frac{1}{2}\theta\cdot\theta F_j\right). \tag{6.54}$$

On the right-hand side there are the following terms:

$$\text{independent of } \theta\text{: } \phi_i\phi_j; \tag{6.55}$$

$$\text{linear in } \theta\text{: } \theta\cdot(\chi_i\phi_j + \chi_j\phi_i); \tag{6.56}$$

$$\text{bilinear in } \theta\text{: } \frac{1}{2}\theta\cdot\theta(\phi_i F_j + \phi_j F_i) + \theta\cdot\chi_i\,\theta\cdot\chi_j. \tag{6.57}$$

In the second term of (6.57) we use the result given in Exercise 6.5 above to write it as

$$\begin{aligned}\theta\cdot\chi_i\,\theta\cdot\chi_j &= \theta^a\chi_{ia}\theta^b\chi_{jb} = -\theta^a\theta^b\chi_{ia}\chi_{jb}\\ &= \frac{1}{2}\epsilon^{ab}\theta\cdot\theta\chi_{ia}\chi_{jb} = \frac{1}{2}\theta\cdot\theta(\chi_{i1}\chi_{j2} - \chi_{i2}\chi_{j1})\\ &= -\frac{1}{2}\theta\cdot\theta\chi_i\cdot\chi_j.\end{aligned} \tag{6.58}$$

Hence the term in the product (6.54) which is bilinear in θ is

$$\frac{1}{2}\theta\cdot\theta(\phi_i F_j + \phi_j F_i - \chi_i\cdot\chi_j). \tag{6.59}$$

Exercise 6.7 Show that the terms in the product (6.54) which are cubic and quartic in θ vanish.

Altogether, then, we have shown that if the product (6.54) is itself expanded in component fields via

$$\Phi_i \Phi_j = \phi_{ij} + \theta \cdot \chi_{ij} + \frac{1}{2}\theta \cdot \theta F_{ij}, \tag{6.60}$$

then

$$\phi_{ij} = \phi_i \phi_j, \quad \chi_{ij} = \chi_i \phi_j + \phi_j \chi_i, \quad F_{ij} = \phi_i F_j + \phi_j F_i - \chi_i \cdot \chi_j. \tag{6.61}$$

Suppose now that we introduce a quantity W_{quad} defined by

$$W_{\text{quad}} = \frac{1}{2} M_{ij} \Phi_i \Phi_j \bigg|_F, \tag{6.62}$$

where '$|_F$' means 'the F-component of' (i.e. the coefficient of $\frac{1}{2}\theta \cdot \theta$ in the product). Here M_{ij} is taken to be symmetric in i and j. Then

$$\begin{aligned} W_{\text{quad}} &= \frac{1}{2} M_{ij}(\phi_i F_j + \phi_j F_i - \chi_i \cdot \chi_j) \\ &= M_{ij}\phi_i F_j - \frac{1}{2} M_{ij} \chi_i \cdot \chi_j. \end{aligned} \tag{6.63}$$

Referring back to the italicized comment at the end of the previous subsection, the fact that (6.63) is the F-component of a chiral superfield (which is the product of two other such superfields, in this case), guarantees that the terms in (6.63) provide a SUSY-invariant Action. In fact, they are precisely the terms involving M_{ij} in the W–Z model of Chapter 5: see (5.3) with W_i given by the first term in (5.19), and W_{ij} given by the first term in (5.7). Note also that our W_{quad} has exactly the same form, as a function of Φ_i and Φ_j, as the M_{ij} part of W in (5.9) had, as a function of ϕ_i and ϕ_j.

Thus encouraged, let us go on to consider the product of three chiral superfields:

$$\begin{aligned} \Phi_i \Phi_j \Phi_k = &\left[\phi_i \phi_j + \theta \cdot (\chi_i \phi_j + \chi_j \phi_i) + \frac{1}{2}\theta \cdot \theta(\phi_i F_j + \phi_j F_i - \chi_i \cdot \chi_j)\right] \\ &\times \left[\phi_k + \theta \cdot \chi_k + \frac{1}{2}\theta \cdot \theta F_k\right]. \end{aligned} \tag{6.64}$$

As our interest is confined to obtaining candidates for SUSY-invariant Actions, we shall only be interested in the F component. Inspection of (6.64) yields the obvious terms

$$\phi_i \phi_j F_k + \phi_j \phi_k F_i + \phi_k \phi_i F_j - \chi_i \cdot \chi_j \phi_k. \tag{6.65}$$

In addition, the term $\theta \cdot (\chi_i \phi_j + \chi_j \phi_i)\theta \cdot \chi_k$ can be re-written as in (6.58) to give

$$-\frac{1}{2}\theta \cdot \theta(\chi_i \phi_j + \chi_j \phi_i) \cdot \chi_k. \tag{6.66}$$

So altogether

$$\Phi_i \Phi_j \Phi_k \big|_F = \phi_i \phi_j F_k + \phi_j \phi_k F_i + \phi_k \phi_i F_j - \chi_i \cdot \chi_j \phi_k - \chi_j \cdot \chi_k \phi_i - \chi_i \cdot \chi_k \phi_j. \tag{6.67}$$

Let us now consider the cubic analogue of (6.62), namely

$$W_{\text{cubic}} = \frac{1}{6} y_{ijk} \Phi_i \Phi_j \Phi_k \bigg|_F, \tag{6.68}$$

where the coefficients y_{ijk} are totally symmetric in i, j and k. Then from (6.67) we immediately obtain

$$W_{\text{cubic}} = \frac{1}{2} y_{ijk} \phi_i \phi_j F_k - \frac{1}{2} y_{ijk} \chi_i \cdot \chi_j \phi_k. \tag{6.69}$$

Sure enough, the first term here is precisely the first term in (5.3) with W_i given by the second (y_{ijk}) term in (5.19), while the second term in (6.69) is the second term in (5.3) with W_{ij} given by the y_{ijk} term in (5.7). Note, again, that our W_{cubic} has exactly the same form, as a function of the Φ's, as the y_{ijk} part of the W in (5.9), as a function of the ϕ's.

Thus we have shown that all the interactions found in Chapter 5 can be expressed as F-components of products of superfields, a result which guarantees the SUSY-invariance of the associated Action. Of course, we must also include the hermitian conjugates of the terms considered here. As all the interactions are generated from the superfield products in W_{quad} and W_{cubic}, such W's are called *superpotentials*. The full superpotential for the W–Z model is thus

$$W = \frac{1}{2} M_{ij} \Phi_i \Phi_j + \frac{1}{6} y_{ijk} \Phi_i \Phi_j \Phi_k, \tag{6.70}$$

it being understood that the F-component is to be taken in the Lagrangian.

That understanding is often made explicit by integrating over θ_1 and θ_2. Integrals over such anticommuting variables are defined by the following rules:

$$\int \mathrm{d}\theta_1 1 = 0; \int \mathrm{d}\theta_1\ \theta_1 = 1; \int \mathrm{d}\theta_1 \int \mathrm{d}\theta_2\ \theta_2 \theta_1 = 1 \tag{6.71}$$

(see Appendix O of [7], for example). These rules imply that

$$\int \mathrm{d}\theta_1 \int \mathrm{d}\theta_2\ \frac{1}{2}\theta \cdot \theta = \int \mathrm{d}\theta_1 \int \mathrm{d}\theta_2\ \theta_2 \theta_1 = 1. \tag{6.72}$$

On the other hand, we can write

$$d\theta_1\, d\theta_2 = -d\theta_2\, d\theta_1 = -\frac{1}{2} d\theta \cdot d\theta \equiv d^2\theta. \tag{6.73}$$

It then follows that

$$\int d^2\theta\ W = \text{coefficient of } \tfrac{1}{2}\theta \cdot \theta \text{ in } W \text{ (i.e. the } F \text{ component).} \tag{6.74}$$

Such integrals are commonly used to project out the desired parts of superfield expressions.

As already noted, the functional form of (6.70) is the same as that of (5.9), which is why they are both called W. Note, however, that the W of (6.70) includes, of course, *all* the interactions of the W–Z model, not only those involving the ϕ fields alone. In the MSSM, superpotentials of the form (6.70) describe the non-gauge interactions of the fields – that is, in fact, interactions involving the Higgs supermultiplets; in this case the quadratic and cubic products of the Φ's must be constructed so as to be singlets (invariant) under the gauge groups.

The reader might suspect that, just as the interactions of the W–Z model can be compactly expressed in terms of superfields, so can the terms of the free Lagrangian (5.1). This can certainly be done, but it requires the formalism of the next section.

6.5 A technical annexe: other forms of chiral superfield

The thoughtful reader may be troubled by the following thought. Our development has been based on the form (6.12) for the unitary operator associated with finite SUSY transformations. We could, however, have started, instead, from

$$U_{\text{real}}(x, \theta, \theta^*) = e^{ix\cdot P} e^{i[\theta\cdot Q + \bar{\theta}\cdot\bar{Q}]}, \tag{6.75}$$

and since Q and $Q^\dagger$ do not commute, (6.75) is not the same as (6.12). Indeed, (6.75) might be regarded as more natural, and certainly more in line with the angular momentum case, which also involves non-commuting generators, and where the corresponding unitary operator is $\exp[i\boldsymbol{\alpha} \cdot \boldsymbol{J}]$. In this case, we shall write the superfield as $\Phi_{\text{real}}(x, \theta, \theta^*)$, where (cf. (6.11) and (6.15))

$$\Phi_{\text{real}}(x, \theta, \theta^*) = e^{i[\theta\cdot Q + \bar{\theta}\cdot\bar{Q}]}\Phi(x, 0, 0)e^{-i[\theta\cdot Q + \bar{\theta}\cdot\bar{Q}]}. \tag{6.76}$$

Now note that if $\Phi^\dagger(x, 0, 0) = \Phi(x, 0, 0)$, then $\Phi^\dagger_{\text{real}}(x, \theta, \theta^*) = \Phi_{\text{real}}(x, \theta, \theta^*)$. For this reason a superfield generated in this way is called 'real type' superfield. It is easy to check that an analogous statement is *not* true for the superfield Φ

generated via (6.15): the latter is called a 'type-I' superfield, denoted (if necessary) by $\Phi_{\rm I}(x, \theta, \theta^*)$. Similarly, the U of (6.12) may be denoted by $U_{\rm I}(x, \theta, \theta^*)$.

In the case of (6.75), the induced transformation corresponding to (6.29) is

$$\begin{aligned}
0 &\to \theta \to \theta + \xi \\
0 &\to \theta^* \to \theta^* + \xi^* \\
0 &\to x^\mu \to x^\mu + a^\mu + \frac{1}{2}\mathrm{i}\xi^a(\sigma^\mu)_{ab}\theta^{b*} - \frac{1}{2}\mathrm{i}\theta^a(\sigma^\mu)_{ab}\xi^{b*} \\
&= x^\mu + a^\mu + \frac{1}{2}\mathrm{i}\xi^\dagger\bar{\sigma}^\mu\theta - \frac{1}{2}\mathrm{i}\theta^\dagger\bar{\sigma}^\mu\xi
\end{aligned} \tag{6.77}$$

where the last line follows via Exercise 6.1 above.[2] Note that the quantity $\mathrm{i}(\xi^\dagger\bar{\sigma}^\mu\theta - \theta^\dagger\bar{\sigma}^\mu\xi)$ is real, again in contrast to the analogous shift (6.29) for a type-I superfield. We can again find differential operators representing the SUSY generators by expanding the change in the field up to first order in ξ and ξ^*, as in (6.31), and this will lead to different expressions from those given in (6.37) and (6.41). However, the new operators will be found to satisy the *same* SUSY algebra (4.48).

We could also imagine using

$$U_{\rm II}(x, \theta, \theta^*) = \mathrm{e}^{\mathrm{i}x\cdot P}\mathrm{e}^{\mathrm{i}\bar{\theta}\cdot\bar{Q}}\mathrm{e}^{\mathrm{i}\theta\cdot Q}, \tag{6.78}$$

which is not the same either, and for which the induced transformation is

$$\begin{aligned}
0 &\to \theta \to \theta + \xi \\
0 &\to \theta^* \to \theta^* + \xi^* \\
0 &\to x^\mu \to x^\mu + a^\mu + \mathrm{i}\xi^a(\sigma^\mu)_{ab}\theta^{b*}.
\end{aligned} \tag{6.79}$$

The corresponding superfield is of 'type-II', denoted by $\Phi_{\rm II}(x, \theta, \theta^*)$. Yet a third set of (differential operator) generators will be found, but again they'll satisfy the same SUSY algebra (4.48).

The three types of superfield are related to each other in a simple way. We have

$$\begin{aligned}
\Phi_{\rm real}(x, \theta, \theta^*) &= \mathrm{e}^{\mathrm{i}(\theta\cdot Q + \bar{\theta}\cdot\bar{Q})}\Phi(x, 0, 0)\mathrm{e}^{-\mathrm{i}(\theta\cdot Q + \bar{\theta}\cdot\bar{Q})} \\
&= \mathrm{e}^{-\mathrm{i}B\cdot P}\mathrm{e}^{\mathrm{i}\theta\cdot Q}\mathrm{e}^{\mathrm{i}\bar{\theta}\cdot\bar{Q}}\Phi(x, 0, 0)\mathrm{e}^{-\mathrm{i}\bar{\theta}\cdot\bar{Q}}\mathrm{e}^{-\mathrm{i}\theta\cdot Q}\mathrm{e}^{\mathrm{i}B\cdot P}
\end{aligned} \tag{6.80}$$

where we have used (6.22) and (6.23) with $\xi \to \theta$ so that

$$B^\mu = \frac{1}{2}\mathrm{i}\theta^a(\sigma^\mu)_{ab}\theta^{b*}. \tag{6.81}$$

[2] The x^μ transformation is essentially the one introduced by Volkov and Akulov [27].

But the second line of (6.80) can be written as

$$e^{-iB\cdot P}\Phi_{\rm I}(x,\theta,\theta^*)e^{iB\cdot P} = \Phi_{\rm I}\left(x^\mu - \frac{1}{2}i\theta^a(\sigma^\mu)_{ab}\theta^{b*},\theta,\theta^*\right). \tag{6.82}$$

Similar steps can be followed for $\Phi_{\rm II}$, and we obtain

$$\begin{aligned}\Phi_{\rm real}(x,\theta,\theta^*) &= \Phi_{\rm I}\left(x^\mu - \frac{1}{2}i\theta^a(\sigma^\mu)_{ab}\theta^{b*},\theta,\theta^*\right)\\ &= \Phi_{\rm II}\left(x^\mu + \frac{1}{2}i\theta^a(\sigma^\mu)_{ab}\theta^{b*},\theta,\theta^*\right).\end{aligned} \tag{6.83}$$

Any of the three superfields $\Phi_{\rm real}(x,\theta,\theta^*)$, $\Phi_{\rm I}(x,\theta,\theta^*)$, $\Phi_{\rm II}(x,\theta,\theta^*)$ can be expanded as a power series in θ and θ^*, just as we did for $\Phi(x,\theta)$. But such an expansion will contain a lot more terms than (6.43), and will involve more component fields than ϕ, χ and F. These general superfields (depending on both θ and θ^*) will provide a representation of the SUSY algebra, but it will be a *reducible* one, in the sense that we'd find that we could pick out sets of components that only transformed among themselves – such as those in a chiral supermultiplet, for example. The *irreducible* sets of fields can be selected out from the beginning by applying a suitable constraint. For example, we got straight to the irreducible left chiral supermultiplet by starting with what we now call $\Phi_{\rm I}(x,\theta,\theta^*)$ and requiring it not to depend on θ^*. That is to say, we required

$$\frac{\partial}{\partial\theta^*_a}\Phi_{\rm I}(x,\theta,\theta^*) = 0. \tag{6.84}$$

The reason that this works is that the operator $\partial/\partial\theta^*_a$ commutes with the SUSY transformation (6.31): that is,

$$\frac{\partial}{\partial\theta^*_a}(\delta\Phi_{\rm I}) = \delta\left(\frac{\partial}{\partial\theta^*_a}\Phi_{\rm I}\right). \tag{6.85}$$

Hence if $\Phi_{\rm I}$ does not depend on θ^*, neither does $\delta\Phi_{\rm I}$, which means that the surviving components form a representation by themselves.

We know that the components of $\Phi_{\rm I}(x,\theta)$ are precisely those of the L-chiral multiplet. A natural question to ask is: how is an L-chiral multiplet described by a real superfield $\Phi_{\rm real}(x,\theta,\theta^*)$? The answer is provided by (6.83), namely

$$\Phi^{\rm L}_{\rm real}(x,\theta,\theta^*) = \Phi_{\rm I}\left(x^\mu - \frac{1}{2}i\theta^a(\sigma^\mu)_{ab}\theta^{b*},\theta\right) \tag{6.86}$$

where

$$\Phi_{\rm I}(x,\theta) = \phi(x) + \theta\cdot\chi(x) + \frac{1}{2}\theta\cdot\theta F(x). \tag{6.87}$$

Hence

$$\Phi^{\mathrm{L}}_{\mathrm{real}}(x,\theta,\theta^*) = \phi\left(x^\mu - \frac{1}{2}\mathrm{i}\theta^a(\sigma^\mu)_{ab}\theta^{b*}\right) + \theta\cdot\chi\left(x^\mu - \frac{1}{2}\mathrm{i}\theta^a(\sigma^\mu)_{ab}\theta^{b*}\right)$$
$$+ \frac{1}{2}\theta\cdot\theta F\left(x^\mu - \frac{1}{2}\mathrm{i}\theta^a(\sigma^\mu)_{ab}\theta^{b*}\right). \tag{6.88}$$

The fields on the right-hand side of (6.88) may be expanded as a Taylor series about the point x, and we obtain

$$\Phi^{\mathrm{L}}_{\mathrm{real}}(x,\theta,\theta^*) = \phi(x) + \theta\cdot\chi(x) + \frac{1}{2}\theta\cdot\theta F(x) - \frac{1}{2}\mathrm{i}\theta^a(\sigma^\mu)_{ab}\theta^{b*}\partial_\mu\phi$$
$$-\frac{1}{2}\mathrm{i}\theta\cdot\partial_\mu\chi\,\theta^a(\sigma^\mu)_{ab}\theta^{b*} - \frac{1}{8}\theta^a(\sigma^\mu)_{ab}\theta^{b*}\theta^c(\sigma^\nu)_{cd}\theta^{d*}\partial_\mu\partial_\nu\phi, \tag{6.89}$$

since terms of higher degree than the second in θ or θ^* vanish. Using equation (6.58), the penultimate term can be written as

$$+\frac{1}{4}\mathrm{i}\theta\cdot\theta\partial_\mu\chi^a(\sigma^\mu)_{ab}\theta^{b*}. \tag{6.90}$$

The last term can be simplified as follows:

$$\begin{aligned}
\theta^a(\sigma^\mu)_{ab}\theta^{b*}\theta^c(\sigma^\nu)_{cd}\theta^{d*} &= -\theta^a\theta^c\theta^{b*}\theta^{d*}(\sigma^\mu)_{ab}(\sigma^\nu)_{cd}\\
&= -\left(-\frac{1}{2}\epsilon^{ac}\theta\cdot\theta\right)\left(+\frac{1}{2}\epsilon^{bd}\bar\theta\cdot\bar\theta\right)(\sigma^\mu)_{ab}(\sigma^\nu)_{cd}\\
&= -\frac{1}{4}\theta\cdot\theta\,\bar\theta\cdot\bar\theta\epsilon^{ca}(\sigma^\mu)_{ab}\epsilon^{bd}(\sigma^\nu)_{cd}\\
&= -\frac{1}{4}\theta\cdot\theta\,\bar\theta\cdot\bar\theta(-\bar\sigma^{\mu\mathrm{T}})^{cd}(\sigma^\nu)_{cd} \quad \text{using (4.30)}\\
&= \frac{1}{4}\theta\cdot\theta\,\bar\theta\cdot\bar\theta\,\mathrm{Tr}(\bar\sigma^\mu\sigma^\nu) = \frac{1}{2}\theta\cdot\theta\,\bar\theta\cdot\bar\theta g^{\mu\nu}.
\end{aligned} \tag{6.91}$$

Finally therefore

$$\Phi^{\mathrm{L}}_{\mathrm{real}}(x,\theta,\theta^*) = \phi(x) + \theta\cdot\chi(x) + \frac{1}{2}\theta\cdot\theta\,F(x) - \frac{1}{2}\mathrm{i}\theta^a(\sigma^\mu)_{ab}\theta^{b*}\partial_\mu\phi$$
$$+\frac{1}{4}\mathrm{i}\theta\cdot\theta\,\partial_\mu\chi^a(\sigma^\mu)_{ab}\theta^{b*} - \frac{1}{16}\theta\cdot\theta\,\bar\theta\cdot\bar\theta\,\partial^2\phi. \tag{6.92}$$

It turns out that a similar story can be told for the R-chiral field, using $\Phi_{\mathrm{II}}(x,\theta,\theta^*)$ restricted to be independent of θ. Indeed we have

$$\begin{aligned}
\Phi^\dagger_{\mathrm{II}}(x,\theta,\theta^*) &= \left[\mathrm{e}^{\mathrm{i}\bar\theta\cdot\bar Q}\mathrm{e}^{\mathrm{i}\theta\cdot Q}\Phi(x,0,0)\mathrm{e}^{-\mathrm{i}\theta\cdot Q}\mathrm{e}^{-\mathrm{i}\bar\theta\cdot\bar Q}\right]^\dagger\\
&= \mathrm{e}^{\mathrm{i}\theta\cdot Q}\mathrm{e}^{\mathrm{i}\bar\theta\cdot\bar Q}\Phi^\dagger(x,0,0)\mathrm{e}^{-\mathrm{i}\bar\theta\cdot\bar Q}\mathrm{e}^{-\mathrm{i}\theta\cdot Q}
\end{aligned} \tag{6.93}$$

which is a type-I superfield built on $\Phi^\dagger(x,0,0)$, whereas $\Phi_{\rm I}(x,\theta,\theta^*)$ was built on $\Phi(x,0,0)$. In a sense, type-I and type-II fields are conjugates of each other, and the simplest description of an R-chiral field is via the conjugate of $\Phi_{\rm I}^{\rm L}(x,\theta)$:

$$\Phi_{\rm II}^{\rm R}(x,\theta^*) = \phi^\dagger(x) + \bar\theta\cdot\bar\chi(x) + \frac{1}{2}\bar\theta\cdot\bar\theta F^\dagger(x). \tag{6.94}$$

$\bar\chi$ is of course an R-chiral (dotted spinor) field (see Section 2.3).

We can now return to the question of representing the free Lagrangian (5.1) in terms of superfields. A glance at (6.92) suggests that the desired terms may be contained in the product

$$\left(\Phi_{\rm real}^{\rm L}(x,\theta,\theta^*)\right)^\dagger \Phi_{\rm real}^{\rm L}(x,\theta,\theta^*). \tag{6.95}$$

The essential point is that, in such a product, the field of highest dimension must transform as a total derivative. In the expansion of $\Phi_{\rm I}(x,\theta)\equiv\Phi(x,\theta)$ this is the coefficient of $\theta\cdot\theta$, namely the field F. Similarly, in the product $\Phi_i(x,\theta)\Phi_j(x,\theta)$ it is the 'F-component'. In the case of the product (6.95) it is the coefficient of $\theta\cdot\theta\,\bar\theta\cdot\bar\theta$, which is called the '$D$-component' (the terminology is taken from the superfield formalism for vector supermultiplets; see [42] Chapter 3). Writing out the product (6.95), the terms which contribute to the D-term are (dropping the subscripts on the component fields)

$$\left[-\frac{1}{16}\phi^\dagger\partial^2\phi - \frac{1}{16}\partial^2\phi\,\phi^\dagger + \frac{1}{4}F^\dagger F\right]\theta\cdot\theta\,\bar\theta\cdot\bar\theta \tag{6.96}$$

$$+\frac{1}{4}{\rm i}\bar\chi\cdot\bar\theta\theta\cdot\theta\,\partial_\mu\chi^a(\sigma^\mu)_{ab}\theta^{b*} - \frac{1}{4}{\rm i}\theta^a(\sigma^\mu)_{ab}\partial_\mu\chi^{b*}\bar\theta\cdot\bar\theta\,\theta\cdot\chi \tag{6.97}$$

$$+\frac{1}{4}\partial\phi^\dagger\theta^a(\sigma^\mu)_{ab}\theta^{b*}\theta^c(\sigma^\nu)_{cd}\theta^{d*}\partial_\nu\phi. \tag{6.98}$$

The first two terms of (6.96) are equivalent to

$$+\frac{1}{8}\partial^\mu\phi^\dagger\partial_\mu\phi\,\theta\cdot\theta\,\bar\theta\cdot\bar\theta \tag{6.99}$$

by partial integrations. The first term of (6.97) can be written as

$$-\frac{1}{4}{\rm i}\bar\theta\cdot\bar\chi\,\theta\cdot\theta\,\bar\theta\bar\sigma^\mu\partial_\mu\chi \tag{6.100}$$

using the result of Exercise 6.1. The expression (6.100) can be further reduced by using the formula

$$\bar\theta\cdot\bar\chi_i\,\bar\theta\cdot\bar\chi_j = -\frac{1}{2}\bar\theta\cdot\bar\theta\,\bar\chi_i\cdot\bar\chi_j \tag{6.101}$$

which is analogous to (6.58); (6.100) becomes

$$+\frac{1}{8}\mathrm{i}\bar{\chi}\bar{\sigma}^{\mu}\partial_{\mu}\chi\,\theta\cdot\theta\,\bar{\theta}\cdot\bar{\theta}. \tag{6.102}$$

Similarly, the second term of (6.97) can be reduced to

$$-\frac{1}{8}\mathrm{i}\partial_{\mu}\bar{\chi}\bar{\sigma}^{\mu}\chi\,\theta\cdot\theta\,\bar{\theta}\cdot\bar{\theta}. \tag{6.103}$$

This is equivalent to (6.102) by a partial integration. Finally using (6.91) the term (6.98) becomes

$$\frac{1}{8}\partial_{\mu}\phi^{\dagger}\partial^{\mu}\phi\,\theta\cdot\theta\,\bar{\theta}\cdot\bar{\theta}. \tag{6.104}$$

Putting together the above results we see that indeed the free part of the W–Z Lagrangian can be written as

$$4\,\Phi^{\mathrm{L}\dagger}_{\mathrm{real}}\Phi^{\mathrm{L}}_{\mathrm{real}}\Big|_{D}\,. \tag{6.105}$$

The D-component of a superfield may be projected out by a Grassmann integration analogous to the one used in (6.74) to project the F-component. We define (compare (6.73))

$$\mathrm{d}^2\bar{\theta}\equiv-\frac{1}{2}\mathrm{d}\bar{\theta}\cdot\mathrm{d}\bar{\theta}=\mathrm{d}\bar{\theta}^{\dot{2}}\mathrm{d}\bar{\theta}^{\dot{1}}, \tag{6.106}$$

from which it follows (compare (6.72)) that

$$\int\mathrm{d}^2\bar{\theta}\,\frac{1}{2}\bar{\theta}\cdot\bar{\theta}=1. \tag{6.107}$$

Then combining (6.72) and (6.107) and defining

$$\mathrm{d}^4\theta\equiv\mathrm{d}^2\bar{\theta}\mathrm{d}^2\theta, \tag{6.108}$$

the free part of the W–Z Lagrangian may be written as

$$\int\mathrm{d}^4\theta\;\Phi^{\mathrm{L}\dagger}_{\mathrm{real}}\Phi^{\mathrm{L}}_{\mathrm{real}}. \tag{6.109}$$

It is time to consider other supermultiplets, in particular ones containing gauge fields, with a view to supersymmetrizing the gauge interactions of the SM.

7

Vector (or gauge) supermultiplets

Having developed a certain amount of superfield formalism, it might seem sensible to use it now to discuss supermultiplets containing vector (gauge) fields. But although this is of course perfectly possible (see for example [42], Chapter 3), it is actually fairly complicated, and we prefer the 'try it and see' approach that we used in Section 3.1, which (as before) establishes the appropriate SUSY transformations more intuitively. We begin with a simple example, a kind of vector analogue of the model of Section 3.1.

7.1 The free Abelian gauge supermultiplet

Consider a simple massless U(1) gauge field $A^{\mu}(x)$, like that of the photon. The spin of such a field is 1, but on-shell it contains only two (rather than three) degrees of freedom, both transverse to the direction of propagation. As we saw in Section 4.4, we expect that SUSY will partner this field with a spin-1/2 field, also with two on-shell degrees of freedom. Such a fermionic partner of a gauge field is called generically a 'gaugino'. This one is a photino, and we'll denote its field by λ, and take it to be L-type. Being in the same multiplet as the photon, it must have the same 'internal' quantum numbers as the photon, in particular it must be electrically neutral. So it doesn't have any coupling to the photon. The photino must also have the same mass as the photon, namely zero. The Lagrangian is therefore just a sum of the Maxwell term for the photon, and the appropriate free massless spinor term for the photino

$$\mathcal{L}_{\gamma\lambda} = -\frac{1}{4}F_{\mu\nu}F^{\mu\nu} + \mathrm{i}\lambda^{\dagger}\bar{\sigma}^{\mu}\partial_{\mu}\lambda, \tag{7.1}$$

where as usual $F^{\mu\nu} = \partial^{\mu}A^{\nu} - \partial^{\nu}A^{\mu}$. We now set about investigating what might be the SUSY transformations between A^{μ} and λ, such that the Lagrangian (7.1) (or the corresponding Action) is invariant.

We anticipate that, as with the chiral supermultiplet, we shall not be able consistently to ignore the off-shell degree of freedom of the gauge field but we shall start by doing so. First, consider $\delta_\xi A^\mu$. This has to be a 4-vector, and also a real rather than complex quantity, linear in ξ and ξ^*. We try (recalling the 4-vector combination from Section 2.2)

$$\delta_\xi A^\mu = \xi^\dagger \bar{\sigma}^\mu \lambda + \lambda^\dagger \bar{\sigma}^\mu \xi, \tag{7.2}$$

where ξ is also an L-type spinor, but has dimension $M^{-1/2}$ as in (3.7). The spinor field λ has dimension $M^{3/2}$, so (7.2) is consistent with A^μ having the desired dimension M^1.

What about $\delta_\xi \lambda$? This must presumably be proportional to A^μ, or better, since λ is gauge-invariant, to the gauge-invariant quantity $F^{\mu\nu}$, so we try

$$\delta_\xi \lambda \sim \xi F^{\mu\nu}. \tag{7.3}$$

Since the dimension of $F^{\mu\nu}$ is M^2, we see that the dimensions already balance on both sides of (7.3), so there is no need to introduce any derivatives. We do, however, need to absorb the two Lorentz indices μ and ν on the right-hand side, and leave ourselves with something transforming correctly as an L-type spinor. This can be neatly done by recalling (Section 2.2) that the quantity $\bar{\sigma}^\nu \xi$ transforms as an R-type spinor ψ, while $\sigma^\mu \psi$ transforms as an L-type spinor. So we try

$$\delta_\xi \lambda = C \sigma^\mu \bar{\sigma}^\nu \xi F_{\mu\nu}, \tag{7.4}$$

where C is a constant to be determined. Then we also have

$$\delta_\xi \lambda^\dagger = C^* \xi^\dagger \bar{\sigma}^\nu \sigma^\mu F_{\mu\nu}. \tag{7.5}$$

Consider the SUSY variation of the Maxwell term in (7.1). Using the antisymmetry of $F^{\mu\nu}$ we have

$$\begin{aligned} \delta_\xi \left(-\frac{1}{4} F_{\mu\nu} F^{\mu\nu} \right) &= -\frac{1}{2} F_{\mu\nu} (\partial^\mu \delta_\xi A^\nu - \partial^\nu \delta_\xi A^\mu) \\ &= -F_{\mu\nu} \partial^\mu \delta_\xi A^\nu \\ &= -F_{\mu\nu} \partial^\mu (\xi^\dagger \bar{\sigma}^\nu \lambda + \lambda^\dagger \bar{\sigma}^\nu \xi). \end{aligned} \tag{7.6}$$

The variation of the spinor term is

$$\begin{aligned} &\mathrm{i}(\delta_\xi \lambda^\dagger) \bar{\sigma}^\mu \partial_\mu \lambda + \mathrm{i} \lambda^\dagger \bar{\sigma}^\mu \partial_\mu (\delta_\xi \lambda) \\ &= \mathrm{i}(C^* \xi^\dagger \bar{\sigma}^\nu \sigma^\mu F_{\mu\nu}) \bar{\sigma}^\rho \partial_\rho \lambda + \mathrm{i} C \lambda^\dagger \bar{\sigma}^\rho \partial_\rho (\sigma^\mu \bar{\sigma}^\nu \xi F_{\mu\nu}). \end{aligned} \tag{7.7}$$

The ξ part of (7.6) must cancel the ξ part of (7.7) (or else their sum must be expressible as a total derivative), and the same is true of the $\xi^\dagger$ parts. So consider the $\xi^\dagger$ part of (7.7). It is

$$\mathrm{i} C^* \xi^\dagger \bar{\sigma}^\nu \sigma^\mu \bar{\sigma}^\rho F_{\mu\nu} \partial_\rho \lambda = -\mathrm{i} C^* \xi^\dagger \bar{\sigma}^\mu \sigma^\nu \bar{\sigma}^\rho F_{\mu\nu} \partial_\rho \lambda. \tag{7.8}$$

Now the σ's are just Pauli matrices, together with the identity matrix, and we know that products of two Pauli matrices will give either the identity matrix or a third Pauli matrix. Hence products of three σ's as in (7.8) must be expressible as a linear combination of σ's. The identity we need is

$$\bar{\sigma}^\mu \sigma^\nu \bar{\sigma}^\rho = g^{\mu\nu}\bar{\sigma}^\rho - g^{\mu\rho}\bar{\sigma}^\nu + g^{\nu\rho}\bar{\sigma}^\mu - \mathrm{i}\epsilon^{\mu\nu\rho\delta}\bar{\sigma}_\delta. \tag{7.9}$$

When (7.9) is inserted into (7.8), some simplifications occur. First, the term involving $\ldots g^{\mu\nu} \ldots F_{\mu\nu}$ vanishes, because $g^{\mu\nu}$ is symmetric in its indices while $F_{\mu\nu}$ is antisymmetric. Next, we can do a partial integration to re-write $F_{\mu\nu}\partial_\rho\lambda$ as $-(\partial_\rho F_{\mu\nu})\lambda = -(\partial_\rho\partial_\mu A_\nu - \partial_\rho\partial_\nu A_\mu)\lambda$. The first of these two terms is symmetric under interchange of ρ and μ, and the second is symmetric under interchange of ρ and ν. However, they are both contracted with $\epsilon^{\mu\nu\rho\delta}$, which is antisymmetric under the interchange of either of these pairs of indices. Hence the term in ϵ vanishes, and (7.8) becomes

$$-\mathrm{i}C^*\xi^\dagger F_{\mu\nu}[-\bar{\sigma}^\nu\partial^\mu\lambda + \bar{\sigma}^\mu\partial^\nu\lambda]. \tag{7.10}$$

In the second term here, if you interchange the indices μ and ν throughout, and then use the antisymmetry of $F_{\nu\mu}$ you will find that the second term equals the first, so that this '$\xi^\dagger$' part of the variation of the fermionic part of $\mathcal{L}_{\gamma\lambda}$ is

$$2\mathrm{i}C^*\xi^\dagger\bar{\sigma}^\nu F_{\mu\nu}\partial^\mu\lambda. \tag{7.11}$$

This will cancel the $\xi^\dagger$ part of (7.6) if $C = \mathrm{i}/2$, and the ξ part of (7.6) will then also cancel. So the required SUSY transformations are (7.2) and

$$\delta_\xi\lambda = \frac{1}{2}\mathrm{i}\sigma^\mu\bar{\sigma}^\nu\xi F_{\mu\nu}, \tag{7.12}$$

$$\delta_\xi\lambda^\dagger = -\frac{1}{2}\mathrm{i}\xi^\dagger\bar{\sigma}^\nu\sigma^\mu F_{\mu\nu}. \tag{7.13}$$

However, if we try to calculate (as in Section 4.5) $\delta_\eta\delta_\xi - \delta_\xi\delta_\eta$ as applied to the fields A^μ and λ, we shall find that consistent results are not obtained unless the free-field equations of motion are assumed to hold, which is not satisfactory. Off-shell, A^μ has a third degree of freedom, and so we expect to have to introduce one more auxiliary field, call it $D(x)$, which is a real scalar field with one degree of freedom. We add to $\mathcal{L}_{\gamma\lambda}$ the extra (non-propagating) term

$$\mathcal{L}_D = \frac{1}{2}D^2. \tag{7.14}$$

We now have to consider SUSY transformations including D.

First note that the dimension of D is M^2, the same as for F. This suggests that D transforms in a similar way to F, as given by (4.109). However, D is a real field, so we modify (4.109) by adding the hermitian conjugate term, arriving at

$$\delta_\xi D = -\mathrm{i}(\xi^\dagger \bar{\sigma}^\mu \partial_\mu \lambda - (\partial_\mu \lambda)^\dagger \bar{\sigma}^\mu \xi). \tag{7.15}$$

As in the case of $\delta_\xi F$, this is also a total derivative. Analogously to (4.113) and (4.115), we expect to modify (7.12) and (7.13) so as to include additional terms

$$\delta_\xi \lambda = \xi D, \quad \delta_\xi \lambda^\dagger = \xi^\dagger D. \tag{7.16}$$

The variation of $\mathcal{L}_D$ is then

$$\delta_\xi \left(\frac{1}{2} D^2 \right) = D \delta_\xi D = -\mathrm{i} D (\xi^\dagger \bar{\sigma}^\mu \partial_\mu \lambda - (\partial_\mu \lambda)^\dagger \bar{\sigma}^\mu \xi), \tag{7.17}$$

and the variation of the fermionic part of $\mathcal{L}_{\gamma\lambda}$ gets an additional contribution which is

$$\mathrm{i}\xi^\dagger \bar{\sigma}^\mu \partial_\mu \lambda D + \mathrm{i}\lambda^\dagger \bar{\sigma}^\mu \partial_\mu \xi D. \tag{7.18}$$

The first term of (7.18) cancels the first term of (7.17), and the second terms also cancel after either one has been integrated by parts.

7.2 Non-Abelian gauge supermultiplets

The preceding example is clearly unrealistic physically, but it will help us in guessing the SUSY transformations in the physically relevant non-Abelian case. For definiteness, we will mostly consider an SU(2) gauge theory, such as occurs in the electroweak sector of the SM. We begin by recalling some necessary facts about non-Abelian gauge theories.

For an SU(2) gauge theory, the Maxwell field strength tensor $F_{\mu\nu}$ of U(1) is generalized to (see, for example, [7] Chapter 13)

$$F^\alpha_{\mu\nu} = \partial_\mu W^\alpha_\nu - \partial_\nu W^\alpha_\mu - g\epsilon^{\alpha\beta\gamma} W^\beta_\mu W^\gamma_\nu, \tag{7.19}$$

where α, β and γ have the values 1, 2 and 3, the gauge field $\boldsymbol{W}_\mu = (W^1_\mu, W^2_\mu, W^3_\mu)$ is an SU(2) triplet (or 'vector', thinking of it in SO(3) terms), and g is the gauge coupling constant. We write the SU(2) indices as superscripts rather than subscripts, but this has no mathematical significance; rather, it is to avoid confusion, later, with the spinor index of the gaugino field λ^α_a. Equation (7.19) can alternatively be written in 'vector' notation as

$$\boldsymbol{F}_{\mu\nu} = \partial_\mu \boldsymbol{W}_\nu - \partial_\nu \boldsymbol{W}_\mu - g \boldsymbol{W}_\mu \times \boldsymbol{W}_\nu. \tag{7.20}$$

If the gauge group was SU(3) there would be eight gauge fields (gluons, in the QCD case), and in general for SU(N) there are $N^2 - 1$. Gauge fields always belong to a particular representation of the gauge group, namely the *regular* or *adjoint* one, which has as many components as there are generators of the group (see pages 400–401 of [7]).

An infinitesimal gauge transformation on the gauge fields W_μ^α takes the form

$$W_\mu^{\prime\alpha}(x) = W_\mu^\alpha(x) - \partial_\mu \epsilon^\alpha(x) - g\epsilon^{\alpha\beta\gamma}\epsilon^\beta(x) W_\mu^\gamma(x), \tag{7.21}$$

where we have here indicated the x-dependence explicitly, to emphasize the fact that this is a local transformation, in which the three infinitesimal parameters $\epsilon^\alpha(x)$ depend on x. In U(1) we would have only one such $\epsilon(x)$, the second term in (7.21) would be absent, and the field strength tensor $F_{\mu\nu}$ would be gauge-invariant. In SU(2), the corresponding tensor (7.20) transforms by

$$F_{\mu\nu}^{\alpha\prime}(x) = F_{\mu\nu}^\alpha(x) - g\epsilon^{\alpha\beta\gamma}\epsilon^\beta(x) F_{\mu\nu}^\gamma(x), \tag{7.22}$$

which is nothing but the statement that $\boldsymbol{F}_{\mu\nu}$ transforms as an SU(2) triplet. Note that (7.22) involves no derivative of $\epsilon(x)$, such as appears in (7.21), even though the transformations being considered are local ones. This fact shows that the simple generalization of the Maxwell Lagrangian in terms of $\boldsymbol{F}_{\mu\nu}$,

$$-\frac{1}{4}\boldsymbol{F}_{\mu\nu} \cdot \boldsymbol{F}^{\mu\nu} = -\frac{1}{4} F_{\mu\nu}^\alpha F^{\mu\nu\alpha}, \tag{7.23}$$

is invariant under local SU(2) transformations; i.e. is SU(2) gauge-invariant.

We now need to generalize the simple U(1) SUSY model of the previous subsection. Clearly the first step is to introduce an SU(2) triplet of gauginos, $\boldsymbol{\lambda} = (\lambda^1, \lambda^2, \lambda^3)$, to partner the triplet of gauge fields. Under an infinitesimal SU(2) gauge transformation, λ^α transforms as in (7.22):

$$\lambda^{\alpha\prime}(x) = \lambda^\alpha(x) - g\epsilon^{\alpha\beta\gamma}\epsilon^\beta(x)\lambda^\gamma(x). \tag{7.24}$$

The gauginos are of course not gauge fields and so their transformation does not include any derivative of $\epsilon(x)$. So the straightforward generalization of (7.1) would be

$$\mathcal{L}_{W\lambda} = -\frac{1}{4} F_{\mu\nu}^\alpha F^{\mu\nu\alpha} + \mathrm{i}\lambda^{\alpha\dagger}\bar{\sigma}^\mu \partial_\mu \lambda^\alpha. \tag{7.25}$$

However, although the first term of (7.25) is SU(2) gauge-invariant, the second is not, because the gradient will act on the x-dependent parameters $\epsilon^\beta(x)$ in (7.24) to leave uncancelled $\partial_\mu \epsilon^\beta(x)$ terms after the gauge transformation. The way to make this term gauge-invariant is to replace the ordinary gradient in it by the appropriate *covariant derivative*; for instance see [7], page 47. The general recipe is

$$\partial_\mu \to D_\mu \equiv \partial_\mu + \mathrm{i}g\mathbf{T}^{(t)} \cdot \boldsymbol{W}_\mu, \tag{7.26}$$

where the three matrices $T^{(t)\alpha}$, $\alpha = 1, 2, 3$, are of dimension $2t + 1 \times 2t + 1$ and represent the generators of SU(2) when acting on a $2t + 1$-component field, which is in the representation of SU(2) characterized by the 'isospin' t (see [7], Section M.5). In the present case, the λ^α's belong in the triplet ($t = 1$) representation, for which the three 3×3 matrices $T^{(1)\alpha}$ are given by (see [7], equation (M.70))

$$\left(T^{(1)\alpha}\right)_{\beta\gamma \text{ element}} = -\mathrm{i}\epsilon^{\alpha\beta\gamma}. \tag{7.27}$$

Thus, in (7.25), we need to make the replacement

$$\begin{aligned}\partial_\mu \lambda^\alpha \to (D_\mu \lambda)^\alpha &= \partial_\mu \lambda^\alpha + \mathrm{i}g\left(\mathbf{T}^{(1)} \cdot \mathbf{W}_\mu\right)_{\alpha\beta \text{ element}} \lambda^\beta \\ &= \partial_\mu \lambda^\alpha + \mathrm{i}g\left(-\mathrm{i}\epsilon^{\gamma\alpha\beta} W_\mu^\gamma\right)\lambda^\beta \\ &= \partial_\mu \lambda^\alpha + g\epsilon^{\gamma\alpha\beta} W_\mu^\gamma \lambda^\beta \\ &= \partial_\mu \lambda^\alpha - g\epsilon^{\alpha\beta\gamma} W_\mu^\beta \lambda^\gamma.\end{aligned} \tag{7.28}$$

With this replacement for $\partial_\mu \lambda^\alpha$ in (7.25), the resulting $\mathcal{L}_{W\lambda}$ is SU(2) gauge-invariant.

What about making it also invariant under SUSY transformations? From the experience of the U(1) case in the previous subsection, we expect that we will need to introduce the analogue of the auxiliary field D. In this case, we need a triplet of D's, D^α, balancing the third off-shell degree of freedom for each W_μ^α. So our shot at a SUSY- and gauge-invariant Lagrangian for an SU(2) gauge supermultiplet is

$$\mathcal{L}_{\text{gauge}} = -\frac{1}{4} F_{\mu\nu}^\alpha F^{\mu\nu\alpha} + \mathrm{i}\lambda^{\alpha\dagger} \bar{\sigma}^\mu (D_\mu \lambda)^\alpha + \frac{1}{2} D^\alpha D^\alpha. \tag{7.29}$$

Confusion must be avoided as between the covariant derivative and the auxiliary field!

What are reasonable guesses for the relevant SUSY transformations? We try the obvious generalizations of the U(1) case:

$$\begin{aligned}\delta_\xi W^{\mu\alpha} &= \xi^\dagger \bar{\sigma}^\mu \lambda^\alpha + \lambda^{\alpha\dagger} \bar{\sigma}^\mu \xi, \\ \delta_\xi \lambda^\alpha &= \frac{1}{2} \mathrm{i}\sigma^\mu \bar{\sigma}^\nu \xi F_{\mu\nu}^\alpha + \xi D^\alpha \\ \delta_\xi D^\alpha &= -\mathrm{i}(\xi^\dagger \bar{\sigma}^\mu (D_\mu \lambda)^\alpha - (D_\mu \lambda)^{\alpha\dagger} \bar{\sigma}^\mu \xi);\end{aligned} \tag{7.30}$$

note that in the last equation we have replaced the '∂_μ' of (7.15) by 'D_μ', so as to maintain gauge-invariance. This in fact works, just as it is! Quite remarkably, the Action for (7.29) is invariant under the transformations (7.30), and $(\delta_\eta \delta_\xi - \delta_\xi \delta_\eta)$ can be consistently applied to all the fields W_μ^α, λ and D^α in this gauge supermultiplet. This supersymmetric gauge theory therefore has two sorts of interactions: (i) the usual self-interactions among the W fields as generated by the term (7.23); and (ii) interactions between the W's and the λ's generated by the covariant derivative coupling in (7.29). We stress again that the supersymmetry requires the gaugino partners to belong to the same representation of the gauge group as the gauge bosons themselves; i.e. to the regular, or adjoint, representation.

We are getting closer to the MSSM at last. The next stage is to build Lagrangians containing both chiral and gauge supermultiplets, in such a way that they (or the Actions) are invariant under both SUSY and gauge transformations.

7.3 Combining chiral and gauge supermultiplets

We do this in two steps. First we introduce, via appropriate covariant derivatives, the couplings of the gauge fields to the scalars and fermions ('matter fields') in the chiral supermultiplets. This will account for the interactions between the gauge fields of the vector supermultiplets and the matter fields of the chiral supermultiplets. However, there are also gaugino and D fields in the vector supermultiplets, and we need to consider whether there are any possible renormalizable interactions between the matter fields and gaugino and D fields, which are both gauge- and SUSY-invariant. Including such interactions is the second step in the programme of combining the two kinds of supermultiplets.

The essential points in such a construction are contained in the simplest case, namely that of a single U(1) (Abelian) vector supermultiplet and a single free chiral supermultiplet, the combination of which we shall now consider.

7.3.1 Combining one U(1) vector supermultiplet and one free chiral supermultiplet

The first step is accomplished by taking the Lagrangian of (5.1), for only a single supermultiplet, replacing ∂_μ by D_μ where (compare (7.26))

$$D_\mu = \partial_\mu + \mathrm{i}qA_\mu, \tag{7.31}$$

where q is the U(1) coupling constant (or charge), and adding on the Lagrangian for the U(1) vector supermultiplet (i.e. (7.1) together with (7.14)). This produces the Lagrangian

$$\mathcal{L} = (D_\mu\phi)^\dagger(D^\mu\phi) + \mathrm{i}\chi^\dagger\bar{\sigma}^\mu D_\mu\chi + F^\dagger F - \frac{1}{4}F_{\mu\nu}F^{\mu\nu} + \mathrm{i}\lambda^\dagger\bar{\sigma}^\mu\partial_\mu\lambda + \frac{1}{2}D^2. \tag{7.32}$$

We now have to consider possible interactions between the matter fields ϕ and χ, and the other fields λ and D in the vector supermultiplet. Any such interaction terms must certainly be Lorentz-invariant, renormalizable (i.e. have mass dimension less than or equal to 4), and gauge-invariant. Given some terms with these characteristics, we shall then have to examine whether we can include them in a SUSY-preserving way.

Since the fields λ and D are neutral, any gauge-invariant couplings between them and the charged fields ϕ and χ must involve neutral bilinear combinations of the latter fields, namely $\phi^\dagger\phi$, $\phi^\dagger\chi$, $\chi^\dagger\phi$ and $\chi^\dagger\chi$. These have mass dimension 2, 5/2, 5/2 and 3 respectively. They have to be coupled to the fields λ and D which have dimension 3/2 and 2 respectively, so as to make quantities with dimension no greater than 4. This rules out the bilinear $\chi^\dagger\chi$, and allows just three possible Lorentz- and gauge-invariant renormalizable couplings: $(\phi^\dagger\chi)\cdot\lambda$, $\lambda^\dagger\cdot(\chi^\dagger\phi)$, and $\phi^\dagger\phi D$. In the first of these the Lorentz invariant is formed as the '$\cdot$' product of the L-type quantity $\phi^\dagger\chi$ and the L-type spinor λ, while in the second it is formed as a '$\lambda^\dagger\cdot\chi^\dagger$'-type product. We take the sum of the first two couplings to obtain a hermitian interaction, and arrive at the possible allowed interaction terms

$$Aq[(\phi^\dagger\chi)\cdot\lambda+\lambda^\dagger\cdot(\chi^\dagger\phi)]+Bq\phi^\dagger\phi D. \tag{7.33}$$

The coefficients A and B are now to be determined by requiring that the complete Lagrangian of (7.32) together with (7.33) is SUSY-invariant (note that for convenience we have extracted an explicit factor of q from A and B).

To implement this programme we need to specify the SUSY transformations of the fields. At first sight, this seems straightforward enough: we use (7.2), (7.12), (7.13) and (7.15) for the fields in the vector supermultiplet, and we 'covariantize' the transformations used for the chiral supermultiplet. For the latter, then, we provisionally assume

$$\delta_\xi\phi=\xi\cdot\chi,\quad \delta_\xi\chi=-\mathrm{i}\sigma^\mu(\mathrm{i}\sigma_2)\xi^{\dagger\mathrm{T}}D_\mu\phi+\xi F,\quad \delta_\xi F=-\mathrm{i}\xi^\dagger\bar{\sigma}^\mu D_\mu\chi, \tag{7.34}$$

together with the analogous transformations for the hermitian conjugate fields. As we shall see, however, there is no choice we can make for A and B in (7.33) such that the complete Lagrangian is invariant under these transformations. One may not be too surprised by this: after all, the transformations for the chiral supermultiplet were found for the case $q=0$, and it is quite possible, one might think, that one or more of the transformations in (7.34) have to be modified by pieces proportional to q. Indeed, we shall find that the transformation for F does need to be so modified. There is, however, a more important reason for the 'failure' to find a suitable A and B. The transformations of (7.2), (7.12), (7.13) and (7.15), on the one hand, and those of (7.34) on the other, certainly do ensure the SUSY-invariance of the gauge and chiral parts of (7.32) respectively, in the limit $q=0$. But there is no *a priori* reason, at least in our 'brute-force' approach, why the 'ξ' parameter in one set of transformations should be exactly the same as that in the other. Either 'ξ' can be rescaled by a constant multiple, and the relevant sub-Lagrangian will remain invariant. However, when we *combine* the Lagrangians and include (7.33), for the case $q\neq 0$, we will see that the requirement of overall SUSY-invariance fixes the relative scale of the two 'ξ's' (up to a sign), and without a rescaling in one or the other transformation we cannot get a SUSY-invariant theory. For definiteness we shall keep the 'ξ' in (7.34) unmodified, and introduce a real scale parameter α into the transformations for the vector supermultiplet, so that they now become

$$\delta_\xi A^\mu = \alpha(\xi^\dagger\bar{\sigma}^\mu\lambda + \lambda^\dagger\bar{\sigma}^\mu\xi) \tag{7.35}$$

$$\delta_\xi \lambda = \frac{\alpha \mathrm{i}}{2}(\sigma^\mu\bar{\sigma}^\nu\xi)F_{\mu\nu} + \alpha\xi D \tag{7.36}$$

$$\delta_\xi \lambda^\dagger = -\frac{\alpha \mathrm{i}}{2}(\xi^\dagger\bar{\sigma}^\nu\sigma^\mu)F_{\mu\nu} + \alpha\xi^\dagger D \tag{7.37}$$

$$\delta_\xi D = -\alpha\mathrm{i}(\xi^\dagger\bar{\sigma}^\mu\partial_\mu\lambda - (\partial_\mu\lambda^\dagger)\bar{\sigma}^\mu\xi). \tag{7.38}$$

Consider first the SUSY variation of the 'A' part of (7.33). This is

$$\begin{aligned} &Aq[(\delta_\xi\phi^\dagger)\chi\cdot\lambda + \phi^\dagger(\delta_\xi\chi)\cdot\lambda + \phi^\dagger\chi\cdot(\delta_\xi\lambda) \\ &+ (\delta_\xi\lambda^\dagger)\cdot\chi^\dagger\phi + \lambda^\dagger\cdot(\delta_\xi\chi^\dagger)\phi + \lambda^\dagger\cdot\chi^\dagger(\delta_\xi\phi)]. \end{aligned} \tag{7.39}$$

Among these terms there are two which are linear in q and D, arising from $\phi^\dagger\chi\cdot(\delta_\xi\lambda)$ and its hermitian conjugate, namely

$$Aq[\alpha\phi^\dagger\chi\cdot\xi D + \alpha\xi^\dagger\cdot\chi^\dagger D\phi]. \tag{7.40}$$

Similarly, the variation of the 'B' part is

$$\begin{aligned} &Bq[(\delta_\xi\phi^\dagger)\phi D + \phi^\dagger(\delta_\xi\phi)D + \phi^\dagger\phi(\delta_\xi D)] \\ &= Bq[\chi^\dagger\cdot\xi^\dagger\phi D + \phi^\dagger\xi\cdot\chi D + \phi^\dagger\phi(-\alpha\mathrm{i})(\xi^\dagger\bar{\sigma}^\mu\partial_\mu\lambda - (\partial_\mu\lambda^\dagger)\bar{\sigma}^\mu\xi)]. \end{aligned} \tag{7.41}$$

The 'D' part of (7.41) will cancel the term (7.40) if (using $\chi^\dagger\cdot\xi^\dagger = \xi^\dagger\cdot\chi^\dagger$ and $\xi\cdot\chi = \chi\cdot\xi$)

$$A\alpha = -B. \tag{7.42}$$

Next, note that the first and last terms of (7.39) produce the changes

$$Aq[\chi^\dagger\cdot\xi^\dagger\chi\cdot\lambda + \lambda^\dagger\cdot\chi^\dagger\xi\cdot\chi]. \tag{7.43}$$

Meanwhile, there is a corresponding change coming from the variation of the term $-q\chi^\dagger\bar{\sigma}^\mu\chi A_\mu$, namely

$$-q\chi^\dagger\bar{\sigma}^\mu\chi(\delta_\xi A_\mu) = -q\alpha\chi^\dagger\bar{\sigma}^\mu\chi(\xi^\dagger\bar{\sigma}_\mu\lambda + \lambda^\dagger\bar{\sigma}_\mu\xi). \tag{7.44}$$

This can be simplified with the help of Exercise 7.1.

Exercise 7.1 Show that

$$(\chi^\dagger\bar{\sigma}^\mu\chi)(\lambda^\dagger\bar{\sigma}_\mu\xi) = 2(\chi^\dagger\cdot\lambda^\dagger)(\chi\cdot\xi). \tag{7.45}$$

So (7.44) becomes

$$-2q\alpha[\chi^\dagger \cdot \xi^\dagger \chi \cdot \lambda + \chi^\dagger \cdot \lambda^\dagger \chi \cdot \xi], \tag{7.46}$$

which will cancel (7.43) if (again using $\chi \cdot \xi = \xi \cdot \chi$ and $\chi^\dagger \cdot \lambda^\dagger = \lambda^\dagger \cdot \chi^\dagger$)

$$A = 2\alpha. \tag{7.47}$$

So far, there is nothing to prevent us from choosing $\alpha = 1$, say, in (7.42) and (7.47). However, a constraint on α arises when we consider the variation of the $A^\mu - \phi$ interaction term in (7.32), namely

$$-\mathrm{i}q\delta_\xi(A^\mu \phi^\dagger \partial_\mu \phi - (\partial_\mu \phi)^\dagger A^\mu \phi). \tag{7.48}$$

The terms in $\delta_\xi A^\mu$ yield a change

$$\mathrm{i}q\alpha[(\partial_\mu \phi^\dagger)(\xi^\dagger \bar{\sigma}^\mu \lambda + \lambda^\dagger \bar{\sigma}^\mu \xi)\phi - (\xi^\dagger \bar{\sigma}^\mu \lambda + \lambda^\dagger \bar{\sigma}^\mu \xi)\phi^\dagger \partial_\mu \phi]. \tag{7.49}$$

A similar change arises from the terms $Aq[\phi^\dagger(\delta_\xi \chi) \cdot \lambda + \lambda^\dagger \cdot (\delta_\xi \chi^\dagger)\phi]$ in (7.39), namely

$$Aq[\phi^\dagger(-\mathrm{i}\sigma^\mu \mathrm{i}\sigma_2 \xi^{\dagger\mathrm{T}} \partial_\mu \phi) \cdot \lambda + \lambda^\dagger \cdot \partial_\mu \phi^\dagger \xi^{\mathrm{T}}(-\mathrm{i}\sigma_2 \mathrm{i}\sigma^\mu \phi)]. \tag{7.50}$$

The first spinor dot product is

$$\xi^\dagger(-\mathrm{i}\sigma_2)(-\mathrm{i}\sigma^{\mu\mathrm{T}})(-\mathrm{i}\sigma_2)\lambda = \mathrm{i}\xi^\dagger \bar{\sigma}^\mu \lambda, \tag{7.51}$$

using (4.30). The second spinor product is the hermitian conjugate of this, so that (7.50) yields a change

$$Aq\mathrm{i}[\phi^\dagger(\partial_\mu \phi)\xi^\dagger \bar{\sigma}^\mu \lambda - (\partial_\mu \phi^\dagger)\phi\, \lambda^\dagger \bar{\sigma}^\mu \xi]. \tag{7.52}$$

Along with (7.49) and (7.52) we must also group the last two terms in (7.41), which we write out again here for convenience

$$Bq[\phi^\dagger \phi(-\alpha \mathrm{i})(\xi^\dagger \bar{\sigma}^\mu \partial_\mu \lambda - (\partial_\mu \lambda^\dagger)\bar{\sigma}^\mu \xi)], \tag{7.53}$$

and integrate by parts to yield

$$\alpha \mathrm{i}Bq\{[(\partial_\mu \phi^\dagger)\phi + \phi^\dagger \partial_\mu \phi](\xi^\dagger \bar{\sigma}^\mu \lambda) - [(\partial_\mu \phi^\dagger)\phi + \phi^\dagger \partial_\mu \phi](\lambda^\dagger \bar{\sigma}^\mu \xi)\}. \tag{7.54}$$

Consider now the terms involving the quantity $\xi^\dagger \bar{\sigma}^\mu \lambda$ in (7.49), (7.52) and (7.54), which are

$$\mathrm{i}q\alpha[(\partial_\mu \phi^\dagger)\phi - \phi^\dagger \partial_\mu \phi] + Aq\mathrm{i}\phi^\dagger \partial_\mu \phi + \alpha \mathrm{i}Bq[(\partial_\mu \phi^\dagger)\phi + \phi^\dagger \partial_\mu \phi]. \tag{7.55}$$

These will all cancel if the condition (7.47) holds, and if in addition

$$B = -1. \tag{7.56}$$

From (7.47) and (7.42) it now follows that

$$\alpha^2 = \frac{1}{2}. \tag{7.57}$$

We conclude that, as promised, the combined Lagrangian will not be SUSY-invariant unless we modify the scale of the transformations of the gauge supermultiplet, relative to those of the chiral supermultiplet, by a non-trivial factor, which we choose (in agreement with what seems to be the usual convention) to be

$$\alpha = -\frac{1}{\sqrt{2}}. \tag{7.58}$$

With this choice, the coefficient A is determined to be

$$A = -\sqrt{2}, \tag{7.59}$$

and our combined Lagrangian is fixed.

We have, of course, not given a complete analysis of all the terms in the SUSY variation of our Lagrangian, an exercise we leave to the dedicated reader, who will find that (with one more adjustment to the SUSY transformations) all the variations do indeed vanish (after partial integrations in some cases, as usual). The need for the adjustment appears when we consider the variation associated with the terms $Aq[\phi^\dagger(\delta_\xi \chi)\cdot\lambda + \lambda^\dagger\cdot(\delta_\xi\chi^\dagger)\phi]$ in (7.39), which includes the term

$$Aq[\phi^\dagger \xi\cdot\lambda F + \lambda^\dagger\cdot\xi^\dagger F^\dagger\phi]. \tag{7.60}$$

This cannot be cancelled by any other variation, and we therefore have to modify the transformation for F and $F^\dagger$ so as to generate a cancelling term from the variation of $F^\dagger F$ in the Lagrangian. This requires

$$\delta_\xi F = -\sqrt{2}q\lambda^\dagger\cdot\xi^\dagger\phi + \text{previous transformation} \tag{7.61}$$

and

$$\delta_\xi F^\dagger = -\sqrt{2}q\xi\cdot\lambda\phi^\dagger + \text{previous transformation}, \tag{7.62}$$

where we have now inserted the known value of A.

In summary then, our SUSY-invariant combined chiral and U(1) gauge supermultiplet Lagrangian is

$$\begin{aligned}\mathcal{L} = {} & (D_\mu\phi)^\dagger(D^\mu\phi) + \mathrm{i}\chi^\dagger\bar{\sigma}^\mu D_\mu\chi + F^\dagger F - \frac{1}{4}F_{\mu\nu}F^{\mu\nu} + \mathrm{i}\lambda^\dagger\bar{\sigma}^\mu\partial_\mu\lambda + \frac{1}{2}D^2 \\ & - \sqrt{2}q[(\phi^\dagger\chi)\cdot\lambda + \lambda^\dagger\cdot(\chi^\dagger\phi)] - q\phi^\dagger\phi D.\end{aligned} \tag{7.63}$$

Note that the terms in the last line of (7.63) are interactions whose strengths are fixed by SUSY to be proportional to the gauge-coupling constant q, even though they do not have the form of ordinary gauge interactions; the terms coupling the photino λ to the matter fields may be thought of as arising from supersymmetrizing the usual coupling of the gauge field to the matter fields.

The equation of motion for the field D is

$$D = q\phi^\dagger\phi. \tag{7.64}$$

Since no derivatives of D enter, we may (as in the W–Z case for F_i and $F_i^\dagger$, cf. equations (5.21) and (5.22)) eliminate the auxiliary field D from the Lagrangian by using (7.64). The effect of this is clearly to replace the two terms involving D in (7.63) by the single term

$$-\frac{1}{2}q^2(\phi^\dagger\phi)^2. \tag{7.65}$$

This is a '$(\phi^\dagger\phi)^2$' type of interaction, just as in the Higgs potential (1.4), but here appearing with a coupling constant, which is not an unknown parameter but is determined by the gauge coupling q. In the next section we shall see that the same feature persists in the more realistic non-Abelian case. Since the Higgs mass is (for a fixed vev of the Higgs field) determined by the $(\phi^\dagger\phi)^2$ coupling (see (1.3)), it follows that there is likely to be less arbitrariness in the mass of the Higgs in the MSSM than in the SM. We shall see in Chapter 10, when we examine the Higgs sector of the MSSM, that this is indeed the case.

7.3.2 The non-Abelian case

Once again, we proceed in two steps. We start from the W–Z Lagrangian for a collection of chiral supermultiplets labelled by i, and including the superpotential terms:

$$\partial_\mu\phi_i^\dagger\partial^\mu\phi_i + \chi_i^\dagger \mathrm{i}\bar{\sigma}^\mu\partial_\mu\chi_i + F_i^\dagger F_i + \left[\frac{\partial W}{\partial\phi_i}F_i - \frac{1}{2}\frac{\partial^2 W}{\partial\phi_i\phi_j}\chi_i\cdot\chi_j + \text{h.c.}\right] \tag{7.66}$$

into which we introduce the gauge couplings via the covariant derivatives

$$\partial_\mu\phi_i \to D_\mu\phi_i = \partial_\mu\phi_i + \mathrm{i}gA_\mu^\alpha(T^\alpha\phi)_i \tag{7.67}$$

$$\partial\chi_i \to D_\mu\chi_i = \partial_\mu\chi_i + \mathrm{i}gA_\mu^\alpha(T^\alpha\chi)_i, \tag{7.68}$$

where g and A_μ^α are the gauge coupling constant and gauge fields (for example, g_s and gluon fields for QCD), and the T^α are the hermitian matrices representing the generators of the gauge group in the representation to which, for given i, ϕ_i and χ_i belong (for example, if ϕ_i and χ_i are SU(2) doublets, the T^α's would be the $\tau^\alpha/2$, with α running from 1 to 3). Recall that SUSY requires that ϕ_i, χ_i and F_i must all be in the same representation of the relevant gauge group. Of course, if, as is the case in the SM, some matter fields interact with more than one gauge field, then all the gauge couplings must be included in the covariant derivatives. There is no covariant derivative for the auxiliary fields F_i, because their ordinary derivatives do not appear in (7.66). To (7.68) we need to add the Lagrangian for the gauge supermultiplet(s), equation (7.29), and then (in the second step) additional 'mixed' interactions as in (7.33).

We therefore need to construct all possible Lorentz- and gauge-invariant renormalizable interactions between the matter fields and the gaugino (λ^α) and auxiliary (D^α) fields, as in the U(1) case. We have the specific particle content of the SM in mind, so we need only consider the cases in which the matter fields are either singlets under the gauge group (for example, the R parts of quark and lepton fields), or belong to the fundamental representation of the gauge group (that is, the triplet for SU(3) and the doublet for SU(2)). For matter fields in singlet representations, there is no possible gauge-invariant coupling between them and λ^α or D^α, which are in the regular representation. For matter fields in the fundamental representation, however, we can form bilinear combinations of them that transform according to the regular representation, and these bilinears can be 'dotted' into λ^α and D^α to give gauge singlets (i.e. gauge-invariant couplings). We must also arrange the couplings to be Lorentz invariant, of course.

The bilinear combinations of the ϕ_i and χ_i which transform as the regular representation are (see, for example, [7] Sections 12.1.3 and 12.2)

$$\phi_i^\dagger T^\alpha \phi_i, \ \phi_i^\dagger T^\alpha \chi_i, \ \chi_i^\dagger T^\alpha \phi_i, \text{ and } \chi_i^\dagger T^\alpha \chi_i, \tag{7.69}$$

where, for example, $T^\alpha = \tau^\alpha/2$ in the case of SU(2), and where the τ^α, $(\alpha = 1, 2, 3)$ are the usual Pauli matrices used in the isospin context. These bilinears are the obvious analogues of the ones considered in the U(1) case; in particular they have the same dimension. Following the same reasoning, then, the allowed additional interaction terms are

$$Ag[(\phi_i^\dagger T^\alpha \chi_i) \cdot \lambda^\alpha + \lambda^{\alpha\dagger} \cdot (\chi_i^\dagger T^\alpha \phi_i)] + Bg(\phi_i^\dagger T^\alpha \phi_i) D^\alpha, \tag{7.70}$$

where A and B are coefficients to be determined by the requirement of SUSY-invariance.

In fact, however, a consideration of the SUSY transformations in this case shows that they are essentially the same as in the U(1) case (apart from straightforward changes involving the matrices T^α). The upshot is that, just as in the U(1) case, we need to change the SUSY transformations of (7.30) by replacing ξ by $-\xi/\sqrt{2}$, and by modifying the transformation of $F_i^\dagger$ to

$$\delta_\xi F_i^\dagger = -\sqrt{2} g \phi_i^\dagger T^\alpha \xi \cdot \lambda^\alpha + \text{previous transformation}, \tag{7.71}$$

and similarly for $\delta_\xi F_i$. The coefficients A and B in (7.70) are then $-\sqrt{2}$ and -1, respectively, as in the U(1) case, and the combined SUSY-invariant Lagrangian is

$$\begin{aligned}\mathcal{L}_{\text{gauge + chiral}} &= \mathcal{L}_{\text{gauge}}(\text{equation (7.29)})\\ &+ \mathcal{L}_{\text{W-Z, covariantized}}(\text{equation (7.66), with } \partial_\mu \to D_\mu \text{ as in (7.67) and (7.68)})\\ &- \sqrt{2}g[(\phi_i^\dagger T^\alpha \chi_i)\cdot\lambda^\alpha + \lambda^{\alpha\dagger}\cdot(\chi_i^\dagger T^\alpha \phi_i)] - g(\phi_i^\dagger T^\alpha \phi_i)D^\alpha. \qquad (7.72)\end{aligned}$$

We draw attention to an important consequence of the terms $-\sqrt{2}g[\ldots]$ in (7.72), for the case in which the chiral multiplets (ϕ_i, χ_i) are the two Higgs supermultiplets H_u and H_d, containing Higgs and Higgsino fields (see Table 8.1 in the next chapter). When the scalar Higgs fields H_u^0 and H_d^0 acquire vevs, these terms will be bilinear in the Higgsino and gaugino fields, implying that mixing will occur among these fields as a consequence of electroweak symmetry breaking. We shall discuss this in Section 11.2.

The equation of motion for the field D^α is

$$D^\alpha = g\sum_i(\phi_i^\dagger T^\alpha \phi_i), \qquad (7.73)$$

where the sum over i (labelling a given chiral supermultiplet) has been re-instated explicitly. As before, we may eliminate these auxiliary fields from the Lagrangian by using (7.73). The complete scalar potential (as in '$\mathcal{L} = \mathcal{T} - \mathcal{V}$') is then

$$\mathcal{V}(\phi_i, \phi_i^\dagger) = |W_i|^2 + \frac{1}{2}\sum_\mathrm{G}\sum_\alpha\sum_{i,j} g_\mathrm{G}^2(\phi_i^\dagger T_\mathrm{G}^\alpha \phi_i)(\phi_j^\dagger T_\mathrm{G}^\alpha \phi_j), \qquad (7.74)$$

where in the summation we have recalled that more than one gauge group G will enter, in general, given the SU(3)×SU(2)×U(1) structure of the SM, with different couplings g_G and generators T_G. The first term in (7.74) is called the 'F-term', for obvious reasons; it is determined by the fermion mass terms M_{ij} and Yukawa couplings (see (5.19)). The second term is called the 'D-term', and is determined by the gauge interactions. There is no room for any other scalar potential, *independent* of these parameters appearing in other parts of the Lagrangian. It is worth emphasizing that $\mathcal{V}$ is a sum of squares, and is hence always greater than or equal to zero for every field configuration. We shall see in Section 10.2 how the form of the D-term allows an important bound to be put on the mass of the lightest Higgs boson in the MSSM.

8
The MSSM

8.1 Specification of the superpotential

We have now introduced all the interactions appearing in the MSSM, apart from specifying the superpotential W. We already assigned the SM fields to supermultiplets in Sections 3.2 and 4.4; let us begin by reviewing those assignments once more.

All the SM fermions, i.e. the quarks and the leptons, have the property that their L ('χ') parts are $\mathrm{SU(2)_L}$ doublets, whereas their R ('ψ') parts are $\mathrm{SU(2)_L}$ singlets. So these weak gauge group properties suggest that we should treat the L and R parts separately, rather than together as in a Dirac 4-component spinor. The basic 'building block' is therefore the chiral supermultiplet, suitably 'gauged'.

We developed chiral supermultiplets in terms of L-type spinors χ: this is clearly fine for e_L^-, μ_L^-, u_L, d_L, etc., but what about e_R^-, μ_R^-, etc.? These *R-type particle fields* can be accommodated within the 'L-type' convention for chiral supermultiplets by regarding them as the charge conjugates of *L-type antiparticle fields*, which we use instead. Charge conjugation was mentioned in Section 2.3; see also Section 20.5 of [7] (but note that we are here using $C_0 = -\mathrm{i}\gamma_2$). If (as is often done) we denote the field by the particle name, then we have $e_R^- \equiv \psi_{e^-}$, while $e_L^+ \equiv \chi_{e^+}$. On the other hand, if we regard e_R^- as the charge conjugate of e_L^+, then (compare equation (2.94))

$$e_R^- \equiv \psi_{e^-} = (e_L^+)^c \equiv \mathrm{i}\sigma_2 \chi_{e^+}^{\dagger T}. \tag{8.1}$$

To remind ourselves of how this works (see also Section 2.4), consider a Dirac mass term for the electron:

$$\begin{aligned} \bar{\Psi}^{(e^-)}\Psi^{(e^-)} &= \psi_{e^-}^\dagger \chi_{e^-} + \chi_{e^-}^\dagger \psi_{e^-} = \left(\mathrm{i}\sigma_2 \chi_{e^+}^{\dagger T}\right)^\dagger \chi_{e^-} + \chi_{e^-}^\dagger \mathrm{i}\sigma_2 \chi_{e^+}^{\dagger T} \\ &= \chi_{e^+}^T(-\mathrm{i}\sigma_2)\chi_{e^-} + \chi_{e^-}^\dagger \mathrm{i}\sigma_2 \chi_{e^+}^{\dagger T} \\ &= \chi_{e^+} \cdot \chi_{e^-} + \chi_{e^-}^\dagger \cdot \chi_{e^+}^\dagger. \end{aligned} \tag{8.2}$$

Table 8.1 *Chiral supermultiplet fields in the MSSM.*

Names		spin 0	spin 1/2	$SU(3)_c$, $SU(2)_L$, $U(1)_y$
squarks, quarks (× 3 families)	Q	$(\tilde{u}_L, \tilde{d}_L)$	(u_L, d_L) or (χ_u, χ_d)	**3**, **2**, 1/3
	$\bar{u}$	$\tilde{\bar{u}}_L = \tilde{u}_R^\dagger$	$\bar{u}_L = (u_R)^c$ or $\chi_{\bar{u}} = \psi_u^c$	$\bar{\mathbf{3}}$, **1**, −4/3
	$\bar{d}$	$\tilde{\bar{d}}_L = \tilde{d}_R^\dagger$	$\bar{d}_L = (d_R)^c$ or $\chi_{\bar{d}} = \psi_d^c$	$\bar{\mathbf{3}}$, **1**, 2/3
sleptons, leptons (× 3 families)	L	$(\tilde{\nu}_{eL}, \tilde{e}_L)$	(ν_{eL}, e_L) or (χ_{ν_e}, χ_e)	**1**, **2**, −1
	$\bar{e}$	$\tilde{\bar{e}}_L = \tilde{e}_R^\dagger$	$\bar{e}_L = (e_R)^c$ or $\chi_{\bar{e}} = \psi_e^c$	**1**, **1**, 2
Higgs, Higgsinos	H_u	(H_u^+, H_u^0)	$(\tilde{H}_u^+, \tilde{H}_u^0)$	**1**, **2**, 1
	H_d	(H_d^0, H_d^-)	$(\tilde{H}_d^0, \tilde{H}_d^-)$	**1**, **2**, −1

Table 8.2 *Gauge supermultiplet fields in the MSSM.*

Names	spin 1/2	spin 1	$SU(3)_c$, $SU(2)_L$, $U(1)_y$
gluinos, gluons	$\tilde{g}$	g	**8**, **1**, 0
winos, W bosons	$\widetilde{W}^\pm$, $\widetilde{W}^0$	$W^\pm$, W^0	**1**, **3**, 0
bino, B boson	$\tilde{B}$	B	**1**, **1**, 0

So it is all expressed in terms of χ's. It is also useful to note that

$$\bar{\Psi}^{(e^-)}\gamma_5\Psi^{(e^-)} = -\chi_{e^+}\cdot\chi_{e^-} + \chi_{e^-}^\dagger\cdot\chi_{e^+}^\dagger. \tag{8.3}$$

The notation for the squark and slepton fields was explained in Section 4.4, following equation (4.97).

In Table 8.1 we list the chiral supermultiplets appearing in the MSSM (our y is twice that of [46], following the convention of [7], Chapter 22). Note that the 'bar' on the fields in this table is merely a label, signifying 'antiparticle', not (for example) Dirac conjugation: thus χ_{e^-} and χ_{e^+}, for instance, now become χ_e and $\chi_{\bar{e}}$. The subscript $_i$ can be added to the names to signify the family index: for example, $u_{1L} = u_L$, $u_{2L} = c_L$, $u_{3L} = t_L$, and similarly for leptons. In Table 8.2, similarly, we list the gauge supermultiplets of the MSSM. After electroweak symmetry breaking, the W^0 and the B fields mix to produce the physical Z^0 and γ fields, while the corresponding 's'-fields mix to produce a zino ($\tilde{Z}^0$) degenerate with the Z^0, and a massless photino $\tilde{\gamma}$.

So, knowing the gauge groups, the particle content, and the gauge transformation properties, all we need to do to specify any proposed model is to give the superpotential W. The MSSM is specified by the choice

$$W = y_{\rm u}^{ij}\bar{u}_i Q_j \cdot H_{\rm u} - y_{\rm d}^{ij}\bar{d}_i Q_j \cdot H_{\rm d} - y_{\rm e}^{ij}\bar{e}_i L_j \cdot H_{\rm d} + \mu H_{\rm u} \cdot H_{\rm d}. \qquad (8.4)$$

The fields appearing in (8.4) are the chiral superfields indicated under the 'Names' column of Table 8.1. In this formulation, we recall from Section 6.4 that the F-component of W is to be taken in the Lagrangian. We can alternatively think of W as being the same function of the scalar fields in each chiral supermultiplet, as explained in Section 6.4. In that case, the W_i of (5.17) and W_{ij} of (5.8) generate the interaction terms in the Lagrangian via (5.3). In either case, the y's are 3×3 matrices in family (or generation) space, *and are exactly the same Yukawa couplings as those which enter the SM* (see, for example, Section 22.7 of [7]).[1] In particular, the terms in (8.4) are all invariant under the SM gauge transformations. The '$\cdot$' notation means that SU(2)-invariant coupling of two doublets;[2] also, colour indices have been suppressed, so that '$\bar{u}_i Q_j$', for example, is really $\bar{u}_{\alpha i} Q_j^\alpha$, where the upstairs $\alpha = 1, 2, 3$ is a colour $\mathbf{3}$ (triplet) index, and the downstairs α is a colour $\bar{\mathbf{3}}$ (antitriplet) index. These couplings give masses to the quarks and leptons when the Higgs fields $H_{\rm u}^0$ and $H_{\rm d}^0$ acquire vacuum expectation values: there are no 'Lagrangian' masses for the fermions, since these would explicitly break the $\mathrm{SU(2)_L}$ gauge symmetry.

In summary, then, at the cost of only one new parameter μ, we have got an exactly supersymmetric extension of the SM. This is, of course, *not* the same 'μ' as appeared in the Higgs potential of the SM, equation (1.4). We follow a common notation, although others are in use which avoid this possible confusion.

[In parenthesis, we note a possibly confusing aspect of the labelling adopted for the Higgs fields. In the conventional formulation of the SM, the Higgs field $\phi = \begin{pmatrix} \phi^+ \\ \phi^0 \end{pmatrix}$ generates mass for the 'down' quark, say, via a Yukawa interaction of the form (suppressing family labels)

$$g_{\rm d}\, \bar{q}_{\rm L} \phi d_{\rm R} + \text{h.c.} \qquad (8.5)$$

[1] However, we stress once again – see Section 3.2 and footnote 1 of Chapter 5, page 72 – that whereas in the SM we can use one Higgs doublet and its charge conjugate doublet (see Section 22.6 of [7]), this is not allowed in SUSY, because W cannot depend on both a complex scalar field ϕ and its Hermitian conjugate $\phi^\dagger$ (which would appear in the charge conjugate via (3.40)). By convention, the MSSM does not include Dirac-type neutrino mass terms, neutrino masses being generally regarded as 'beyond the SM' physics.

[2] To take an elementary example: consider the isospin part of the deuteron's wavefunction. It has $I = 0$; i.e. it is the SU(2)-invariant coupling of the two doublets $N^{(1)} = \begin{pmatrix} p^{(1)} \\ n^{(1)} \end{pmatrix}$, $N^{(2)} = \begin{pmatrix} p^{(2)} \\ n^{(2)} \end{pmatrix}$. This $I = 0$ wavefunction is, as usual, $\frac{1}{\sqrt{2}}(p^{(1)}n^{(2)} - n^{(1)}p^{(2)})$, which (dropping the $1/\sqrt{2}$) we may write as $N^{(1)\mathrm{T}} \mathrm{i}\tau_2 N^{(2)} \equiv N^{(1)} \cdot N^{(2)}$. Clearly this isospin-invariant coupling is basically the same as the Lorentz-invariant spinor coupling '$\chi^{(1)} \cdot \chi^{(2)}$' (see (2.47)–(2.49)), which is why we use the same '$\cdot$' notation for both, we hope without causing confusion.

where $q_L = \begin{pmatrix} u_L \\ d_L \end{pmatrix}$. In this case, the SU(2) dot product is simply $q_L^\dagger \phi$, which is plainly invariant under $q_L \to U q_L$, $\phi \to U\phi$. Now $q_L^\dagger \phi = u_L^\dagger \phi^+ + d_L^\dagger \phi^0$; so when ϕ^0 develops a vev, (8.5) contributes

$$g_d \langle \phi^0 \rangle \bar{d}_L d_R + \text{h.c.} \tag{8.6}$$

which is a d-quark mass. Why, then, do we label our Higgs field $\begin{pmatrix} H_u^+ \\ H_u^0 \end{pmatrix}$ with a subscript 'u' rather than 'd'? The point is that, in the SUSY version (8.4), the SU(2) dot product involving the superfield H_u is taken with the superfield Q which has the quantum numbers of the quark doublet q_L rather than the antiquark doublet $q_L^\dagger$. If we revert for the moment to the procedure of Section 5.1, and write W just in terms of the corresponding scalar fields, the first term in (8.4) is

$$y_u^{ij} \tilde{\bar{u}}_{Li} (\tilde{u}_{Lj} H_u^0 - \tilde{d}_{Lj} H_u^+). \tag{8.7}$$

The first term here will, via (5.8) and (5.3), generate a term in the Lagrangian (cf. (5.45))

$$-\frac{1}{2} y_u^{ij} (\chi_{\bar{u}_{Li}} \cdot \chi_{u_{Lj}} + \chi_{u_{Li}} \cdot \chi_{\bar{u}_{Lj}}) H_u^0 + \text{h.c.} = -y_u^{ij} (\chi_{\bar{u}_{Li}} \cdot \chi_{u_{Lj}}) H_u^0 + \text{h.c.} \tag{8.8}$$

When H_u^0 develops a (real) vacuum value v_u (see Section 10.1), this will become a Dirac-type mass term for the u-quark (cf. (8.2)):

$$-(m_{uij} \chi_{\bar{u}_{Li}} \cdot \chi_{u_{Lj}} + \text{h.c.}) \tag{8.9}$$

where

$$m_{uij} = v_u y_u^{ij}. \tag{8.10}$$

Transforming to the basis which diagonalizes the mass matrices then leads to flavour mixing exactly as in the SM (see Section 22.7 of [7], for example).]

The fermion masses are evidently proportional to the relevant y parameter, so since the top, bottom and tau are the heaviest fermions in the SM, it is sometimes useful to consider an approximation in which the only non-zero y's are

$$y_u^{33} = y_t; \quad y_d^{33} = y_b; \quad y_e^{33} = y_\tau. \tag{8.11}$$

Writing W now in terms of the scalar fields, and omitting the μ term, this gives

$$W \approx y_t\big[\bar{\tilde{t}}_L\big(\tilde{t}_L H_u^0 - \tilde{b}_L H_u^+\big)\big] - y_b\big[\bar{\tilde{b}}_L\big(\tilde{t}_L H_d^- - \tilde{b}_L H_d^0\big)\big] - y_\tau\big[\bar{\tilde{\tau}}_L\big(\tilde{\nu}_{\tau L} H_d^- - \tilde{\tau}_L H_d^0\big)\big]. \tag{8.12}$$

The minus signs in W have been chosen so that the terms $y_t\bar{\tilde{t}}_L\tilde{t}_L$, $y_b\bar{\tilde{b}}_L\tilde{b}_L$ and $y_\tau\bar{\tilde{\tau}}_L\tilde{\tau}_L$ have the correct sign to generate mass terms for the top, bottom and tau when $\langle H_u^0\rangle \neq 0$ and $\langle H_d^0\rangle \neq 0$. Note that $\bar{\tilde{t}}_L$, $\bar{\tilde{b}}_L$ and $\bar{\tilde{\tau}}_L$ could equally well have been written as $\tilde{t}_R^\dagger$, $\tilde{b}_R^\dagger$ and $\tilde{\tau}_R^\dagger$.

It is worth recalling that in such a SUSY theory, in addition to the Yukawa couplings of the SM, which couple the Higgs fields to quarks and to leptons, there must also be similar couplings between Higgsinos, squarks and quarks, and between Higgsinos, sleptons and leptons (i.e. we change two ordinary particles into their superpartners). There are also scalar quartic interactions with strength proportional to y_t^2, as noted in Section 5.1, arising from the term '$|W_i|^2$' in the scalar potential (7.74). In addition, there are scalar quartic interactions proportional to the squares of the gauge couplings g and g' coming from the 'D-term' in (7.74). These include quartic Higgs couplings such as are postulated in the SM, but now appearing with coefficients that are determined in terms of the parameters g and g' already present in the model. The important phenomenological consequences of this will be discussed in Chapter 10.

Although there are no conventional mass terms in (8.4), there is one term which is quadratic in the fields, the so-called 'μ term', which is the SU(2)-invariant coupling of the two different Higgs superfield doublets:

$$W(\mu \text{ term}) = \mu H_u \cdot H_d = \mu(H_{u1}H_{d2} - H_{u2}H_{d1}), \tag{8.13}$$

where the subscripts 1 and 2 denote the isospinor component. This is the only such bilinear coupling of the Higgs fields allowed in W, because the other possibilities $H_u^\dagger \cdot H_u$ and $H_d^\dagger \cdot H_d$ involve hermitian conjugate fields, which would violate SUSY. As always, we need the F-component of (8.13), which is (see (6.61))

$$\mu\big[\big(H_u^+ F_d^- + H_d^- F_u^+ - H_u^0 F_d^0 - H_d^0 F_u^0\big) - \big(\tilde{H}_u^+ \cdot \tilde{H}_d^- - \tilde{H}_u^0 \cdot \tilde{H}_d^0\big)\big], \tag{8.14}$$

and we must include also the hermitian conjugate of (8.14). The second term in (8.14) will contribute to (off-diagonal) Higgsino mass terms. The first term has the general form '$W_i F_i$' of Section 5.1, and hence (see (5.22)) it leads to the following term in the scalar potential, involving the Higgs fields:

$$|\mu|^2\big(|H_u^+|^2 + |H_d^-|^2 + \big|H_u^0\big|^2 + \big|H_d^0\big|^2\big). \tag{8.15}$$

But these terms all have the (positive) sign appropriate to a standard '$m^2\phi^\dagger\phi$' bosonic mass term, *not* the negative sign needed for electroweak symmetry breaking via the Higgs mechanism (recall the discussion following equation (1.4)). This means that our SUSY-invariant Lagrangian cannot accommodate electroweak symmetry breaking.

Of course, SUSY itself – in the MSSM application we are considering – cannot be an exact symmetry, since we have not yet observed the s-partners of the SM fields. We shall discuss SUSY breaking briefly in Chapter 9, but it is clear from the above that some SUSY-breaking terms will be needed in the Higgs potential, in order to allow electroweak symmetry breaking. This very fact even suggests that a common mechanism might be responsible for both symmetry breakings.

The 'μ term' actually poses something of a puzzle [53]. The parameter μ should presumably lie roughly in the range 100 GeV–1 TeV, or else we'd need delicate cancellations between the positive $|\mu|^2$ terms in (8.15) and the negative SUSY-breaking terms necessary for electroweak symmetry breaking (see a similar argument in Section 1.1). We saw in Section 1.1 that the general 'no fine-tuning' argument suggested that SUSY-breaking masses should not be much greater than 1 TeV. But the μ term does not break SUSY! We are faced with an apparent difficulty: where does this low scale for the SUSY-respecting parameter μ come from? References to some proposed solutions to this 'μ problem' are given in [46] Section 5.1, where some further discussion is also given of the various interactions present in the MSSM; see also [47] Section 4.2, and particularly the review of the μ problem in [54].

8.2 The SM interactions in the MSSM

By now we seem to have travelled a long way from the Standard Model, and it may be helpful, before continuing with features of the MSSM which go beyond the SM, to take a slight backwards detour and reassure ourselves that the familiar SM interactions are indeed contained (in possibly unfamiliar notation) in the MSSM.

We start with the QCD interactions of the SM quarks and gluons. First of all, the 3- and 4-gluon interactions are as usual contained in the $\mathrm{SU(3)_c}$ field strength tensor $-\frac{1}{4}F_{a\mu\nu}F_a^{\mu\nu}$ (cf. (7.29)), where the colour index 'a' runs from 1 to 8; see, for example, Section 14.2.3 of [7]. Next, consider the $\mathrm{SU(3)_c}$ triplet of 'up' quarks, described by the 4-component Dirac field

$$\Psi_{\mathrm{u}} = \begin{pmatrix} \psi_{\mathrm{u}} \\ \chi_{\mathrm{u}} \end{pmatrix}. \tag{8.16}$$

We shall not indicate the colour labels explicitly on the spinor fields. The covariant derivative (7.68) is (see, for example, Section 13.4 of [7])

$$D_\mu = \partial_\mu + \frac{1}{2}\mathrm{i}g_{\mathrm{s}}\boldsymbol{\lambda}\cdot\boldsymbol{A}_\mu \tag{8.17}$$

where g_s is the strong interaction coupling constant, and $\boldsymbol{A}_\mu = (A_{1\mu}, A_{2\mu}, \ldots, A_{8\mu})$ are the eight gluon fields. Then the gauge-kinetic term for the field χ_u yields the interaction

$$-\frac{1}{2} g_s \chi_u^\dagger \bar{\sigma}^\mu \boldsymbol{\lambda} \cdot \boldsymbol{A}_\mu \chi_u. \tag{8.18}$$

In (8.18) the 3×3 $\boldsymbol{\lambda}$ matrices act on the colour labels of χ_u, while the 2×2 $\bar{\sigma}^\mu$ matrices act on the spinor labels. As regards the R-part ψ_u, we write it as the charge conjugate of the L-type field for $\bar{u}$ (cf. (2.93) and (2.94))

$$\psi_u = \chi_{\bar{u}}^c = i\sigma_2 \chi_{\bar{u}}^*. \tag{8.19}$$

We now need the interaction term for the field $\chi_{\bar{u}}$. Antiquarks belong to the $\bar{\mathbf{3}}$ representation of SU(3), and the matrices representing the generators in this representation are $-\boldsymbol{\lambda}^*/2$ (see [7], page 21). Hence the QCD interaction term for the L-chiral multiplet containing $\chi_{\bar{u}}$ is

$$-\frac{1}{2} g_s \chi_{\bar{u}}^\dagger \bar{\sigma}^\mu (-\boldsymbol{\lambda}^*) \cdot \boldsymbol{A}_\mu \chi_{\bar{u}}. \tag{8.20}$$

We can rewrite (8.20) in terms of the field $\chi_{\bar{u}}^c$ which appears in Ψ_u by inverting (8.19):

$$-\frac{1}{2} g_s \big(-i\sigma_2 \chi_{\bar{u}}^{c*} \big)^\dagger \bar{\sigma}^\mu (-\boldsymbol{\lambda}^*) \cdot \boldsymbol{A}_\mu \big(-i\sigma_2 \chi_{\bar{u}}^{c*} \big) = \frac{1}{2} g_s \chi_{\bar{u}}^{cT} \sigma_2 \bar{\sigma}^\mu \sigma_2 \boldsymbol{\lambda}^* \cdot \boldsymbol{A}_\mu \chi_{\bar{u}}^{c*}. \tag{8.21}$$

Now take the transpose of (8.21), remembering the minus sign from interchanging fermion operators, and use (2.83) together with the fact that the $\boldsymbol{\lambda}$ matrices are hermitian; this converts (8.21) to

$$-\frac{1}{2} g_s \chi_{\bar{u}}^{c\dagger} \sigma^\mu \boldsymbol{\lambda} \cdot \boldsymbol{A}_\mu \chi_{\bar{u}}^c. \tag{8.22}$$

On the other hand, the QCD interaction for the Dirac field

$$\Psi_u = \begin{pmatrix} \chi_{\bar{u}}^c \\ \chi_u \end{pmatrix} \tag{8.23}$$

is

$$-\frac{1}{2} g_s \bar{\Psi}_u \gamma^\mu \boldsymbol{\lambda} \cdot \boldsymbol{A}_\mu \Psi_u = -\frac{1}{2} g_s \big[\chi_{\bar{u}}^{c\dagger} \sigma^\mu \boldsymbol{\lambda} \cdot \boldsymbol{A}_\mu \chi_{\bar{u}}^c + \chi_u^\dagger \bar{\sigma}^\mu \boldsymbol{\lambda} \cdot \boldsymbol{A}_\mu \chi_u \big]. \tag{8.24}$$

We see that (8.24) is recovered as the sum of (8.18) and (8.22).

It may be useful, in passing, to show the analogous steps in the Majorana formalism. As explained in Section 2.5.1, we need two Majorana fields to represent the degees of freedom in $\Psi_{\rm u}$, namely

$$\Psi_{\rm M}^{\chi_{\rm u}} = \begin{pmatrix} \chi_{\rm u}^{\rm c} = {\rm i}\sigma_2\chi_{\rm u}^* \\ \chi_{\rm u} \end{pmatrix} \tag{8.25}$$

and

$$\Psi_{\rm M}^{\psi_{\rm u}} = \begin{pmatrix} \psi_{\rm u} \\ \psi_{\rm u}^{\rm c} = -{\rm i}\sigma_2\psi_{\rm u}^* \end{pmatrix}, \tag{8.26}$$

so that

$$\Psi_{\rm u} = P_{\rm R}\Psi_{\rm M}^{\psi_{\rm u}} + P_{\rm L}\Psi_{\rm M}^{\chi_{\rm u}}. \tag{8.27}$$

Now, we already know from Section 2.4 that a Weyl kinetic energy term of the form $\chi^\dagger {\rm i}\bar{\sigma}^\mu\partial_\mu\chi$ is equivalent to the Majorana expression $\frac{1}{2}\bar{\Psi}_{\rm M}^{\chi}{\rm i}\gamma^\mu\partial_\mu\Psi_{\rm M}^{\chi}$, and similarly for $\psi^\dagger{\rm i}\sigma^\mu\partial_\mu\psi$. Thus the QCD interactions for (8.25) and (8.26) are contained in

$$\frac{1}{2}\bar{\Psi}_{\rm M}^{\chi_{\rm u}}{\rm i}\gamma^\mu D_\mu\Psi_{\rm M}^{\chi_{\rm u}} + \frac{1}{2}\bar{\Psi}_{\rm M}^{\psi_{\rm u}}{\rm i}\gamma^\mu D_\mu\Psi_{\rm M}^{\psi_{\rm u}}. \tag{8.28}$$

In evaluating (8.28) we must remember that although the R-part of $\Psi_{\rm M}^{\psi_{\rm u}}$ and the L-part of $\Psi_{\rm M}^{\chi_{\rm u}}$ transform as **3**'s of SU(3), the L-part of $\Psi_{\rm M}^{\psi_{\rm u}}$ and the R-part of $\Psi_{\rm M}^{\chi_{\rm u}}$ transform as $\bar{\mathbf{3}}$'s. The interaction part of the first term of (8.28) is therefore

$$\begin{aligned} &-\frac{1}{4}\bar{\Psi}_{\rm M}^{\chi_{\rm u}}\gamma^\mu[\boldsymbol{\lambda}\cdot\boldsymbol{A}_\mu P_{\rm L} - \boldsymbol{\lambda}^*\cdot\boldsymbol{A}_\mu P_{\rm R}]\Psi_{\rm M}^{\chi_{\rm u}} \\ &= -\frac{1}{4}\Psi_{\rm M}^{\chi_{\rm u}\dagger}\begin{pmatrix} -\sigma^\mu\boldsymbol{\lambda}^*\cdot\boldsymbol{A}_\mu & 0 \\ 0 & \bar{\sigma}^\mu\boldsymbol{\lambda}\cdot\boldsymbol{A}_\mu \end{pmatrix}\Psi_{\rm M}^{\chi_{\rm u}} \\ &= -\frac{1}{4}\left[\chi_{\rm u}^{\rm T}(-{\rm i}\sigma_2)(-\sigma^\mu\boldsymbol{\lambda}^*\cdot\boldsymbol{A}_\mu){\rm i}\sigma_2\chi_{\rm u}^* + \chi_{\rm u}^\dagger\bar{\sigma}^\mu\boldsymbol{\lambda}\cdot\boldsymbol{A}_\mu\chi_{\rm u}\right]. \end{aligned} \tag{8.29}$$

Exercise 8.1 Show that the first term of (8.29) is equal to the second, and hence that the interaction part of the first term in (8.28) is

$$-\frac{1}{2}g_{\rm s}\chi_{\rm u}^\dagger\bar{\sigma}^\mu\boldsymbol{\lambda}\cdot\boldsymbol{A}_\mu\chi_{\rm u}, \tag{8.30}$$

as in (8.18).

In a similar way the interaction part of the second term of (8.28) is just

$$-\frac{1}{2}g_{\rm s}\psi_{\rm u}^\dagger\sigma^\mu\boldsymbol{\lambda}\cdot\boldsymbol{A}_\mu\psi_{\rm u}, \tag{8.31}$$

and we have again recovered the full (Dirac) QCD term (this time using $\psi_{\rm u}$ rather than $\chi_{\bar{\rm u}}^{\rm c}$).

The electroweak interactions of the SM particles emerge very simply. The trilinear and quadrilinear self-interactions of the weak gauge bosons are contained in the

$\mathrm{SU(2)_L} \times U(1)_y$ field strength tensors. Consider the interaction of the left-handed electron-type doublet

$$\begin{pmatrix} \chi_{\nu_e} \\ \chi_e \end{pmatrix} \tag{8.32}$$

with the SU(2) gauge field $\boldsymbol{W}^\mu$, which is given by

$$-\frac{1}{2} g \left(\chi^\dagger_{\nu_e} \chi^\dagger_e\right) \bar{\sigma}^\mu \boldsymbol{\tau} \cdot \boldsymbol{W}_\mu \begin{pmatrix} \chi_{\nu_e} \\ \chi_e \end{pmatrix}. \tag{8.33}$$

Here the $\boldsymbol{\tau}$'s act in the two-dimensional 'ν_e–e' space, while $\bar{\sigma}^\mu$ acts on the spinor components of χ_{ν_e} and χ_e. On the other hand, in 4-component Dirac notation the interaction is

$$-\frac{1}{2} g \left(\bar{\Psi}_{\nu_e \mathrm{L}} \bar{\Psi}_{\mathrm{eL}}\right) \gamma^\mu \boldsymbol{\tau} \cdot \boldsymbol{W}_\mu \begin{pmatrix} \Psi_{\nu_e \mathrm{L}} \\ \Psi_{\mathrm{eL}} \end{pmatrix}, \tag{8.34}$$

where

$$\Psi_{\mathrm{eL}} = \left(\frac{1-\gamma_5}{2}\right) \Psi_e = \begin{pmatrix} 0 \\ \chi_e \end{pmatrix} \tag{8.35}$$

and similarly for Ψ_{ν_e}. Now for any two Dirac fields Ψ_1 and Ψ_2 we have

$$\bar{\Psi}_{2\mathrm{L}} \gamma^\mu \Psi_{1\mathrm{L}} = (0\ \chi_2^\dagger) \begin{pmatrix} 0 & 1 \\ 1 & 0 \end{pmatrix} \begin{pmatrix} 0 & \bar{\sigma}^\mu \\ \sigma^\mu & 0 \end{pmatrix} \begin{pmatrix} 0 \\ \chi_1 \end{pmatrix} = \chi_2^\dagger \bar{\sigma}^\mu \chi_1. \tag{8.36}$$

It is therefore clear that (8.33) is the same as (8.34).

The $\mathrm{U(1)}_y$ covariant derivative takes the form

$$\partial_\mu + \frac{1}{2} \mathrm{i} g' y_{\mathrm{L\,f}} B_\mu \ \text{ on 'L' SU(2) doublets} \tag{8.37}$$

and

$$\partial_\mu + \frac{1}{2} \mathrm{i} g' y_{\mathrm{R\,f}} B_\mu \ \text{ on 'R' SU(2) singlets,} \tag{8.38}$$

where for example $y_{\mathrm{L\,e}} = -1$ and $y_{\mathrm{R\,e}} = -2$. For the doublet (8.32), therefore, the $\mathrm{U(1)}_y$ interaction is

$$-\frac{1}{2} g' y_{\mathrm{L\,e}} \left(\chi^\dagger_{\nu_e} \chi^\dagger_e\right) \bar{\sigma}^\mu B_\mu \begin{pmatrix} \chi_{\nu_e} \\ \chi_e \end{pmatrix}, \tag{8.39}$$

which is the same as the 4-component version

$$-\frac{1}{2} g' y_{\mathrm{L\,e}} \left(\bar{\Psi}_{\nu_e \mathrm{L}} \bar{\Psi}_{\mathrm{eL}}\right) \gamma^\mu B_\mu \begin{pmatrix} \Psi_{\nu_e \mathrm{L}} \\ \Psi_{\mathrm{e\,L}} \end{pmatrix}. \tag{8.40}$$

We replace the singlet ψ_{e} by $\chi_{\bar{\mathrm{e}}}^{\mathrm{c}}$ as before. If we denote the y-value of $\chi_{\bar{\mathrm{e}}}$ by $y_{\mathrm{L}\,\bar{\mathrm{e}}}$, we have

$$y_{\mathrm{L}\,\bar{\mathrm{e}}} = -y_{\mathrm{R}\,\mathrm{e}}. \tag{8.41}$$

The $\mathrm{U}(1)_y$ interaction for $\chi_{\bar{\mathrm{u}}}$ is therefore

$$-\frac{1}{2}g'(-y_{\mathrm{R}\,\mathrm{e}})\chi_{\bar{\mathrm{e}}}^{\dagger}\bar{\sigma}^{\mu}B_{\mu}\chi_{\bar{\mathrm{e}}}. \tag{8.42}$$

Performing the same steps as in (8.20)–(8.22) we find that (8.42) is the same as

$$-\frac{1}{2}g' y_{\mathrm{R}\,\mathrm{e}}\chi_{\bar{\mathrm{e}}}^{\mathrm{c}}\sigma^{\mu}B_{\mu}\chi_{\bar{\mathrm{e}}}^{\mathrm{c}} \tag{8.43}$$

which is the correct interaction for the R-part of the electron field (2.93).

In the quark sector, electroweak interactions will be complicated by the usual intergenerational mixing, but no new point of principle arises; the simple examples we have considered are sufficient for our purpose. We now proceed to discuss one of the key predictions of the MSSM.

8.3 Gauge coupling unification in the MSSM

As mentioned in Section 1.2(b), the idea [55] that the three scale-dependent ('running') SU(3) × SU(2) × U(1) gauge couplings of the SM should converge to a common value – or *unify* – at some very high energy scale does not, in fact, prove to be the case for the SM itself, but it does work very convincingly in the MSSM [56]. The evolution of the gauge couplings is determined by the numbers and types of the gauge and matter multiplets present in the theory, which we have just now given for the MSSM; we can therefore proceed to describe this celebrated result.

The couplings α_3 and α_2 are defined by

$$\alpha_3 = g_{\mathrm{s}}^2/4\pi, \qquad \alpha_2 = g^2/4\pi \tag{8.44}$$

where g_{s} is the $\mathrm{SU(3)_c}$ gauge coupling of QCD and g is that of the electroweak $\mathrm{SU(2)_L}$. The definition of the third coupling α_1 is a little more complicated. It obviously has to be related in some way to g'^2, where g' is the gauge coupling of the $\mathrm{U}(1)_y$ of the SM. The constants g and g' appear in the $\mathrm{SU(2)_L}$ covariant derivative (see equation (22.21) of [7] for example)

$$D_{\mu} = \partial_{\mu} + \mathrm{i}g(\boldsymbol{\tau}/2)\cdot\boldsymbol{W}_{\mu} + \mathrm{i}g'(y/2)B_{\mu}. \tag{8.45}$$

The problem is that, strictly within the SM framework, the scale of 'g'' is arbitrary: we could multiply the weak hypercharge generator y by an arbitrary constant c, and divide g' by c, and nothing would change. In contrast to this, the normalization of whatever couplings multiply the three generators τ^1, τ^2 and τ^3 in (8.45) is fixed by the normalization of the τ's:

$$\mathrm{Tr}\left(\frac{\tau^\alpha}{2}\frac{\tau^\beta}{2}\right) = \frac{1}{2}\delta_{\alpha\beta}. \tag{8.46}$$

Since each generator is normalized to the same value, the same constant g must multiply each one; no relative rescalings are possible. Within a 'unified' framework, therefore, we hypothesize that some multiple of y, say $Y = c(y/2)$, is one of the generators of a larger group (SU(5) for instance), which also includes the generators of $\mathrm{SU(3)_c}$ and $\mathrm{SU(2)_L}$, all being subject to a common normalization condition; there is then only one (unified) gauge coupling. The quarks and leptons of one family will all belong to a single representation of the larger group, although this need not necessarily be the fundamental representation. All that matters is that the generators all have a common normalization. For example, we can demand the condition

$$\mathrm{Tr}(c^2(y/2)^2) = \mathrm{Tr}(t_3)^2 \tag{8.47}$$

say, where t_3 is the third $\mathrm{SU(2)_L}$ generator (any generator will give the same result), and the Trace is over all states in the representation: here, u, d, ν_e and e^-. The Traces are simply the sums of the squares of the eigenvalues. On the right-hand side of (8.47) we obtain

$$3\left(\frac{1}{4}+\frac{1}{4}\right)+\frac{1}{4}+\frac{1}{4}=2, \tag{8.48}$$

where the '3' comes from colour, while on the left we find from Table 8.1

$$c^2\left(\frac{3}{36}+\frac{3}{36}+\frac{3.4}{9}+\frac{3.1}{9}+\frac{1}{4}+1+\frac{1}{4}\right)=c^2\frac{20}{6}. \tag{8.49}$$

It follows that

$$c=\sqrt{\frac{3}{5}}, \tag{8.50}$$

so that the correctly normalized generator is

$$Y=\sqrt{\frac{3}{5}}\,y/2. \tag{8.51}$$

The B_μ term in (8.45) is then

$$\mathrm{i}g'\sqrt{\frac{5}{3}}\,Y\,B_\mu, \tag{8.52}$$

indicating that the correctly normalized α_1 is

$$\alpha_1 = \frac{5}{3}\frac{g'^2}{4\pi} \equiv \frac{g_1^2}{4\pi}. \tag{8.53}$$

Equation (8.53) can also be interpreted as a prediction for the weak angle θ_W at the unification scale: since $g \tan\theta_W = g' = \sqrt{3/5}g_1$ and $g = g_1$ at unification, we have $\tan\theta_W = \sqrt{3/5}$, or

$$\sin^2\theta_W(\text{unification scale}) = \frac{3}{8}. \tag{8.54}$$

We are now ready to consider the running of the couplings α_i. To one loop order, the renormalization group equation (RGE) has the form (for an introduction, see Chapter 15 of [7], for example)

$$\frac{d\alpha_i}{dt} = -\frac{b_i}{2\pi}\alpha_i^2 \tag{8.55}$$

where $t = \ln Q$ and Q is the 'running' energy scale, and the coefficients b_i are determined by the gauge group and the matter multiplets to which the gauge bosons couple. For SU(N) gauge theories with matter (scalars and fermions) in multiplets belonging to the fundamental representation, we have (see [50], for example)

$$b_N = \frac{11}{3}N - \frac{1}{3}n_f - \frac{1}{6}n_s \tag{8.56}$$

where n_f is the number of fermion multiplets (counting the two chirality states separately), and n_s is the number of (complex) scalar multiplets, which couple to the gauge bosons. For a U(1)$_Y$ gauge theory in which the fermionic matter particles have charges Y_f and the scalars have charges Y_s, the corresponding formula is

$$b_1 = -\frac{2}{3}\sum_f Y_f^2 - \frac{1}{3}\sum_s Y_s^2. \tag{8.57}$$

To examine unification, it is convenient to rewrite (8.55) as

$$\frac{d}{dt}(\alpha_i^{-1}) = \frac{b_i}{2\pi}, \tag{8.58}$$

which can be immediately integrated to give

$$\alpha_i^{-1}(Q) = \alpha_i^{-1}(Q_0) + \frac{b_i}{2\pi}\ln(Q/Q_0), \tag{8.59}$$

where Q_0 is the scale at which running commences. We see that the inverse couplings run linearly with $\ln Q$. Q_0 is taken to be m_Z, where the couplings are well measured. 'Unification' is then the hypothesis that, at some higher scale $Q_U = m_U$, the couplings are equal:

$$\alpha_1(m_U) = \alpha_2(m_U) = \alpha_3(m_U) \equiv \alpha_U. \tag{8.60}$$

This implies that the three equations (8.59), for $i = 1, 2, 3$, become

$$\alpha_U^{-1} = \alpha_3^{-1}(m_Z) + \frac{b_3}{2\pi}\ln(m_U/m_Z) \tag{8.61}$$

$$\alpha_U^{-1} = \alpha_2^{-1}(m_Z) + \frac{b_2}{2\pi}\ln(m_U/m_Z) \tag{8.62}$$

$$\alpha_U^{-1} = \alpha_1^{-1}(m_Z) + \frac{b_1}{2\pi}\ln(m_U/m_Z). \tag{8.63}$$

Eliminating α_U and $\ln(m_U/m_Z)$ from these equations gives one condition relating the measured constants $\alpha_i^{-1}(m_Z)$ and the calculated numbers b_i, which is

$$\frac{\alpha_3^{-1}(m_Z) - \alpha_2^{-1}(m_Z)}{\alpha_2^{-1}(m_Z) - \alpha_1^{-1}(m_Z)} = \frac{b_2 - b_3}{b_1 - b_2}. \tag{8.64}$$

Checking the truth of (8.64) is one simple way [57] of testing unification quantitatively (at least, at this one-loop level).

Let us call the left-hand side of (8.64) B_{exp}, and the right-hand side B_{th}. For B_{exp}, we use the data

$$\sin^2\theta_W(m_Z) = 0.231 \tag{8.65}$$

$$\alpha_3(m_Z) = 0.119, \quad \text{or} \quad \alpha_3^{-1}(m_Z) = 8.40 \tag{8.66}$$

$$\alpha_{\text{em}}^{-1}(m_Z) = 128. \tag{8.67}$$

We are not going to bother with errors here, but the uncertainty in $\alpha_3(m_Z)$ is about 2%, and that in $\sin^2\theta_W(m_Z)$ and $\alpha_{\text{em}}(m_Z)$ is much less. Here α_{em} is defined by $\alpha_{\text{em}} = e^2/4\pi$, where $e = g\sin\theta_W$. Hence

$$\alpha_2^{-1}(m_Z) = \alpha_{\text{em}}^{-1}(m_Z)\sin^2\theta_W(m_Z) = 29.6. \tag{8.68}$$

Finally,

$$g'^2 = g^2\tan^2\theta_W \tag{8.69}$$

and hence

$$\alpha_1^{-1}(m_Z) = \frac{3}{5}\alpha'^{-1}(m_Z) = \frac{3}{5}\alpha_2^{-1}(m_Z)\cot^2\theta_W(m_Z) = 59.12. \tag{8.70}$$

From these values we obtain

$$B_{\text{exp}} = 0.72. \tag{8.71}$$

Now let us look at B_{th}. First, consider the SM. For $SU(3)_c$ we have

$$b_3^{\text{SM}} = 11 - \frac{1}{3}12 = 7. \tag{8.72}$$

For $SU(2)_L$ we have

$$b_2^{\text{SM}} = \frac{22}{3} - 4 - \frac{1}{6} = \frac{19}{6}, \tag{8.73}$$

and for $U(1)_Y$ we have

$$b_1^{\text{SM}} = -\frac{2}{3}\frac{3}{5}\sum_{\text{f}}(y_{\text{f}}/2)^2 - \frac{1}{3}\frac{3}{5}\sum_{\text{s}}(y_{\text{s}}/2)^2 \tag{8.74}$$

$$= -\frac{2}{5}3\frac{20}{6} - \frac{1}{5}\frac{1}{2} = -\frac{41}{10}. \tag{8.75}$$

Hence, in the SM, the right-hand side of (8.64) gives

$$B_{\text{th}} = \frac{115}{218} = 0.528, \tag{8.76}$$

which is in very poor accord with (8.71).

What about the MSSM? Expression (8.56) must be modified to take account of the fact that, in each SU(N), the gauge bosons are accompanied by gauginos in the regular representation of the group. Their contribution to b_N is $-2N/3$. In addition, we have to include the scalar partners of the quarks and of the leptons, in the fundamental representations of SU(3) and SU(2); and we must not forget that we have two Higgs doublets, both accompanied by Higgsinos, all in the fundamental representation of SU(2). These changes give

$$b_3^{\text{MSSM}} = 7 - 2 - \frac{1}{6}12 = 3, \tag{8.77}$$

and

$$b_2^{\text{MSSM}} = \frac{19}{6} - \frac{4}{3} - \frac{1}{6}12 - \frac{1}{3}2 - \frac{1}{6} = -1. \tag{8.78}$$

It is interesting that the sign of b_2 has been reversed. For b_1^{MSSM}, there is no contribution from the gauge bosons or their fermionic partners. The left-handed fermions contribute as in (8.74), and are each accompanied by corresponding scalars, so that

$$b_1^{\text{MSSM}}(\text{fermions and sfermions}) = -\frac{3}{5}10 = -6. \tag{8.79}$$

The Higgs and Higginos contribute

$$b_1^{\text{MSSM}}(\text{Higgs and Higgsinos}) = -\frac{3}{5}4\frac{1}{4} = -\frac{3}{5}. \tag{8.80}$$

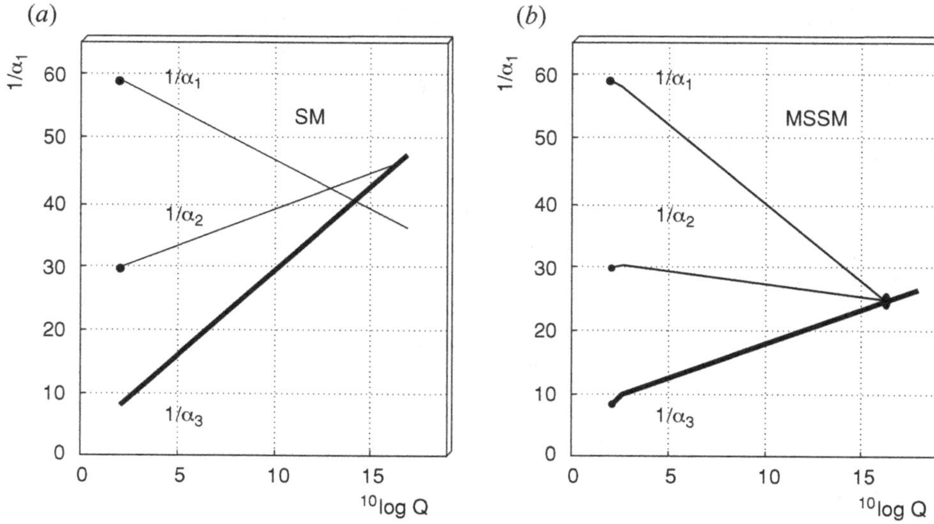

Figure 8.1 (a) Failure of the SM couplings to unify. (b) Gauge coupling unification in the MSSM. The blob represents model-dependent threshold corrections at the GUT scale. [Figure reprinted with permission from the review of Grand Unified Theories by S. Raby, Section 15 in *The Review of Particle Physics*, W.-M. Yao *et al. Journal of Physics* **G33**: 1–1232 (2006), p. 175, IOP Publishing Limited.]

In total, therefore,

$$b_1^{\text{MSSM}} = -\frac{33}{5}. \tag{8.81}$$

From (8.77), (8.78) and (8.81) we obtain [57]

$$B_{\text{th}}^{\text{MSSM}} = \frac{5}{7} = 0.714 \tag{8.82}$$

which is in excellent agreement with (8.71).

This has been by no means a 'professional' calculation. One should consider two-loop corrections. Furthermore, SUSY must be broken, presumably at a scale of order 1 TeV or less, and the resulting mass differences between the particles and their s-partners will lead to 'threshold' corrections. Similarly, the details of the theory at the high scale (in particular, its breaking) may be expected to lead to (high-energy) threshold corrections. A recent analysis by Pokorski [58] concludes that the present data are in good agreement with the predictions of supersymmetric unification, for reasonable estimates of such corrections. Figure 8.1, which is taken from Raby's review of grand unified theories [59], illustrates the situation.

Returning to (8.62) and (8.63), and inserting the values of $\alpha_2^{-1}(m_Z)$ and $\alpha_1^{-1}(m_Z)$, we can obtain an estimate of the unification scale m_U. We find

$$\ln(m_U/m_Z) = \frac{10\pi}{28}\left[\alpha_1^{-1}(m_Z) - \alpha_2^{-1}(m_Z)\right] \simeq 33.1, \tag{8.83}$$

which implies

$$m_{\mathrm{U}} \simeq 2.2 \times 10^{16}\ \mathrm{GeV}. \tag{8.84}$$

The first estimates of m_{U} with essentially the field content of the MSSM were made by Dimopoulos *et al.* [60], Dimopoulos and Georgi [20] and Sakai [21]; see also Ibáñez and Ross [61] and Einhorn and Jones [62].

Of course, one can make up any number of models yielding the experimental value B_{exp}, but there is no denying that the prediction (8.82) is an unforced consequence simply of the matter content of the MSSM, and agreement with (8.71) was clearly not inevitable. It does seem to provide support both for the inclusion of supersymmetric particles in the RGE, and for gauge unification.

8.4 *R*-parity

As stated in Section 8.1, the 'minimal' supersymmetric extension of the SM is specified by the choice (8.4) for the superpotential. There are, however, other gauge-invariant and renormalizable terms which could also be included in the superpotential, namely ([46] Section 5.2)

$$W_{\Delta L=1} = \lambda_e^{ijk} L_i \cdot L_j \bar{e}_k + \lambda_L^{ijk} L_i \cdot Q_j \bar{d}_k + \mu_L^{\prime i} L_i \cdot H_{\mathrm{u}} \tag{8.85}$$

and

$$W_{\Delta B=1} = \lambda_B^{ijk} \bar{u}_i \bar{d}_j \bar{d}_k. \tag{8.86}$$

The superfields Q_i carry baryon number $B = 1/3$ and $\bar{u}, \bar{d}$ carry $B = -1/3$, while L_i carries lepton number $L = 1$ and $\bar{e}$ carries $L = -1$. Hence the terms in (8.85) violate lepton number conservation by one unit of L, and those in (8.86) violate baryon number conservation by one unit of B. Now, B- and L-violating processes have never been seen experimentally. If both the couplings λ_L and λ_B were present, the proton could decay via channels such as $\mathrm{e}^+\pi^0$, $\mu^+\pi^0, \ldots,$ etc. The non-observance of such decays places strong limits on the strengths of such couplings, which would have to be extraordinarily small (being renormalizable, the couplings are dimensionless, and there is no natural suppression by a high scale such as would occur in a non-renormalizable term). It is noteworthy that in the SM, there are no possible renormalizable terms in the Lagrangian which violate B or L; this is indeed a nice bonus provided by the SM. We could of course just impose B and L conservation as a principle, thus forbidding (8.85) and (8.86), but in fact both are known to be violated by non-perturbative electroweak effects, which are negligible at ordinary energies but which might be relevant in the early universe. Neither B nor L can therefore be regarded as a fundamental symmetry. Instead, an alternative symmetry is required, which forbids (8.85) and (8.86), while allowing all the interactions of the MSSM.

This symmetry is called R-parity [33, 34], which is multiplicatively conserved, and is defined by

$$R = (-)^{3B+L+2s} \tag{8.87}$$

where s is the spin of the particle. One quickly finds that R is $+1$ for all conventional matter particles, and (because of the $(-)^{2s}$ factor) -1 for all their s-partners ('sparticles'). Since the product of $(-)^{2s}$ is $+1$ for the particles involved in any interaction vertex, by angular momentum conservation, it is clear that both (8.85) and (8.86) do not conserve R-parity, while the terms in (8.4) do. In fact, every interaction vertex in (8.4) contains an even number of $R = -1$ sparticles, which has important phenomenological consequences:

- The lightest sparticle ('LSP') is absolutely stable, and if electrically uncharged it could be an attractive candidate for non-baryonic dark matter.
- The decay products of all other sparticles must contain an odd number of LSP's.
- In accelerator experiments, sparticles can only be produced in pairs.

In the context of the MSSM, the LSP must lack electromagnetic and strong interactions; otherwise, LSP's surviving from the Big Bang era would have bound to nuclei forming objects with highly unusual charge to mass ratios, but searches for such exotics have excluded all models with stable charged or strongly interacting particles unless their mass exceeds several TeV, which is unacceptably high for the LSP. An important implication is that in collider experiments LSP's will carry away energy and momentum while escaping detection. Since all sparticles will decay into at least one LSP (plus SM particles), and since in the MSSM sparticles are pair-produced, it follows that at least $2m_{\tilde{\chi}_1^0}$ missing energy will be associated with each SUSY event, where $m_{\tilde{\chi}_1^0}$ is the mass of the LSP (often taken to be a neutralino; see Section 11.2). In e^-e^+ machines, the total visible energy and momentum can be well measured, and the beams have very small spread, so that the missing energy and momentum can be well correlated with the energy and momentum of the LSP's. In hadron colliders, the distribution of energy and longitudinal momentum of the partons (i.e. quarks and gluons) is very broad, so in practice only the missing transverse momentum (or missing transverse energy $\not{E}_T$) is useful.

Further aspects of R-parity, and of R-parity violation, are discussed in [46–49].

9
SUSY breaking

Since SUSY is manifestly not an exact symmetry of the known particle spectrum, the issue of SUSY breaking must be addressed before the MSSM can be applied phenomenologically. We know only two ways in which a symmetry can be broken: either (a) by explicit symmetry-breaking terms in the Lagrangian, or (b) by spontaneous symmetry breaking, such as occurs in the case of the chiral symmetry of QCD, and is hypothesized to occur for the electroweak symmetry of the SM via the Higgs mechanism. In the electroweak case, the introduction of explicit symmetry-breaking (gauge non-invariant) mass terms for the fermions and massive gauge bosons would spoil renormalizability, which is why in this case spontaneous symmetry breaking (which preserves renormalizability) is preferred theoretically – and indeed is strongly indicated by experiment, via the precision measurement of finite radiative corrections. We shall give a brief introduction to spontaneous SUSY breaking, since it presents some novel features as compared, say, to the more 'standard' examples of the spontaneous breaking of chiral symmetry in QCD, and of gauge symmetry in the electroweak theory. But in fact there is no consensus on how 'best' to break SUSY spontaneously, and in practice one is reduced to introducing explicit SUSY-breaking terms as in approach (a) after all, which parametrize the low-energy effects of the unknown breaking mechanism presumed (usually) to operate at some high mass scale. We shall see in Section 9.2 that these SUSY-breaking terms (which are gauge invariant and super-renormalizable) are quite constrained by the requirement that they do not re-introduce quadratic divergences which would spoil the SUSY solution to the SM fine-tuning problem of Section 1.1; nevertheless, over 100 parameters are needed to characterize them.

9.1 Breaking SUSY spontaneously

The fundamental requirement for a symmetry in field theory to be spontaneously broken (see, for example, [7] Part 7) is that a field, which is not invariant under the symmetry, should have a non-vanishing vacuum expectation value. That is, if the field in question is denoted by ϕ', then we require $\langle 0|\phi'(x)|0\rangle \neq 0$. Since ϕ' is not invariant, it must belong to a symmetry multiplet of some kind, along with other fields, and it must be possible to express ϕ' as

$$\phi'(x) = \mathrm{i}[Q, \phi(x)], \tag{9.1}$$

where Q is a hermitian generator of the symmetry group, and ϕ is a suitable field in the multiplet to which ϕ' belongs. So then we have

$$\langle 0|\phi'|0\rangle = \langle 0|\mathrm{i}[Q, \phi]|0\rangle = \langle 0|\mathrm{i}Q\phi - \mathrm{i}\phi Q|0\rangle \neq 0. \tag{9.2}$$

Now the vacuum state $|0\rangle$ is usually assumed to be such that $Q|0\rangle = 0$, since this implies that $|0\rangle$ is invariant under the transformation generated by Q, but if we take $Q|0\rangle = 0$, we violate (9.2). Hence for spontaneous symmetry breaking we have to assume $Q|0\rangle \neq 0$.

It is tempting to infer, in the latter case, that the application of Q to a vacuum state $|0\rangle$ gives, not zero, but another vacuum state $|0\rangle'$, leading to the physically suggestive idea of 'degenerate vacua'. But this notion is not mathematically correct. There are, in fact, only two alternatives: either $Q|0\rangle = 0$, or the state $Q|0\rangle$ has infinite norm, and hence cannot be regarded as a legitimate state. A fuller discussion is provided on pages 197–8 of [7], for example.

In the case of SUSY, there is a remarkable connection between the symmetry generators Q_a, $Q_b^\dagger$ of Chapter 3, and the Hamiltonian. The SUSY algebra (4.48) is

$$\{Q_a, Q_b^\dagger\} = (\sigma^\mu)_{ab} P_\mu. \tag{9.3}$$

So we have

$$\begin{aligned} Q_1 Q_1^\dagger + Q_1^\dagger Q_1 &= (\sigma^\mu)_{11} P_\mu = P_0 + P_3 \\ Q_2 Q_2^\dagger + Q_2^\dagger Q_2 &= (\sigma^\mu)_{22} P_\mu = P_0 - P_3, \end{aligned} \tag{9.4}$$

and consequently

$$H \equiv P_0 = \frac{1}{2}(Q_1 Q_1^\dagger + Q_1^\dagger Q_1 + Q_2 Q_2^\dagger + Q_2^\dagger Q_2), \tag{9.5}$$

where H is the Hamiltonian of the theory considered. Hence we find

$$\begin{aligned} \langle 0|H|0\rangle &= \frac{1}{2}(\langle 0|Q_1 Q_1^\dagger|0\rangle + \langle 0|Q_1^\dagger Q_1|0\rangle + \cdots) \\ &= \frac{1}{2}(\,|(Q_1^\dagger|0\rangle)|^2 + |(Q_1|0\rangle)|^2 + \cdots). \end{aligned} \tag{9.6}$$

It follows that the vacuum energy of a SUSY-invariant theory is zero.

Now we may assume that the kinetic energy parts of the Hamiltonian density do not contribute to the vacuum energy. On the other hand, the SUSY-invariant potential energy density $\mathcal{V}$ is given by (7.74) (which could equally well be written in terms of the auxiliary fields F_i and D^α). We remarked at the end of Section 7.3 that the form (7.74) implies that $\mathcal{V}$ is always greater than or equal to zero – and we now see that $\mathcal{V} = 0$ corresponds to the SUSY-invariant case.

For SUSY to be spontaneously broken, therefore, $\mathcal{V}$ must have no SUSY-invariant minimum: for, if it did, such a configuration would necessarily have zero energy, and since this is the minimum value of $\mathcal{V}$, SUSY breaking will simply not happen, on energy grounds. In the spontaneously broken case, when some field develops a non-zero vev, the minimum value of $\mathcal{V}$ will be a positive constant, and the vacuum energy will diverge. This is consistent with (9.6) and the infinite norm (in this case) of $Q_a|0\rangle$.

What kinds of field ϕ' could have a non-zero value in the SUSY case? Returning to (9.1), with Q now a SUSY generator, we consider all such possible commutation relations, beginning with those for the chiral supermultiplet. The commutation relations of the Q's with the fields are determined by the SUSY transformations, which are

$$\begin{aligned}
\delta_\xi \phi &= \mathrm{i}[\xi \cdot Q, \phi] = \xi \cdot \chi \\
\delta_\xi \chi &= \mathrm{i}[\xi \cdot Q, \chi] = -\mathrm{i}\sigma^\mu \mathrm{i}\sigma_2 \xi^* \partial_\mu \phi + \xi F \\
\delta_\xi F &= \mathrm{i}[\xi \cdot Q, F] = -\mathrm{i}\xi^\dagger \bar{\sigma}^\mu \partial_\mu \chi.
\end{aligned} \tag{9.7}$$

Now Lorentz invariance implies that only scalar fields may acquire vevs, since only such vevs are invariant under Lorentz transformations. Considering the terms on the right-hand side of each of the three relations in (9.7), we see that the only possibility for a symmetry-breaking vev is

$$\langle 0|F|0\rangle \neq 0. \tag{9.8}$$

This is called 'F-type SUSY breaking', since it is the auxiliary field F which acquires a vev.

Recall now that in W–Z models, with superpotentials of the form (5.9) such as are used in the MSSM, we had

$$F_i = -\left(\frac{\partial W}{\partial \phi_i}\right)^\dagger = -\left(M_{ij}\phi_j + \frac{1}{2}y_{ijk}\phi_j\phi_k\right)^\dagger, \tag{9.9}$$

and $\mathcal{V}(\phi) = |F_i|^2$, which has an obvious minimum when all the ϕ's are zero. Hence with this form of W, SUSY can not be spontaneously broken. To get spontaneous SUSY breaking, we must add a constant to F_i, that is a linear term in W (see footnote 2 of Chapter 5, page 72). Even this is tricky, and it needs ingenuity to produce a simple working model. One (due to O'Raifeartaigh [63]) employs three chiral supermultiplets, and takes W to be

$$W = m\phi_1\phi_3 + g\phi_2(\phi_3^2 - M^2), \tag{9.10}$$

where m and g may be chosen to be real and positive, and M is real. This produces

$$-F_1^\dagger = m\phi_3, \quad -F_2^\dagger = g\left(\phi_3^2 - M^2\right), \quad -F_3^\dagger = m\phi_1 + 2g\phi_2\phi_3. \tag{9.11}$$

Hence

$$\begin{aligned}\mathcal{V} &= |F_1|^2 + |F_2|^2 + |F_3|^2 \\ &= m^2|\phi_3|^2 + g^2\left|\phi_3^2 - M^2\right|^2 + |m\phi_1 + 2g\phi_2\phi_3|^2.\end{aligned} \tag{9.12}$$

The first two terms in (9.12) cannot both vanish at once, and so there is no possible field configuration giving $\mathcal{V} = 0$, which is the SUSY-preserving case. On the other hand, the third term in (9.12) can always be made to vanish by a suitable choice of ϕ_1, given ϕ_2 and ϕ_3. Hence to find the (SUSY breaking) minimum of $\mathcal{V}$ it suffices to examine just the first two terms of (9.12), which depend only on ϕ_3. Introducing the real and imaginary parts of ϕ_3 via

$$\phi_3 = (A + \mathrm{i}B)/\sqrt{2}, \tag{9.13}$$

these terms are

$$\mathcal{V}_3 = \frac{1}{2}(m^2 - 2g^2M^2)A^2 + \frac{1}{2}(m^2 + 2g^2M^2)B^2 + \frac{g^2}{4}(A^2 + B^2)^2 + g^2M^4. \tag{9.14}$$

The details of the further analysis depend on the sign of the coefficient $(m^2 - 2g^2M^2)$. We shall consider the case

$$m^2 > 2g^2M^2; \tag{9.15}$$

the reader may pursue the alternative one (or consult Section 7.2.2 of [49]). Assuming (9.15) holds, $\mathcal{V}_3$ clearly has a (SUSY breaking) minimum at

$$A = B = 0 \quad \text{i.e.} \quad \phi_3 = 0, \tag{9.16}$$

which implies from (9.12) that we also require

$$\phi_1 = 0. \tag{9.17}$$

Conditions (9.16) and (9.17) are interpreted as the corresponding vevs. Note, however, that ϕ_2 is left undetermined (a so-called 'flat' direction in field space). This solution therefore gives

$$\langle 0|F_1^\dagger|0\rangle = \langle 0|F_3^\dagger|0\rangle = 0, \tag{9.18}$$

but

$$\langle 0|F_2^\dagger|0\rangle = gM^2. \tag{9.19}$$

The minimum value of $\mathcal{V}$ is g^2M^4, which is strictly positive, as expected. The parameter M does indeed have the dimensions of a mass: it can be understood as signifying the scale of spontaneous SUSY breaking, via $\langle 0|F_2^\dagger|0\rangle \neq 0$, much as the Higgs vev sets the scale of electroweak symmetry breaking.

Now all the terms in W must be gauge-invariant, in particular the term linear in ϕ_2 in (9.10), but there is no field in the SM which is itself gauge-invariant (i.e. all its gauge quantum numbers are zero, often called a 'gauge singlet'). Hence in the MSSM we cannot have a linear term in W, and must look beyond this model if we want to pursue this form of SUSY breaking.

Nevertheless, it is worth considering some further aspects of F-type SUSY breaking. We evidently have

$$0 \neq \langle 0|[Q, \chi(x)]|0\rangle = \sum_n \langle 0|Q|n\rangle\langle n|\chi(x)|0\rangle - \langle 0|\chi(x)|n\rangle\langle n|Q|0\rangle, \tag{9.20}$$

where $|n\rangle$ is a complete set of states. It can be shown that (9.20) implies that there must exist among the states $|n\rangle$ a *massless* state $|\tilde{g}\rangle$ which couples to the vacuum via the generator Q: $\langle 0|Q|\tilde{g}\rangle \neq 0$. This is the SUSY version of Goldstone's theorem (see, for example, Section 17.4 of [7]). The theorem states that when a symmetry is spontaneously broken, one or more massless particles must be present, which couple to the vacuum via a symmetry generator. In the non-SUSY case, they are (Goldstone) bosons; in the SUSY case, since the generators are fermionic, they are fermions, 'Goldstinos'.[1] You can check that the fermion spectrum in the above model contains a massless field χ_2 – it is in fact in a supermultiplet along with F_2, the auxiliary field which gained a vev, and the scalar field ϕ_2, where ϕ_2 is the field direction along which the potential is 'flat' – a situation analogous to that for the standard Goldstone 'wine-bottle' potential, where the massless mode is associated with excitations round the flat rim of the bottle.

Exercise 9.1 Show that the mass spectrum of the O'Raifeartaigh model consists of (a) six real scalar fields with tree-level squared masses 0, 0 (the real and imaginary parts of the complex field ϕ_2) m^2, m^2 (ditto for the complex field ϕ_1) and $m^2 - 2g^2M^2, m^2 + 2g^2M^2$ (the no longer degenerate real and imaginary parts of the complex field ϕ_3); (b) three L-type fermions with masses 0 (the Goldstino χ_2), m, m (linear combinations of the fields χ_1 and χ_3). (Hint: for the scalar masses, take $\langle\phi_2\rangle = 0$ for convenience, expand the potential about the point $\phi_1 = \phi_2 = \phi_3 = 0$, and examine the quadratic terms. For the fermions, the mass matrix of (5.22) is $W_{13} = W_{31} = m$, all other components vanishing; diagonalize the mass term

[1] Note the (conventionally) different use of the '-ino' suffix here: the Goldstino is not the fermionic superpartner of a scalar Goldstone mode, but is itself the (fermionic) Goldstone mode. In general, the Goldstino is the fermionic component of the supermultiplet whose auxiliary field develops a vev.

by introducing the linear combinations $\chi_- = (\chi_1 - \chi_3)/\sqrt{2}$, $\chi_+ = (\chi_1 + \chi_3)/\sqrt{2}$. See also [49] Section 7.2.1.)

In the absence of SUSY breaking, a single massive chiral supermultiplet consists (as in the W–Z model of Chapter 5) of a complex scalar field (or equivalently two real scalar fields) degenerate in mass with an L-type spin-1/2 field. It is interesting that in the O'Raifeartaigh model the masses of the '3' supermultiplet, after SUSY breaking, obey the relation

$$(m^2 - 2g^2M^2) + (m^2 + 2g^2M^2) = 2m^2 = 2m^2_{\chi_3}, \tag{9.21}$$

which is evidently a generalization of the relation that would hold in the SUSY-preserving case $g = 0$. Indeed, there is a general sum rule for the tree-level (mass)^2 values of complex scalars and chiral fermions in theories with spontaneous SUSY breaking [64]:

$$\sum m^2_{\text{real scalars}} = 2\sum m^2_{\text{chiral fermions}}, \tag{9.22}$$

where it is understood that the sums are over sectors with the same electric charge, colour charge, baryon number and lepton number. Unfortunately, (9.22) implies that this kind of SUSY breaking cannot be phenomenologically viable, since it requires the existence of (for example) light scalar partners of the light SM fermions – and this is excluded experimentally.

We also need to consider possible SUSY breaking via terms in a gauge supermultiplet. This time the transformations are

$$\begin{aligned}
\delta_\xi W^{\mu\alpha} &= \mathrm{i}[\xi \cdot Q, W^{\mu\alpha}] = -\frac{1}{\sqrt{2}}(\xi^\dagger \bar{\sigma}^\mu \lambda^\alpha + \lambda^{\alpha\dagger}\bar{\sigma}^\mu \xi) \\
\delta_\xi \lambda^\alpha &= \mathrm{i}[\xi \cdot Q, \lambda^\alpha] = -\frac{\mathrm{i}}{2\sqrt{2}}\sigma^\mu \bar{\sigma}^\nu \xi F^\alpha_{\mu\nu} + \frac{1}{\sqrt{2}}\xi D^\alpha \\
\delta_\xi D^\alpha &= \mathrm{i}[\xi \cdot Q, D^\alpha] = \frac{\mathrm{i}}{\sqrt{2}}(\xi^\dagger \bar{\sigma}^\mu (D_\mu \lambda)^\alpha - (D_\mu\lambda)^{\alpha\dagger}\bar{\sigma}^\mu \xi).
\end{aligned} \tag{9.23}$$

Inspection of (9.23) shows that, as for the chiral supermultiplet, only the auxiliary fields can have a non-zero vev:

$$\langle 0|D^\alpha|0\rangle \neq 0, \tag{9.24}$$

which is called D-type symmetry breaking.

At first sight, however, such a mechanism can not operate in the MSSM, for which the scalar potential is as given in (7.74). 'F-type' SUSY breaking comes from the first term $|W_i|^2$, D-type from the second, and the latter clearly has a SUSY-preserving minimum at $\mathcal{V} = 0$ when all the fields vanish. But there is another possibility, rather like the 'linear term in W' trick used for F-type breaking, which was discovered by Fayet and Iliopoulos [65] for the U(1) gauge case. The auxiliary field D of a U(1) gauge supermultiplet is gauge-invariant, and a term in the Lagrangian proportional to D is SUSY-invariant too, since (see (7.38)) it transforms by a total derivative. Suppose, then, that we add a term $M^2 D$, the *Fayet–Iliopoulos term*, to the Lagrangian (7.72). The part involving D is now

$$\mathcal{L}_D = M^2 D + \frac{1}{2} D^2 - g_1 D \sum_i e_i \phi_i^\dagger \phi_i, \tag{9.25}$$

where e_i are the U(1) charges of the scalar fields ϕ_i in units of g_1, the U(1) coupling constant. Then the equation of motion for D is

$$D = -M^2 + g_1 \sum_i e_i \phi_i^\dagger \phi_i. \tag{9.26}$$

The corresponding potential is now

$$\mathcal{V}_D = \frac{1}{2} \left(-M^2 + g_1 \sum_i e_i \phi_i^\dagger \phi_i \right)^2. \tag{9.27}$$

Consider for simplicity the case of just one scalar field ϕ, with charge eg_1. If $eg_1 > 0$ there will be a SUSY-preserving solution, i.e. with $\mathcal{V}_D = 0$, and $\langle 0|D|0\rangle = 0$, and hence $|\langle 0|\phi|0\rangle| = (M^2/eg_1)^{1/2}$. This is actually a Higgs-type breaking of the U(1) symmetry, and it will also generate a mass for the U(1) gauge field. On the other hand, if $eg_1 < 0$, we find the SUSY-breaking solution $\mathcal{V}_D = \frac{1}{2}M^4$ when $\langle 0|D|0\rangle = -M^2$ and $\langle 0|\phi|0\rangle = 0$, which is U(1)-preserving. In fact, we then have

$$\mathcal{L}_D = -\frac{1}{2} M^4 - |eg_1| M^2 \phi^\dagger \phi + \cdots \tag{9.28}$$

showing that the ϕ field has a mass $M(|eg_1|)^{1/2}$. The gaugino field λ and the gauge field A^μ remain massless, and λ can be interpreted as a Goldstino.

This mechanism can not be used in the non-Abelian case, because no term of the form $M^2 D^\alpha$ can be gauge-invariant (it is in the adjoint representation, not a singlet). Could we have D-term breaking in the $\mathrm{U(1)}_y$ sector of the MSSM? Unfortunately not. What we want is a situation in which the scalar fields in (9.27) do not acquire vevs (for example, because they have large mass terms in the superpotential), so that the minimum of (9.27) forces D to have a non-zero (vacuum) value, thus breaking SUSY. In the MSSM, however, the squark and slepton fields have no superpotential mass terms, and so would not be prevented from acquiring vevs *en route* to minimizing (9.27). However, this would imply the breaking of any symmetry associated with quantum numbers carried by these fields, for example colour, which is not acceptable.

One common viewpoint seems to be that spontaneous SUSY breaking could occur in a sector that is weakly coupled to the chiral supermultiplets of the MSSM. For example, it could be (a) via gravitational interactions – presumably at the Planck scale, so that SUSY-breaking mass terms would enter as (the vev of an F- or D-type field which has dimension M^2, and is a singlet under the SM gauge group)$/M_{\rm P}$, which gives $\sqrt{(\text{vev})} \sim 10^{11}$ GeV, say; or (b) via electroweak gauge interactions. These possibilities are discussed in [46] Section 6. Recent reviews, embracing additional SUSY-breaking mechanisms, are contained in [47] Section 3, [48] Chapters 12 and 13, and [49] Chapter 11.

9.2 Soft SUSY-breaking terms

In any case, however the necessary breaking of SUSY is effected, we can always look for a parametrization of the SUSY-breaking terms which should be present at 'low' energies, and do phenomenology with them. It is a vital point that such phenomenological SUSY-breaking terms in the (now effective) Lagrangian should be 'soft', as the jargon has it – that is, they should have positive mass dimension, for example '$M^2\phi^2$', '$M\phi^3$', '$M\chi \cdot \chi$', etc. The reason is that such terms (which are super-renormalizable) will not introduce new divergences into, for example, the relations between the dimensionless coupling constants which follow from SUSY, and which guarantee the cancellation of quadratic divergences and hence the stability of the mass hierarchy, which was one of the prime motivations for SUSY in the first place. As we saw in Section 1.1, a typical leading one-loop radiative correction to a scalar (mass)^2 is

$$\delta m^2 \sim \left(\lambda_{\rm scalar} - g^2_{\rm fermion}\right)\Lambda^2, \tag{9.29}$$

where Λ is the high-energy cutoff. In SUSY we essentially have $\lambda_{\rm scalar} = g^2_{\rm fermion}$, and the dependence on Λ becomes safely logarithmic. Suppose, on the other hand, that the dimensionless couplings $\lambda_{\rm scalar}$ or $g_{\rm fermion}$ themselves received divergent one-loop corrections, arising from renormalizable (rather than super-renormalizable) SUSY-breaking interactions.[2] Then $\lambda_{\rm scalar}$ and $g_{\rm fermion}$ would differ by terms of order $\ln \Lambda$, with the result that the mass shift (9.29) becomes very large indeed, once more. In general, soft SUSY-breaking terms maintain the cancellations of quadratically divergent radiative corrections to scalar (mass)^2 terms, to all orders in perturbation theory [66]. This means that corrections δm^2 go like

[2] One example of such a renormalizable SUSY-breaking interaction would be the Standard Model Yukawa interaction that generates mass for 'up' fermions and which involves the charge-conjugate of the Higgs doublet that generates mass for the 'down' fermions. The argument being given here implies that we do *not* want to generate 'up' masses this way, but rather via a second, independent, Higgs field.

$m_{\text{soft}}^2 \ln(\Lambda/m_{\text{soft}})$, where m_{soft} is the typical mass scale of the soft SUSY-breaking terms. This is a stable shift in the sense of the SM fine-tuning problem, provided of course that (as remarked in Section 1.1) the new mass scale m_{soft} is not much greater than 1 TeV, say. The origin of this mass scale remains unexplained.

The forms of possible gauge invariant soft SUSY-breaking terms are quite limited. They are as follows.

(a) Gaugino masses for each gauge group:

$$-\frac{1}{2}(M_3 \tilde{g}^a \cdot \tilde{g}^a + M_2 \tilde{W}^a \cdot \tilde{W}^a + M_1 \tilde{B} \cdot \tilde{B} + \text{h.c.}) \tag{9.30}$$

where in the first (gluino) term a runs from 1 to 8 and in the second (wino) term it runs from 1 to 3, the dot here signifying the Lorentz invariant spinor product. The fields $\tilde{g}^a$, $\tilde{W}^a$ and $\tilde{B}$ are all L-type spinors, in a slightly simpler notation than $\chi_{\tilde{g}^a}$ etc. As in the case of the spinor field in the W–Z model (cf. (5.41)), the gaugino masses are given by the absolute values $|M_i|$. For simplicity, we shall assume that the parameters M_i are all real, which implies that they will not introduce any new CP-violation. There is, however, no necessity for the M_i to be positive, and we shall discuss how to deal with the possibility of negative M_i in Section 11.1.1.

(b) Squark (mass)^2 terms:

$$-m_{\tilde{\text{Q}}ij}^2 \tilde{Q}_i^\dagger \cdot \tilde{Q}_j - m_{\tilde{\bar{\text{u}}}ij}^2 \tilde{\bar{u}}_{\text{L}i}^\dagger \tilde{\bar{u}}_{\text{L}j} - m_{\tilde{\bar{\text{d}}}ij}^2 \tilde{\bar{d}}_{\text{L}i}^\dagger \tilde{\bar{d}}_{\text{L}j}, \tag{9.31}$$

where i and j are family labels, colour indices have been suppressed, and the first term is an $\text{SU}(2)_\text{L}$-invariant dot product of scalars in the $\bar{\mathbf{2}}$ and $\mathbf{2}$ representations; for example,

$$\tilde{Q}_1^\dagger \cdot \tilde{Q}_2 = \tilde{u}_\text{L}^\dagger \tilde{c}_\text{L} + \tilde{d}_\text{L}^\dagger \tilde{s}_\text{L}. \tag{9.32}$$

We remind the reader that all fields $\tilde{\bar{f}}_{\text{L}i}$ can equally well be written as $\tilde{f}_{\text{R}i}^\dagger$.

(c) Slepton (mass)^2 terms:

$$-m_{\tilde{\text{L}}ij}^2 \tilde{L}_i^\dagger \cdot \tilde{L}_j - m_{\tilde{\bar{\text{e}}}ij}^2 \tilde{\bar{e}}_{\text{L}i}^\dagger \tilde{\bar{e}}_{\text{L}j}. \tag{9.33}$$

(d) Higgs (mass)^2 terms:

$$-m_{\text{H}_\text{u}}^2 H_\text{u}^\dagger \cdot H_\text{u} - m_{\text{H}_\text{d}}^2 H_\text{d}^\dagger \cdot H_\text{d} - (b H_\text{u} \cdot H_\text{d} + \text{h.c.}) \tag{9.34}$$

where the $\text{SU}(2)_\text{L}$-invariant dot products are

$$H_\text{u}^\dagger \cdot H_\text{u} = |H_\text{u}^+|^2 + |H_\text{u}^0|^2 \tag{9.35}$$

and similarly for $H_\text{d}^\dagger \cdot H_\text{d}$, while

$$H_\text{u} \cdot H_\text{d} = H_\text{u}^+ H_\text{d}^- - H_\text{u}^0 H_\text{d}^0. \tag{9.36}$$

(e) Triple scalar couplings[3]

$$-a_{\rm u}^{ij}\tilde{\bar{u}}_{{\rm L}i}\tilde{Q}_j \cdot H_{\rm u} + a_{\rm d}^{ij}\tilde{\bar{d}}_{{\rm L}i}\tilde{Q}_j \cdot H_{\rm d} + a_{\rm e}^{ij}\tilde{\bar{e}}_{{\rm L}i}\tilde{L}_j \cdot H_{\rm d} + {\rm h.c.} \tag{9.37}$$

The five (mass)^2 matrices are in general complex, but must be hermitian so that the Lagrangian is real. All the terms (9.30)–(9.37) manifestly break SUSY, since they involve only the scalars and gauginos, and omit their respective superpartners.

On the other hand, it is important to emphasize that the terms (9.30)–(9.37) do respect the SM gauge symmetries. The b term in (9.34) and the triple scalar couplings in (9.37) have the same form as the 'μ' and 'Yukawa' couplings in the (gauge-invariant) superpotential (8.4), but here involving just the scalar fields, of course. It is particularly noteworthy that gauge-invariant mass terms are possible for all these superpartners, in marked contrast to the situation for the known SM particles. Consider (9.30) for instance. The gluinos are in the regular (adjoint) representation of a gauge group, like their gauge boson superpartners: for example, in SU(2) the winos are in the $t = 1$ ('vector') representation. In this representation, the transformation matrices can be chosen to be real (the generators are pure imaginary, $(T_i^{(1)})_{jk} = -{\rm i}\epsilon_{ijk}$), which means that they are orthogonal rather than unitary, just like rotation matrices in ordinary three-dimensional space. Thus quantities of the form '$\widetilde{\boldsymbol{W}} \cdot \widetilde{\boldsymbol{W}}$' are invariant under SU(2) transformations, including local (i.e. gauge) ones since no derivatives are involved; similarly for the gluinos and the bino. Coming to (9.31) and (9.33), squark and slepton mass terms of this form are allowed if i and j are family indices, and the m_{ij}^2's are hermitian matrices in family space, since under a gauge transformation $\phi_i \to U\phi_i$, $\phi_j \to U\phi_j$, where $U^\dagger U = 1$, and the ϕ's stand for a squark or slepton flavour multiplet. Higgs mass terms like $-m_{\rm H_u}^2 H_{\rm u}^\dagger H_{\rm u}$ are of course present in the SM already, and (as we saw in Chapter 8 – see the remarks following equation (8.15)) from the perspective of the MSSM we need to include such SUSY-violating terms in order to have any chance of breaking electroweak symmetry spontaneously (the parameter written as '$m_{\rm H_u}^2$' can of course have either sign). The b term in (9.34) is like the SUSY-invariant μ term of (8.13), but it only involves the Higgs, not the Higgsinos, and is hence SUSY breaking. Mass terms for the Higgsinos themselves are forbidden by gauge invariance, but the μ-term of (8.13) is gauge invariant and does, as noted after (8.14), contribute to off-diagonal Higgsino mass terms.

The upshot of these considerations is that mass terms which preserve electroweak symmetry can be written down for all the so-far unobserved particles of the MSSM.

[3] A further set of triple scalar couplings is also possible, namely $-c_{\rm u}^{ij}\tilde{\bar{u}}_{{\rm L}i}Q_j \cdot H_{\rm d}^\dagger + c_{\rm d}^{ij}\tilde{\bar{d}}_{{\rm L}i}\tilde{Q}_j \cdot H_{\rm u}^\dagger + c_{\rm e}^{ij}\tilde{\bar{e}}_{{\rm L}i}\tilde{L}_j \cdot H_{\rm u}^\dagger + {\rm h.c.}$ However, these are generally omitted, because it turns out that they are either absent or small in many models of SUSY-breaking (see [46] Section 4, for example).

By contrast, of course, similar mass terms for the known particles of the SM would all break electroweak symmetry explicitly, which is unacceptable (as leading to non-renormalizability, or unitarity violations; see, for example, [7] Sections 21.3, 21.4 and 22.6): the masses of the known SM particles must all arise via spontaneous breaking of electroweak symmetry. Thus it could be argued that, from the viewpoint of the MSSM, it is natural that the known particles have been found, since they are 'light', with a scale associated with electroweak symmetry breaking. The masses of the undiscovered particles, on the other hand, can be significantly higher.[4] As against this, it must be repeated that electroweak symmetry breaking is not possible while preserving SUSY: the Yukawa-like terms in (8.4) do respect SUSY, but will not generate fermion masses unless some Higgs fields have a non-zero vev, and this will not happen with a potential of the form (7.74) (see also (8.15)); similarly, the gauge-invariant couplings (7.67) are part of a SUSY-invariant theory, but the electroweak gauge boson masses require a Higgs vev in (7.67). So some, at least, of the SUSY-breaking parameters must have values not too far from the scale of electroweak symmetry breaking, if we don't want fine tuning. From this point of view, then, there seems no very clear distinction between the scales of electroweak and of SUSY breaking.

Unfortunately, although the terms (9.30)–(9.37) are restricted in form, there are nevertheless quite a lot of possible terms in total, when all the fields in the MSSM are considered, and this implies very many new parameters. In fact, Dimopoulos and Sutter [67] counted a total of 105 new parameters describing masses, mixing angles and phases, after allowing for all allowed redefinitions of bases. It is worth emphasizing that this massive increase in parameters is entirely to do with SUSY breaking, the SUSY-invariant (but unphysical) MSSM Lagrangian having only one new parameter (μ) with respect to the SM.

One may well be dismayed by such an apparently huge arbitrariness in the theory, but this impression is in a sense misleading since extensive regions of parameter space are in fact excluded phenomenologically. This is because generic values of most of the new parameters allow flavour changing neutral current (FCNC) processes, or new sources of CP violation, at levels that are excluded by experiment. For example, if the matrix $\mathbf{m}^2_{\tilde{\mathrm{L}}}$ in (9.33) has a non-suppressed off-diagonal term such as

$$\left(m^2_{\tilde{\mathrm{L}}}\right)_{\mathrm{e}\mu} \tilde{e}^\dagger_{\mathrm{L}} \tilde{\mu}_{\mathrm{L}} \tag{9.38}$$

(on the basis in which the lepton masses are diagonal), then unacceptably large lepton flavour changing ($\mu \to \mathrm{e}$) will be generated. We can, for instance, envisage

[4] The Higgs is an interesting special case (taking it to be unobserved as yet). In the SM its mass is arbitrary (though see footnote 3 of Chapter 1, page 11), but in the MSSM the lightest Higgs particle is predicted to be no heavier than about 140 GeV (see Section 10.2).

a loop diagram contributing to $\mu \to \mathrm{e} + \gamma$, in which the μ first decays virtually (via its L-component) to $\tilde{\mu}_\mathrm{L}+$ bino through one of the couplings in (7.72), the $\tilde{\mu}_\mathrm{L}$ then changing to $\tilde{\mathrm{e}}_\mathrm{L}$ via (9.38), followed by $\tilde{\mathrm{e}}_\mathrm{L}$ re-combining with the bino to make an electron (L-component), after emitting a photon. The upper limit on the branching ratio for $\mu \to \mathrm{e} + \gamma$ is 1.2×10^{-11}, and our loop amplitude will be many orders of magnitude larger than this, even for sleptons as heavy as 1 TeV. Similarly, the squark (mass)^2 matrices are tightly constrained both as to flavour mixing and as to CP-violating complex phases by data on $\mathrm{K}^0 - \bar{\mathrm{K}}^0$ mixing, $\mathrm{D}^0 - \bar{\mathrm{D}}^0$ and $\mathrm{B}^0 - \bar{\mathrm{B}}^0$ mixing, and the decay $\mathrm{b} \to \mathrm{s}\gamma$. For a recent survey, with further references, see [47] Section 5.

The existence of these strong constraints on the SUSY-breaking parameters at the SM scale suggests that whatever the actual SUSY-breaking mechanism might be, it should be such as to lead naturally to the suppression of such dangerous off-diagonal terms. One framework which guarantees this is that of supergravity unification [68–70], specifically the 'minimal supergravity (mSUGRA)' theory [69, 70], in which the parameters (9.30)–(9.37) take a particularly simple form at the GUT scale:

$$M_3 = M_2 = M_1 = m_{1/2}; \tag{9.39}$$

$$\mathbf{m}^2_{\tilde{\mathrm{Q}}} = \mathbf{m}^2_{\bar{\tilde{\mathrm{u}}}} = \mathbf{m}^2_{\bar{\tilde{\mathrm{d}}}} = \mathbf{m}^2_{\tilde{\mathrm{L}}} = \mathbf{m}^2_{\bar{\tilde{\mathrm{e}}}} = m_0^2\,\mathbf{1}, \tag{9.40}$$

where '**1**' stands for the unit matrix in family space;

$$m^2_{\mathrm{H_u}} = m^2_{\mathrm{H_d}} = m_0^2; \tag{9.41}$$

and

$$\mathbf{a}_\mathrm{u} = A_0\mathbf{y}_\mathrm{u},\ \mathbf{a}_\mathrm{d} = A_0\mathbf{y}_\mathrm{d},\ \mathbf{a}_\mathrm{e} = A_0\mathbf{y}_\mathrm{e}, \tag{9.42}$$

where the $\mathbf{y}$ matrices are those appearing in (8.4). Relations (9.40) imply that at m_P all squark and sleptons are degenerate in mass (independent of both flavour and family, in fact) and so, in particular, squarks and sleptons with the same electroweak quantum numbers can be freely transformed into each other by unitary transformations. All mixings can then be eliminated, apart from that originating via the triple scalar terms, but conditions (9.42) ensure that only the squarks and sleptons of the (more massive) third family can have large triple scalar couplings. If $m_{1/2}$, A_0 and b of (9.34) all have the same complex phase, the only CP-violating phase in the theory will be the usual Cabibbo–Kobayashi–Maskawa (CKM) one (leaving aside CP-violation in the neutrino sector). Somewhat weaker conditions than (9.39)–(9.42) would also suffice to accommodate the phenomenological constraints. (For completeness, we mention other SUSY-breaking mechanisms that have been proposed: gauge-mediated [71], gaugino-mediated [72] and anomaly-mediated [73] symmetry breaking.)

We must now remember, of course, that if we use this kind of effective Lagrangian to calculate quantities at the electroweak scale, in perturbation theory, the results will involve logarithms of the form[5]

$$\ln[(\text{high scale, for example the unification scale } m_{\text{U}})/\text{low scale } m_{\text{Z}}], \tag{9.43}$$

coming from loop diagrams, which can be large enough to invalidate perturbation theory. As usual, such 'large logarithms' must be re-summed by the renormalization group technique (see Chapter 15 of [7] for example). This amounts to treating all couplings and masses as running parameters, which evolve as the energy scale changes according to RGEs, whose coefficients can be calculated perturbatively. Conditions such as (9.39)–(9.42) are then interpreted as boundary conditions on the parameters at the high scale.

This implies that after evolution to the SM scale the relations (9.39)–(9.42) will no longer hold, in general. However, RG corrections due to gauge interactions will not introduce flavour-mixing or CP-violating phases, while RG corrections due to Yukawa interactions are quite small except for the third family. It seems to be generally the case that if FCNC and CP-violating terms are suppressed at a high Q_0, then supersymmetric contributions to FCNC and CP-violating observables are not in conflict with present bounds, although this may change as the bounds are improved.

9.3 RGE evolution of the parameters in the (softly broken) MSSM

It is fair to say that the apparently successful gauge unification in the MSSM (Section 8.3) encourages us to apply a similar RG analysis to the other MSSM couplings and to the soft parameters (9.39)–(9.42). One-loop RGEs for the MSSM are given in [47] Appendix C.6; see also [46] Section 5.5.

A simple example is provided by the gaugino mass parameters M_i ($i = 1, 2, 3$) whose evolution (at 1-loop order) is determined by an equation very similar to (8.55) for the running of the α_i, namely

$$\frac{\mathrm{d}M_i}{\mathrm{d}t} = -\frac{b_i}{2\pi}\alpha_i M_i. \tag{9.44}$$

[5] Expression (9.43) may be thought of in the context either of running the quantities 'down' in scale; i.e. from a supposedly 'fundamental' high scale $Q_0 \sim m_{\text{U}}$ to a low scale $\sim m_{\text{Z}}$; or, as in (8.61)–(8.63), of running 'up' from a low scale $Q_0 \sim m_{\text{Z}}$ to a high scale $\sim m_{\text{U}}$ (in order, perhaps, to try and infer high-scale physics from weak-scale input). Either way, a crucial hypothesis is, of course, that no new physics intervenes between $\sim m_{\text{Z}}$ and $\sim m_{\text{U}}$.

From (8.55) and (9.44) we obtain

$$\frac{1}{\alpha_i}\frac{\mathrm{d}M_i}{\mathrm{d}t} - M_i\frac{1}{\alpha_i^2}\frac{\mathrm{d}\alpha_i}{\mathrm{d}t} = 0, \tag{9.45}$$

and hence

$$\frac{\mathrm{d}}{\mathrm{d}t}(M_i/\alpha_i) = 0. \tag{9.46}$$

It follows that the three ratios (M_i/α_i) are RG-scale independent at 1-loop order. In mSUGRA-type models, then, we can write

$$\frac{M_i(Q)}{\alpha_i(Q)} = \frac{m_{1/2}}{\alpha_i(m_\mathrm{P})}, \tag{9.47}$$

and since all the α_i's are already unified below M_P we obtain

$$\frac{M_1(Q)}{\alpha_1(Q)} = \frac{M_2(Q)}{\alpha_2(Q)} = \frac{M_3(Q)}{\alpha_3(Q)} \tag{9.48}$$

at any scale Q, up to small 2-loop corrections and possible threshold effects at high scales.

Applying (9.48) at $Q = m_\mathrm{Z}$ we find

$$M_1(m_\mathrm{Z}) = \frac{\alpha_1(m_\mathrm{Z})}{\alpha_2(m_\mathrm{Z})}M_2(m_\mathrm{Z}) = \frac{5}{3}\tan^2\theta_\mathrm{W}(m_\mathrm{Z})M_2(m_\mathrm{Z}) \simeq 0.5M_2(m_\mathrm{Z}) \tag{9.49}$$

and

$$M_3(m_\mathrm{Z}) = \frac{\alpha_3(m_\mathrm{Z})}{\alpha_2(m_\mathrm{Z})}M_2(m_\mathrm{Z}) = \frac{\sin^2\theta_\mathrm{W}(m_\mathrm{Z})}{\alpha_\mathrm{em}(m_\mathrm{Z})}\alpha_3(m_\mathrm{Z})M_2(m_\mathrm{Z}) \simeq 3.5M_2(m_\mathrm{Z}), \tag{9.50}$$

where we have used (8.65)–(8.67). Equations (9.49) and (9.50) may be summarized as

$$M_3(m_\mathrm{Z}) : M_2(m_\mathrm{Z}) : M_1(m_\mathrm{Z}) \simeq 7 : 2 : 1. \tag{9.51}$$

This simple prediction is common to most supersymmetric phenomenology. It implies that the gluino is expected to be heavier than the states associated with the electroweak sector. (The latter are 'neutralinos', which are mixtures of the neutral Higgsinos ($\tilde{H}_\mathrm{u}^0$, $\tilde{H}_\mathrm{d}^0$) and neutral gauginos ($\tilde{B}$, $\tilde{W}^0$), and 'charginos', which are mixtures of the charged Higgsinos ($\tilde{H}_\mathrm{u}^+$, $\tilde{H}_\mathrm{d}^-$) and winos ($\tilde{W}^+$, $\tilde{W}^-$); see Sections 11.2 and 11.3.)

A second significant example concerns the running of the scalar masses. Here the gauginos contribute to the RHS of '$\mathrm{d}m^2/\mathrm{d}t$' with a negative coefficient, which tends to increase the mass as the scale Q is lowered. On the other hand, the contributions from fermion loops have the opposite sign, tending to decrease the mass at low scales. The dominant such contribution is provided by top quark loops since y_t is so much larger than the other Yukawa couplings. If we retain only the top quark Yukawa coupling, the 1-loop evolution equations for $m^2_{\mathrm{H_u}}$, $m^2_{\tilde{\mathrm{Q}}_3}$ and $m^2_{\tilde{\bar{\mathrm{u}}}_3}$ are

$$\frac{\mathrm{d}m^2_{\mathrm{H_u}}}{\mathrm{d}t} = \left[\frac{3X_t}{4\pi} - 6\alpha_2 M_2^2 - \frac{6}{5}\alpha_1 M_1^2\right] \Big/ 4\pi \tag{9.52}$$

$$\frac{\mathrm{d}m^2_{\tilde{\mathrm{Q}}_3}}{\mathrm{d}t} = \left[\frac{X_t}{4\pi} - \frac{32}{3}\alpha_3 M_3^2 - 6\alpha_2 M_2^2 - \frac{2}{15}\alpha_1 M_1^2\right] \Big/ 4\pi \tag{9.53}$$

$$\frac{\mathrm{d}m^2_{\tilde{\bar{\mathrm{u}}}_3}}{\mathrm{d}t} = \left[\frac{2X_t}{4\pi} - \frac{32}{3}\alpha_3 M_3^2 - \frac{32}{15}\alpha_1 M_1^2\right] \Big/ 4\pi, \tag{9.54}$$

where

$$X_t = 2|y_t|^2\left(m^2_{\mathrm{H_u}} + m^2_{\tilde{\mathrm{Q}}_3} + m^2_{\tilde{\bar{\mathrm{u}}}_3} + A_0^2\right) \tag{9.55}$$

and we have used (9.42). In contrast, the corresponding equation for $m^2_{\mathrm{H_d}}$, to which the top quark loop does not contribute, is

$$\frac{\mathrm{d}m^2_{\mathrm{H_d}}}{\mathrm{d}t} = \left[-6\alpha_2 M_2^2 - \frac{6}{5}\alpha_1 M_1^2\right] \Big/ 4\pi. \tag{9.56}$$

Since the quantity X_t is positive, its effect is always to decrease the appropriate (mass)^2 parameter at low scales. From (9.52)–(9.54) we can see that, of the three masses, $m^2_{\mathrm{H_u}}$ is (a) decreased the most because of the factor 3, and (b) increased the least because the gluino contribution (which is larger than those of the other gauginos) is absent. On the other hand, $m^2_{\mathrm{H_d}}$ will always tend to increase at low scales. The possibility then arises that $m^2_{\mathrm{H_u}}$ could run from a positive value at $Q_0 \sim 10^{16}$ GeV to a negative value at the electroweak scale, while all the other (mass)^2 parameters of the scalar particles remain positive.[6] This can indeed happen, thanks to the large value of the top quark mass (or equivalently the large value of y_t): see [74–80]. Such a negative (mass)^2 value would tend to destabilize the point $H_u^0 = 0$, providing a strong (although not conclusive; see Section 10.1) indication that this is the trigger for electroweak symmetry breaking. A representative example of the effect is shown in Figure 9.1 (taken from [80]).

The parameter y_t in (9.52)–(9.54), and the other Yukawa couplings in (8.4), all run too; consideration of the RGEs for these couplings provides some further interesting results. If (for simplicity) we make the approximations that only third-family couplings are significant, and ignore contributions from α_1 and α_2, the 1-loop

[6] Negative values for the squark (mass)^2 parameters would have the undesirable consequence of spontaneously breaking the colour SU(3).

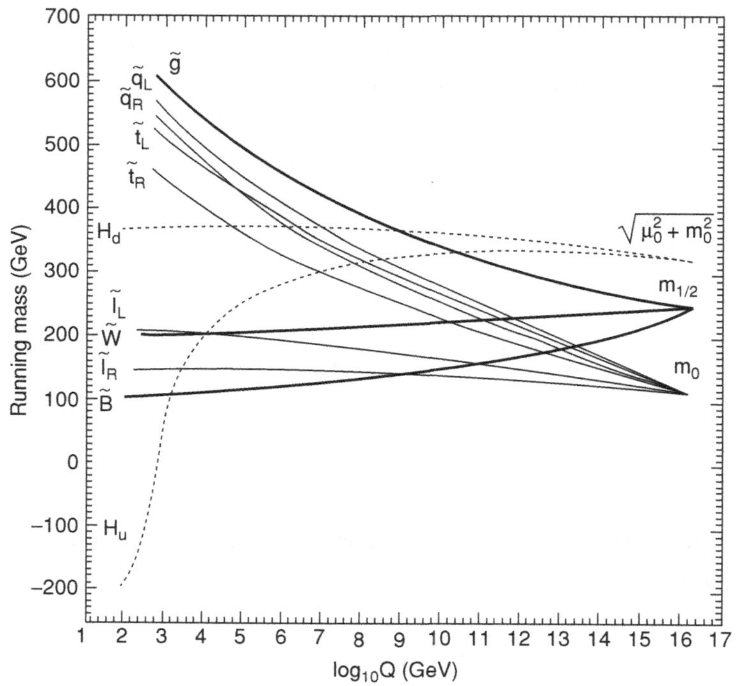

Figure 9.1 The running of the soft MSSM masses from the GUT scale to the electroweak scale, for a sample choice of input parameters (see [80]). The three gaugino masses M_1, M_2 and M_3 are labelled by $\tilde{B}$, $\tilde{W}$ and $\tilde{g}$ respectively. Similarly, the squark and slepton masses are labelled by the corresponding field label. The dashed lines labelled $\mathrm{H_d}$ and $\mathrm{H_d}$ represent the evolution of the masses $m_{\mathrm{H_u}}$ and $m_{\mathrm{H_d}}$. For convenience, negative values of $m^2_{\mathrm{H_u}}$ are shown on this plot as negative values of $m_{\mathrm{H_u}}$: that is, the figure shows $\mathrm{sign}(m^2_{\mathrm{H_u}})\sqrt{|m^2_{\mathrm{H_u}}|}$. [Figure reprinted with permission from G. L. Kane *et al.*, *Phys. Rev.* **D49** page 6183 (1994). Copyright (1994) by the American Physical Society.]

RGEs for the parameters y_t, y_b and y_τ are

$$\frac{\mathrm{d}y_\mathrm{t}}{\mathrm{d}t} = \frac{y_\mathrm{t}}{4\pi}\left[\left(6y_\mathrm{t}^2 + y_\mathrm{b}^2\right)/4\pi - \frac{16}{3}\alpha_\mathrm{s}\right] \tag{9.57}$$

$$\frac{\mathrm{d}y_\mathrm{b}}{\mathrm{d}t} = \frac{y_\mathrm{b}}{4\pi}\left[\left(6y_\mathrm{b}^2 + y_\mathrm{t}^2 + y_\tau^2\right)/4\pi - \frac{16}{3}\alpha_\mathrm{s}\right] \tag{9.58}$$

$$\frac{\mathrm{d}y_\tau}{\mathrm{d}t} = \frac{y_\tau}{16\pi^2}\left[4y_\tau^2 + 3y_\mathrm{b}^2\right]. \tag{9.59}$$

As in equations (9.52)–(9.54) the Yukawa couplings and the gauge coupling α_s enter the right-hand side of (9.57)–(9.59) with opposite signs; the former tend to increase the y's at high scales, while α_s tends to reduce y_t and y_b. It is then conceivable that, starting at low scales with $y_{\mathrm{t}} > y_{\mathrm{b}} > y_{\tau}$, the three y's might unify at or around m_{U}. Indeed, there is some evidence that the condition $y_{\mathrm{b}}(m_{\mathrm{U}}) = y_{\tau}(m_{\mathrm{U}})$, which arises naturally in many GUT models, leads to good low-energy phenomenology [81–84].

Further unification with $y_{\mathrm{t}}(m_{\mathrm{U}})$ must be such as to be consistent with the known top quark mass at low scales. To get a rough idea of how this works, we return to the relation (8.10), and similar ones for $m_{\mathrm{d}ij}$ and $m_{\mathrm{e}ij}$, which in the mass-diagonal basis give

$$y_{\mathrm{t}} = \frac{m_{\mathrm{t}}}{v_{\mathrm{u}}}, \quad y_{\mathrm{b}} = \frac{m_{\mathrm{b}}}{v_{\mathrm{d}}}, \quad y_{\tau} = \frac{m_{\tau}}{v_{\mathrm{d}}}, \tag{9.60}$$

where v_{d} is the vev of the field H_{d}^{0}. It is clear that the viability of $y_{\mathrm{t}} \approx y_{\mathrm{b}}$ will depend on the value of the additional parameter $v_{\mathrm{u}}/v_{\mathrm{d}}$ (denoted by $\tan\beta$; see Section 10.1). It seems that 'Yukawa unification' at m_{U} may work in the parameter regime where $\tan\beta \approx m_{\mathrm{t}}/m_{\mathrm{b}}$ [85–91].

In the following chapter we shall discuss the Higgs sector of the MSSM where, even without assumptions such as (9.39)–(9.42), only a few parameters enter, and one important prediction can be made: namely, an upper bound on the mass of the lightest Higgs boson, which is well within reach of the Large Hadron Collider.

10

The Higgs sector and electroweak symmetry breaking in the MSSM

10.1 The scalar potential and the conditions for electroweak symmetry breaking

We largely follow the treatment in Martin [46], Section 7.1. The first task is to find the potential for the scalar Higgs fields in the MSSM. As frequently emphasized, there are two complex Higgs $\mathrm{SU(2)_L}$ doublets which we are denoting by $H_{\mathrm{u}} = (H_{\mathrm{u}}^+, H_{\mathrm{u}}^0)$ which has weak hypercharge $y = 1$, and $H_{\mathrm{d}} = (H_{\mathrm{d}}^0, H_{\mathrm{d}}^-)$ which has $y = -1$. The classical (tree-level) potential for these scalar fields is made up of several terms. First, quadratic terms arise from the SUSY-invariant ('F-term') contribution (8.15) which involves the μ parameter from (8.4), and also from SUSY-breaking terms of the type (9.34). The latter two contributions are

$$m_{\mathrm{H_u}}^2\left(\left|H_{\mathrm{u}}^+\right|^2 + \left|H_{\mathrm{u}}^0\right|^2\right) + m_{\mathrm{H_d}}^2\left(\left|H_{\mathrm{d}}^0\right|^2 + \left|H_{\mathrm{d}}^-\right|^2\right), \tag{10.1}$$

where despite appearances it must be remembered that the arbitrary parameters '$m_{\mathrm{H_u}}^2$' and '$m_{\mathrm{H_d}}^2$' may have either sign, and

$$b\left(H_{\mathrm{u}}^+ H_{\mathrm{d}}^- - H_{\mathrm{u}}^0 H_{\mathrm{d}}^0\right) + \text{h.c.} \tag{10.2}$$

To these must be added the quartic SUSY-invariant 'D-terms' of (7.74), of the form $(\text{Higgs})^2\,(\text{Higgs})^2$, which we need to evaluate for the electroweak sector of the MSSM.

There are two groups G, $\mathrm{SU(2)_L}$ with coupling g and $\mathrm{U(1)}_y$ with coupling $g'/2$ (in the convention of [7]; see equation (22.21) of that reference). For the first, the matrices T^α are just $\tau^\alpha/2$, and we must evaluate

$$\begin{aligned}
&\sum_\alpha (H_{\mathrm{u}}^\dagger(\tau^\alpha/2)H_{\mathrm{u}} + H_{\mathrm{d}}^\dagger(\tau^\alpha/2)H_{\mathrm{d}})(H_{\mathrm{u}}^\dagger(\tau^\alpha/2)H_{\mathrm{u}} + H_{\mathrm{d}}^\dagger(\tau^\alpha/2)H_{\mathrm{d}}) \\
&\quad = (H_{\mathrm{u}}^\dagger(\boldsymbol{\tau}/2)H_{\mathrm{u}})\cdot(H_{\mathrm{u}}^\dagger(\boldsymbol{\tau}/2)H_{\mathrm{u}}) + (H_{\mathrm{d}}^\dagger(\boldsymbol{\tau}/2)H_{\mathrm{d}})\cdot(H_{\mathrm{d}}^\dagger(\boldsymbol{\tau}/2)H_{\mathrm{d}}) \\
&\qquad + 2(H_{\mathrm{u}}^\dagger(\boldsymbol{\tau}/2)H_{\mathrm{u}})\cdot(H_{\mathrm{d}}^\dagger(\boldsymbol{\tau}/2)H_{\mathrm{d}}).
\end{aligned} \tag{10.3}$$

If we write

$$H_{\mathrm{u}} = \begin{pmatrix} a \\ b \end{pmatrix}, \; H_{\mathrm{d}} = \begin{pmatrix} c \\ d \end{pmatrix}, \tag{10.4}$$

then brute force evaluation of the matrix and dot products in (10.3) yields the result

$$\frac{1}{4}\{[(|a|^2 + |b|^2) - (|c|^2 + |d|^2)]^2 + 4(ac^* + bd^*)(a^*c + b^*d)\}, \tag{10.5}$$

so that the SU(2) contribution is (10.5) multiplied by $g^2/2$. The U(1) contribution is

$$\frac{1}{2}(g'/2)^2[H_{\mathrm{u}}^\dagger H_{\mathrm{u}} - H_{\mathrm{d}}^\dagger H_{\mathrm{d}}]^2 = \frac{g'^2}{8}[(|a|^2 + |b|^2) - (|c|^2 + |d|^2)]^2. \tag{10.6}$$

Re-writing (10.5) and (10.6) in the notation of the fields, and including the quadratic pieces, the complete potential for the scalar fields in the MSSM is

$$\begin{aligned}\mathcal{V} = {} & \left(|\mu|^2 + m_{\mathrm{H_u}}^2\right)\left(|H_{\mathrm{u}}^+|^2 + \left|H_{\mathrm{u}}^0\right|^2\right) + \left(|\mu|^2 + m_{\mathrm{H_d}}^2\right)\left(\left|H_{\mathrm{d}}^0\right|^2 + |H_{\mathrm{d}}^-|^2\right) \\ & + \left[b\left(H_{\mathrm{u}}^+ H_{\mathrm{d}}^- - H_{\mathrm{u}}^0 H_{\mathrm{d}}^0\right) + \text{h.c.}\right] + \frac{(g^2 + g'^2)}{8} \\ & \times \left(|H_{\mathrm{u}}^+|^2 + \left|H_{\mathrm{u}}^0\right|^2 - \left|H_{\mathrm{d}}^0\right|^2 - |H_{\mathrm{d}}^-|^2\right)^2 + \frac{g^2}{2}\left|H_{\mathrm{u}}^+ H_{\mathrm{d}}^{0\dagger} + H_{\mathrm{u}}^0 H_{\mathrm{d}}^{-\dagger}\right|^2. \end{aligned} \tag{10.7}$$

We prefer not to re-write $(|\mu|^2 + m_{\mathrm{H_u}}^2)$ and $(|\mu|^2 + m_{\mathrm{H_d}}^2)$ as m_1^2 and m_2^2, say, so as to retain a memory of the fact that $|\mu|^2$ arises from a SUSY-invariant term, and is necessarily positive, while $m_{\mathrm{H_u}}^2$ and $m_{\mathrm{H_d}}^2$ are SUSY-breaking and of either sign *a priori*.

We must now investigate whether, and if so under what conditions, this potential can have a minimum which (like that of the simple Higgs potential (1.4) of the SM) breaks the $\mathrm{SU(2)_L} \times \mathrm{U(1)}_y$ electroweak symmetry down to $\mathrm{U(1)_{em}}$.

We can use the gauge symmetry to simplify the algebra somewhat. As in the SM (see, for example, Sections 17.6 and 19.6 of [7]) we can reduce a possible vev of one component of either H_{u} or H_{d} to zero by an $\mathrm{SU(2)_L}$ transformation. We choose $H_{\mathrm{u}}^+ = 0$ at the minimum of $\mathcal{V}$. The conditions $H_{\mathrm{u}}^+ = 0$ and $\partial\mathcal{V}/\partial H_{\mathrm{u}}^+ = 0$ then imply that, at the minimum of the potential, either

$$H_{\mathrm{d}}^- = 0 \tag{10.8}$$

or

$$b + \frac{g^2}{2} H_{\mathrm{d}}^{0\dagger} H_{\mathrm{u}}^{0\dagger} = 0. \tag{10.9}$$

The second condition (10.9) implies that the b term in (10.7) becomes

$$g^2\left|H_{\mathrm{u}}^0\right|^2\left|H_{\mathrm{d}}^0\right|^2 \tag{10.10}$$

which is definitely positive, and unfavourable to symmetry-breaking. As we shall see, condition (10.8) leads to a negative b-contribution. Accepting alternative (10.8) then, it follows that neither $H_{\rm u}^+$ nor $H_{\rm d}^-$ acquire a vev, which means (satisfactorily) that electromagnetism is not spontaneously broken. We can now ignore the charged components, and concentrate on the potential for the neutral fields which is

$$\mathcal{V}_{\rm n} = \left(|\mu|^2 + m_{\rm H_u}^2\right)\left|H_{\rm u}^0\right|^2 + \left(|\mu|^2 + m_{\rm H_d}^2\right)\left|H_{\rm d}^0\right|^2 - \left(bH_{\rm u}^0 H_{\rm d}^0 + {\rm h.c.}\right) + \frac{(g^2 + g'^2)}{8}\left(\left|H_{\rm u}^0\right|^2 - \left|H_{\rm d}^0\right|^2\right)^2. \qquad (10.11)$$

This is perhaps an appropriate point to note that the coefficient of the quartic term is *not* a free parameter, but is determined by the known electroweak couplings ($(g^2 + g'^2)/8 \approx 0.065$). This is of course in marked contrast to the case of the SM, where the coefficient $\lambda/4$ in (1.4) is a free parameter. Recalling from (1.3) that, in the SM, the mass of the Higgs boson is proportional to $\sqrt{\lambda}$, for given Higgs vev, this suggests that in the MSSM there should be a relatively light Higgs particle. As we shall see, this is indeed the case, although the larger field content of the Higgs sector in the MSSM makes the analysis more involved.

Consider now the b-term in (10.11), which is the only one that depends on the phases of the fields. Without loss of generality, b may be taken to be real and positive, any possible phase of b being absorbed into the relative phase of $H_{\rm u}^0$ and $H_{\rm d}^0$. For a minimum of $\mathcal{V}_{\rm n}$, the product $H_{\rm u}^0 H_{\rm d}^0$ must be real and positive too, which means that (at the minimum) the vev's of $H_{\rm u}^0$ and $H_{\rm d}^0$ must have equal and opposite phases. Since these fields have equal and opposite hypercharges, we can make a $\mathrm{U}(1)_y$ gauge transformation to reduce both their phases to zero. All vev's and couplings can therefore be chosen to be real, which means that **CP** is not spontaneously broken by the 2-Higgs potential of the MSSM, any more than it is in the 1-Higgs potential of the SM.[1]

The scalar potential now takes the more manageable form

$$\mathcal{V}_{\rm n} = \left(|\mu|^2 + m_{\rm H_u}^2\right)x^2 + \left(|\mu|^2 + m_{\rm H_d}^2\right)y^2 - 2bxy + \frac{(g^2 + g'^2)}{8}(x^2 - y^2)^2, \qquad (10.12)$$

where $x = |H_{\rm u}^0|$, $y = |H_{\rm d}^0|$; it depends on three parameters, $|\mu|^2 + m_{\rm H_u}^2$, $|\mu|^2 + m_{\rm H_d}^2$ and b. We want to identify the conditions required for the stable minimum of $\mathcal{V}_{\rm n}$ to occur at non-zero values of x and y. First note that, along the special ('flat')

[1] While this is true at tree level, **CP** symmetry could be broken significantly by radiative corrections, specifically via loops involving third generation squarks [92]; this would imply that the three neutral Higgs eigenstates would not have well defined **CP** quantum numbers (for the usual, **CP** conserving, case, see comments following equation (10.96) below).

direction $x = y$, the potential will be unbounded from below (no minimum) unless

$$2|\mu|^2 + m^2_{\mathrm{H_u}} + m^2_{\mathrm{H_d}} > 2b > 0. \tag{10.13}$$

Hence $(|\mu|^2 + m^2_{\mathrm{H_u}})$ and $(|\mu|^2 + m^2_{\mathrm{H_d}})$ cannot both be negative. This implies, referring to (10.12), that the point $x = y = 0$ cannot be a maximum of $\mathcal{V}_{\mathrm{n}}$. If $(|\mu|^2 + m^2_{\mathrm{H_u}})$ and $(|\mu|^2 + m^2_{\mathrm{H_d}})$ are both positive, then the origin is a minimum (which would be an unwanted symmetry-preserving solution) unless

$$\left(|\mu|^2 + m^2_{\mathrm{H_u}}\right)\left(|\mu|^2 + m^2_{\mathrm{H_d}}\right) < b^2, \tag{10.14}$$

which is the condition for the origin to be a saddle point. Equation (10.14) is automatically satisfied if either $(|\mu|^2 + m^2_{\mathrm{H_u}})$ or $(|\mu|^2 + m^2_{\mathrm{H_d}})$ is negative.

The b-term favours electroweak symmetry breaking, but it is not required to be non-zero. What can be said about $m^2_{\mathrm{H_u}}$ and $m^2_{\mathrm{H_d}}$? A glance at conditions (10.13) and (10.14) shows that they cannot both be satisfied if $m^2_{\mathrm{H_u}} = m^2_{\mathrm{H_d}}$, a condition that is typically taken to hold at a high scale $\sim 10^{16}$ GeV. However, the parameter $m^2_{\mathrm{H_u}}$ is, in fact, the one whose renormalization group evolution can drive it to negative values at the electroweak scale, as discussed at the end of the previous chapter (see Figure 9.1). It is clear that a negative value of $m^2_{\mathrm{H_u}}$ will tend to help condition (10.14) to be satisfied, but it is neither necessary nor sufficient ($|\mu|$ may be too large or b too small). A 'large' negative value for $m^2_{\mathrm{H_u}}$ is a significant factor, but it falls short of a demonstration that electroweak symmetry breaking *will* occur via this mechanism.

Having established the conditions (10.13) and (10.14) required for $|H^0_{\mathrm{u}}|$ and $|H^0_{\mathrm{d}}|$ to have non-zero vevs, say v_{u} and v_{d}, respectively, we can now proceed to write down the equations determining these vevs which follow from imposing the stationary conditions

$$\frac{\partial \mathcal{V}_{\mathrm{n}}}{\partial x} = \frac{\partial \mathcal{V}_{\mathrm{n}}}{\partial y} = 0. \tag{10.15}$$

Performing the differentiations and setting $x = v_{\mathrm{u}}$ and $y = v_{\mathrm{d}}$ we obtain

$$\left(|\mu|^2 + m^2_{\mathrm{H_u}}\right)v_{\mathrm{u}} = b v_{\mathrm{d}} + \frac{1}{4}(g^2 + g'^2)\left(v^2_{\mathrm{d}} - v^2_{\mathrm{u}}\right)v_{\mathrm{u}} \tag{10.16}$$

$$\left(|\mu|^2 + m^2_{\mathrm{H_d}}\right)v_{\mathrm{d}} = b v_{\mathrm{u}} - \frac{1}{4}(g^2 + g'^2)\left(v^2_{\mathrm{d}} - v^2_{\mathrm{u}}\right)v_{\mathrm{d}}. \tag{10.17}$$

One combination of v_{u} and v_{d} is fixed by experiment, since it determines the mass of the W and Z bosons, just as in the SM. The relevant terms in the electroweak sector are

$$(D_\mu H_{\mathrm{u}})^\dagger(D^\mu H_{\mathrm{u}}) + (D_\mu H_{\mathrm{d}})^\dagger(D^\mu H_{\mathrm{d}}) \tag{10.18}$$

where (see equation (22.21) of [7])

$$D_\mu = \partial_\mu + \mathrm{i}g(\boldsymbol{\tau}/2)\cdot \boldsymbol{W}_\mu + \mathrm{i}(g'/2)yB_\mu. \tag{10.19}$$

The mass terms for the vector particles come (in unitary gauge) from inserting the vevs for H_u and H_d, and defining

$$Z^\mu = \left(-g'B^\mu + gW_3^\mu\right)/(g^2 + g'^2)^{1/2}. \tag{10.20}$$

One finds

$$m_\mathrm{Z}^2 = \frac{1}{2}(g^2 + g'^2)\left(v_\mathrm{u}^2 + v_\mathrm{d}^2\right) \tag{10.21}$$

$$m_\mathrm{W}^2 = \frac{1}{2}g^2\left(v_\mathrm{u}^2 + v_\mathrm{d}^2\right). \tag{10.22}$$

Hence (see equations (22.29)–(22.32) of [7])

$$\left(v_\mathrm{u}^2 + v_\mathrm{d}^2\right)^{1/2} = \left(\frac{2m_\mathrm{W}^2}{g^2}\right)^{1/2} = 174\,\mathrm{GeV}. \tag{10.23}$$

Equations (10.16) and (10.17) may now be written as

$$\left(|\mu|^2 + m_{\mathrm{H_u}}^2\right) = b\cot\beta + \left(m_\mathrm{Z}^2/2\right)\cos 2\beta \tag{10.24}$$

$$\left(|\mu|^2 + m_{\mathrm{H_d}}^2\right) = b\tan\beta - \left(m_\mathrm{Z}^2/2\right)\cos 2\beta, \tag{10.25}$$

where

$$\tan\beta \equiv v_\mathrm{u}/v_\mathrm{d}. \tag{10.26}$$

It is easy to check that (10.24) and (10.25) satisfy the necessary conditions (10.13) and (10.14). We may use (10.24) and (10.25) to eliminate the parameters $|\mu|$ and b in favour of $\tan\beta$, but the phase of μ is not determined. As both v_u and v_d are real and positive, the angle β lies between 0 and $\pi/2$.

We are now ready to calculate the mass spectrum.

10.2 The tree-level masses of the scalar Higgs states in the MSSM

In the SM, there are four real scalar degrees of freedom in the Higgs doublet (1.5); after electroweak symmetry breaking (i.e. given a non-zero Higgs vev), three of them become the longitudinal modes of the massive vector bosons $\mathrm{W}^\pm$ and Z^0, while the fourth is the neutral Higgs boson of the SM, the mass of which is found by considering quadratic deviations away from the symmetry-breaking minimum (see Chapter 19 of [7], for example). In the MSSM, there are eight real scalar degrees of freedom. Three of them are massless, and just as in the SM they get 'swallowed' by the $\mathrm{W}^\pm$ and Z^0. The masses of the other five are again calculated by expanding the potential about the minimum, up to second order in the fields. Although straightforward, the work is complicated by the fact that the quadratic deviations are not diagonal in the fields, so that some diagonalization has to be done before the physical masses can be extracted.

To illustrate the procedure, consider the Lagrangian

$$\mathcal{L}_{12} = \partial_\mu \phi_1 \partial^\mu \phi_1 + \partial_\mu \phi_2 \partial^\mu \phi_2 - V(\phi_1, \phi_2), \tag{10.27}$$

where $V(\phi_1, \phi_2)$ has a minimum at $\phi_1 = v_1, \phi_2 = v_2$. We expand V about the minimum, retaining only quadratic terms, and discarding an irrelevant constant; this yields

$$\begin{aligned}\mathcal{L}_{12,\text{quad}} = {} & \partial_\mu \phi_1 \partial^\mu \phi_1 + \partial_\mu \phi_2 \partial^\mu \phi_2 - \frac{1}{2}\frac{\partial^2 V}{\partial \phi_1^2}(\phi_1 - v_1)^2 \\ & - \frac{1}{2}\frac{\partial^2 V}{\partial \phi_2^2}(\phi_2 - v_2)^2 - \frac{\partial^2 V}{\partial \phi_1 \partial \phi_2}(\phi_1 - v_1)(\phi_2 - v_2)\end{aligned} \tag{10.28}$$

where the derivatives are evaluated at the minimum (v_1, v_2). Defining

$$\tilde{\phi_1} = \sqrt{2}(\phi_1 - v_1), \quad \tilde{\phi_2} = \sqrt{2}(\phi_2 - v_2), \tag{10.29}$$

(10.28) can be written as

$$\mathcal{L}_{12,\text{quad}} = \frac{1}{2}\partial_\mu \tilde{\phi_1} \partial^\mu \tilde{\phi_1} + \frac{1}{2}\partial_\mu \tilde{\phi_2} \partial^\mu \tilde{\phi_2} - \frac{1}{2}(\tilde{\phi_1}\ \tilde{\phi_2})\mathbf{M}^{\text{sq}}\begin{pmatrix}\tilde{\phi_1} \\ \tilde{\phi_2}\end{pmatrix}, \tag{10.30}$$

where the (mass)^2 matrix $\mathbf{M}^{\text{sq}}$ is given by

$$\mathbf{M}^{\text{sq}} = \frac{1}{2}\begin{pmatrix} V''_{11} & V''_{12} \\ V''_{12} & V''_{22}\end{pmatrix}, \tag{10.31}$$

where

$$V''_{ij} = \frac{\partial^2 V}{\partial \phi_i \partial \phi_j}(v_1, v_2). \tag{10.32}$$

The matrix $\mathbf{M}^{\text{sq}}$ is real and symmetric, and can be diagonalized via an orthogonal transformation of the form

$$\begin{pmatrix}\phi_+ \\ \phi_-\end{pmatrix} = \begin{pmatrix}\cos\alpha & -\sin\alpha \\ \sin\alpha & \cos\alpha\end{pmatrix}\begin{pmatrix}\tilde{\phi_1} \\ \tilde{\phi_2}\end{pmatrix}. \tag{10.33}$$

If the eigenvalues of $\mathbf{M}^{\text{sq}}$ are m_+^2 and m_-^2, we see that in the new basis (10.30) becomes

$$\mathcal{L}_{12,\text{quad}} = \frac{1}{2}\partial_\mu \phi_+ \partial^\mu \phi_+ + \frac{1}{2}\partial_\mu \phi_- \partial^\mu \phi_- - \frac{1}{2}(\phi_+)^2 m_+^2 - \frac{1}{2}(\phi_-)^2 m_-^2, \tag{10.34}$$

from which it follows (via the equations of motion for ϕ_+ and ϕ_-) that m_+^2 and m_-^2 are the squared masses of the modes described by ϕ_+ and ϕ_-.

We apply this formalism first to the pair of fields ($\mathrm{Im}H_{\mathrm{u}}^0$, $\mathrm{Im}H_{\mathrm{d}}^0$). The part of our scalar potential involving this pair is

$$\mathcal{V}_{\mathrm{A}} = \left(|\mu|^2 + m_{\mathrm{H_u}}^2\right)\left(\mathrm{Im}H_{\mathrm{u}}^0\right)^2 + \left(|\mu|^2 + m_{\mathrm{H_d}}^2\right)\left(\mathrm{Im}H_{\mathrm{d}}^0\right)^2 + 2b\left(\mathrm{Im}H_{\mathrm{u}}^0\right)\left(\mathrm{Im}H_{\mathrm{d}}^0\right)$$
$$+ \frac{(g^2+g'^2)}{8}\left[\left(\mathrm{Re}H_{\mathrm{u}}^0\right)^2 + \left(\mathrm{Im}H_{\mathrm{u}}^0\right)^2 - \left(\mathrm{Re}H_{\mathrm{d}}^0\right)^2 - \left(\mathrm{Im}H_{\mathrm{d}}^0\right)^2\right]^2. \quad (10.35)$$

Evaluating the second derivatives at the minimum point, we find the elements of the (mass)^2 matrix:

$$M_{11}^{\mathrm{sq}} = |\mu|^2 + m_{\mathrm{H_u}}^2 + \frac{(g^2+g'^2)}{4}\left(v_{\mathrm{u}}^2 - v_{\mathrm{d}}^2\right) = b\cot\beta, \quad (10.36)$$

where we have used (10.16), and similarly

$$M_{12}^{\mathrm{sq}} = b, \quad M_{22}^{\mathrm{sq}} = b\tan\beta. \quad (10.37)$$

The eigenvalues of $\mathbf{M}^{\mathrm{sq}}$ are easily found to be

$$m_+^2 = 0, \quad m_-^2 = 2b/\sin 2\beta. \quad (10.38)$$

The eigenmode corresponding to the massless state is

$$\sqrt{2}\left[\sin\beta\left(\mathrm{Im}H_{\mathrm{u}}^0\right) - \cos\beta\left(\mathrm{Im}H_{\mathrm{d}}^0\right)\right], \quad (10.39)$$

and this will become the longitudinal state of the Z^0. The orthogonal combination

$$\sqrt{2}\left[\cos\beta\left(\mathrm{Im}H_{\mathrm{u}}^0\right) + \sin\beta\left(\mathrm{Im}H_{\mathrm{d}}^0\right)\right] \quad (10.40)$$

is the field of a scalar particle 'A^0', with mass

$$m_{\mathrm{A}^0} = (2b/\sin 2\beta)^{1/2}. \quad (10.41)$$

In discussing the parameter space of the Higgs sector of the MSSM, the pair of parameters (b, $\tan\beta$) is usually replaced by the pair (m_{A^0}, $\tan\beta$).

Next, consider the charged pair (H_{u}^+, $H_{\mathrm{d}}^{-\dagger}$). In this case the relevant part of the Lagrangian is

$$\mathcal{L}_{\mathrm{ch,\,quad}} = (\partial_\mu H_{\mathrm{u}}^+)^\dagger(\partial^\mu H_{\mathrm{u}}^+) + (\partial_\mu H_{\mathrm{d}}^-)^\dagger(\partial^\mu H_{\mathrm{d}}^-) - \frac{\partial^2\mathcal{V}}{\partial H_{\mathrm{u}}^{+\dagger}\partial H_{\mathrm{u}}^+}H_{\mathrm{u}}^{+\dagger}H_{\mathrm{u}}^+$$
$$- \frac{\partial^2\mathcal{V}}{\partial H_{\mathrm{d}}^{-\dagger}\partial H_{\mathrm{d}}^-}H_{\mathrm{d}}^{-\dagger}H_{\mathrm{d}}^- - \frac{\partial^2\mathcal{V}}{\partial H_{\mathrm{u}}^+\partial H_{\mathrm{d}}^-}H_{\mathrm{u}}^+H_{\mathrm{d}}^- - \frac{\partial^2\mathcal{V}}{\partial H_{\mathrm{u}}^{+\dagger}\partial H_{\mathrm{d}}^{-\dagger}}H_{\mathrm{u}}^{+\dagger}H_{\mathrm{d}}^{-\dagger}, \quad (10.42)$$

where we use (10.7) for $\mathcal{V}$, and the derivatives are evaluated at $H_{\rm u}^0 = v_{\rm u}$, $H_{\rm d}^0 = v_{\rm d}$, $H_{\rm u}^+ = H_{\rm d}^- = 0$. We write the potential terms as

$$(H_{\rm u}^{+\dagger}\ H_{\rm d}^-)\mathbf{M}_{\rm ch}^{\rm sq}\begin{pmatrix} H_{\rm u}^+ \\ H_{\rm d}^{-\dagger} \end{pmatrix} \tag{10.43}$$

where

$$\mathbf{M}_{\rm ch}^{\rm sq} = \begin{pmatrix} M_{++}^{\rm sq} & M_{+-}^{\rm sq} \\ M_{-+}^{\rm sq} & M_{--}^{\rm sq} \end{pmatrix} \tag{10.44}$$

with $M_{++}^{\rm sq} = \partial^2\mathcal{V}/\partial H_{\rm u}^{+\dagger}\partial H_{\rm u}^+$ etc. Performing the differentiations and evaluating the results at the minimum, we obtain

$$\mathbf{M}_{\rm ch}^{\rm sq} = \begin{pmatrix} b\cot\beta + g^2 v_{\rm d}^2/2 & b + g^2 v_{\rm u} v_{\rm d}/2 \\ b + g^2 v_{\rm u} v_{\rm d}/2 & b\tan\beta + g^2 v_{\rm u}^2/2 \end{pmatrix}. \tag{10.45}$$

This matrix has eigenvalues 0 and $m_{\rm W}^2 + m_{\rm A^0}^2$. The massless state corresponds to the superposition

$$G^+ = \sin\beta H_{\rm u}^+ - \cos\beta H_{\rm d}^{-\dagger}, \tag{10.46}$$

and it provides the longitudinal mode of the $\rm W^+$ boson. There is a similar state $G^- \equiv (G^+)^\dagger$, which goes into the $\rm W^-$. The massive (orthogonal) state is

$$H^+ = \cos\beta H_{\rm u}^+ + \sin\beta H_{\rm d}^{-\dagger}, \tag{10.47}$$

which has mass $m_{\rm H^+} = (m_{\rm W}^2 + m_{\rm A^0}^2)^{1/2}$, and there is a similar state $H^- \equiv (H^+)^\dagger$. Note that after diagonalization (10.42) becomes

$$(\partial_\mu G^+)^\dagger(\partial^\mu G^+) + (\partial H^+)^\dagger(\partial^\mu H^+) - m_{\rm H^+}^2 H^{+\dagger} H^+ \tag{10.48}$$

and the equation of motion for H^+ shows that $m_{\rm H^+}^2$ is correctly identified with the physical squared mass, without the various factors of 2 that appeared in our example (10.28)–(10.34) of two neutral fields.

Finally, we consider the coupled pair $({\rm Re}H_{\rm u}^0 - v_{\rm u}, {\rm Re}H_{\rm d}^0 - v_{\rm d})$, which is of the same type as our example, and as the pair $({\rm Im}H_{\rm u}^0, {\rm Im}H_{\rm d}^0)$. The (mass)^2 matrix is

$$\mathbf{M}_{\rm h,H}^{\rm sq} = \begin{pmatrix} b\cot\beta + m_{\rm Z}^2\sin^2\beta & -b - \left(m_{\rm Z}^2\sin 2\beta\right)/2 \\ -b - \left(m_{\rm Z}^2\sin 2\beta\right)/2 & b\tan\beta + m_{\rm Z}^2\cos^2\beta \end{pmatrix} \tag{10.49}$$

which has eigenvalues

$$m_{\rm h^0}^2 = \frac{1}{2}\left\{m_{\rm A^0}^2 + m_{\rm Z}^2 - \left[\left(m_{\rm A^0}^2 + m_{\rm Z}^2\right)^2 - 4m_{\rm A^0}^2 m_{\rm Z}^2\cos^2 2\beta\right]^{1/2}\right\} \tag{10.50}$$

and

$$m_{\rm H^0}^2 = \frac{1}{2}\left\{m_{\rm A^0}^2 + m_{\rm Z}^2 + \left[\left(m_{\rm A^0}^2 + m_{\rm Z}^2\right)^2 - 4m_{\rm A^0}^2 m_{\rm Z}^2\cos^2 2\beta\right]^{1/2}\right\}. \tag{10.51}$$

Equations (10.50) and (10.51) display the dependence of m_{h^0} and m_{H^0} on the parameters m_{A^0} and β. The corresponding eigenmodes will be given in Section 10.4.

The crucial point now is that, whereas the masses m_{A^0}, m_{H^0} and $m_{H^\pm}$ are unconstrained (since they all grow as $b/\sin\beta$ which can in principle be arbitrarily large), the mass m_{h^0} is bounded from above. Let us write $x = m^2_{A^0}$, $a = m^2_Z$; then

$$m^2_{h^0} = \frac{1}{2}\{x + a - [(x+a)^2 - 4ax\cos^2 2\beta]^{1/2}\}. \tag{10.52}$$

It is easy to verify that this function has no stationary point for finite values of x. Further, for small x we find

$$m^2_{h^0} \approx x\cos^2 2\beta, \tag{10.53}$$

whereas for large x

$$m^2_{h^0} \to a\cos^2 2\beta - (a^2/4x)\sin^2 4\beta. \tag{10.54}$$

Hence the maximum value of $m^2_{h^0}$, reached as $m^2_{A^0} \to \infty$, is $a\cos^2 2\beta$; that is

$$m_{h^0} \le m_Z|\cos 2\beta| \le m_Z. \tag{10.55}$$

Note that $|\cos 2\beta|$ vanishes when $\tan\beta = 1$, and is maximized for small or large $\tan\beta$ ($\beta \approx 0$ or $\pi/2$).

This is the promised upper bound on the mass of one of the neutral Higgs bosons in the MSSM, and it is surely a remarkable result [93, 94]. The bound (10.55) has, of course, already been exceeded by the current experimental lower bound [95]

$$m_H \ge 114.4 \text{ GeV } (95\% \text{ c.l.}). \tag{10.56}$$

Fortunately for the MSSM, the tree-level mass formulae derived above receive significant 1-loop corrections, particularly in the case of the h^0, whose mass is shifted upwards, possibly by a substantial amount [96–99]. One important contribution to $m^2_{h^0}$ arises from the incomplete cancellation of top quark and top squark loops, which would cancel in the exact SUSY limit (recall the paragraph following equation (1.22)). The magnitude of this contribution depends on the top squark masses, which we shall discuss in Section 11.4. If, for simplicity, we neglect top squark mixing effects (i.e. set the off-diagonal elements of the mass-squared matrix of (11.59) to zero), then the inclusion of this contribution modifies the bound (10.55) to

$$m^2_{h^0} \le m^2_Z + \frac{3m_t^4}{2\pi^2(v_u^2 + v_d^2)}\ln(m_S/m_t) \tag{10.57}$$

where m_t is the top quark mass and $m^2_S = \frac{1}{2}(m^2_{\tilde{t}_1} + m^2_{\tilde{t}_2})$ is the average of the squared masses of the two top squarks. To get an idea of the orders of magnitude involved, let us set $m_S = 500$ GeV, together with $m_t = 174$ GeV $=(v_u^2 + v_d^2)^{1/2}$. Then (10.57) gives

$$m_{h^0}^2 \leq m_Z^2 + (70 \text{ GeV})^2 = (115 \text{ GeV})^2. \tag{10.58}$$

Evidently for this *a priori* not unreasonable value of the top squark mass parameter, the (approximate) 1-loop corrected squared mass $m_{h^0}^2$ just clears the experimental bound (10.56).

These simple considerations indicate that the shift in $m_{h^0}^2$ required for consistency with the bound (10.55) may be attributable to radiative corrections. However, the shift must be almost as large as the tree-level value, so that higher order effects cannot be neglected. More complete treatments (see, for example, [100] and [101]) show that the inclusion of only the 1-loop terms somewhat overestimates the true upper bound on $m_{h^0}^2$. Equivalently, to reach a given value of $m_{h^0}^2$ using the more complete calculation requires a larger value of m_S. For example, if squark mixing is still relatively small, then the bound (10.55) requires $m_S \sim 800$–1200 GeV. This estimate is further increased if a lower value of m_t is used.

The magnitude of these radiative corrections is obviously very sensitive to the value of m_t. It also depends on the quark mixing parameters. The latter may be tuned so as to maximize m_{h^0} for each value of m_{A^0} and $\tan\beta$ (see [102] for example). Typically, an increase of about 15 GeV in m_{h^0} is produced, compared with the no-mixing case. This, in turn, allows the bound (10.56) to be met for a smaller value of m_S: $m_S \sim 400$–500 GeV.

It is natural to wonder how large m_{h^0} can become in the MSSM, keeping $m_S \leq 2$ TeV say. A recent summary [103] which includes leading 2-loop effects and takes the average top squark squared mass to be $(2 \text{ TeV})^2$, concludes that in the '$m_{h^0}^{\text{max}}$' scenario [102], with $m_t = 179.4$ GeV, the bound (10.56) places no constraint on $\tan\beta$, and predicts $m_{h^0} \leq 140$ GeV (with an accuracy of a few GeV). This is still an extremely interesting result. In the words of Drees [104]: "If experiments.... fail to find at least one Higgs boson [in this energy region], the MSSM can be completely excluded, independent of the value of its 100 or so free parameters."

In concluding this section, we should note that, while the bound (10.56) is generally accepted, alternative interpretations of the data do exist. Thus Drees has suggested [105] that the 2.3 σ excess of events around 98 GeV and the 1.7 σ excess around 115 GeV reported by the four LEP experiments [95] might actually be the h^0 and H^0 respectively (see also footnote 4 below, page 170). More recently, Dermíšek and Gunion have proposed [106] that the 98 GeV excess correlates well with there being a Higgs boson of that mass with SM-like ZZh^0 coupling, which decays dominantly via $h^0 \to a\,a$, where 'a' is a CP-odd Higgs boson of the kind present in the 'next to minimal supersymmetric standard model' (NMSSM), and $m_a < 2m_b$. These authors argue further that this scenario is phenomenologically viable for parameter choices in the NMSSM which yield the lowest possible 'fine-tuning '. This is an appropriate moment to take a small detour in that direction.

10.3 The SM fine-tuning problem revisited, in the MSSM

The foregoing discussion has provided estimates of the scale of m_S required to accommodate Higgs mass values in the range 115–140 GeV in the MSSM. At first sight, the indicated superpartner range (500 GeV $\leq m_S \leq$ few TeV) seems pretty much as anticipated from the qualitative account given in Section 1.1 of how a supersymmetric theory could solve the fine-tuning problem present in the SM. On further reflection, however, this range of m_S – particularly the upper end of it – may seem to present something of a problem for the MSSM. For one thing, the natural parameter for bosonic mass terms is (mass)^2, and if indeed $m_S \sim 1$ TeV then m_S^2 may be up to two orders of magnitude larger than the weak scale (given by m_Z^2 or $(v_u^2 + v_d^2)$). While certainly very far from the problem created by the scale of M_P^2 or M_{GUT}^2, this relatively large scale of m_S^2 consitutes, for many physicists, a 'little fine-tuning problem'. Correspondingly, within the specific context of the MSSM, the indicated scale of m_S^2 leads, in the view of many, to an 'MSSM fine-tuning problem'. We shall give a brief outline of these concerns.

Let us begin by formulating more precisely the argument of Section 1.1 regarding the fine-tuning problem in the SM. The SM is viewed as an effective theory, valid below some cut-off Λ_{SM}. At one-loop order, the (mass)^2 parameter in the Higgs potential, which we shall now call $-\mu_H^2$, receives quadratically divergent contributions from Higgs boson, gauge boson and (dominantly) top quark loops, giving a total shift [17, 107]

$$\delta_q(-\mu_H^2) = \frac{3}{16\pi^2 v^2}\left(2M_W^2 + M_Z^2 + M_H^2 - 4m_t^2\right)\Lambda_{SM}^2, \tag{10.59}$$

where $v = 246$ GeV. The one-loop corrected physical value $-\mu_{H\,phys}^2$ is then $-\mu_{H\,phys}^2 = -\mu_H^2 + \delta_q(-\mu_H^2)$, or equivalently

$$\mu_{H\,phys}^2 = \mu_H^2 - \frac{3}{16\pi^2 v^2}\left(2M_W^2 + M_Z^2 + M_H^2 - 4m_t^2\right)\Lambda_{SM}^2 \tag{10.60}$$

(compare equation (1.11)). If delicate cancellations between the two terms on the right-hand side of (10.60) are to be avoided (i.e. no fine-tuning), then neither term should be (say) one order of magnitude greater than the left-hand side. Now we saw in Section 1.1 that the natural scale of $\mu_{H\,phys}$ is of order $v/2$. Hence we require

$$\left|\frac{3}{16\pi^2 v^2}\left(2M_W^2 + M_Z^2 + M_H^2 - 4m_t^2\right)\Lambda_{SM}^2\right| \leq 10\frac{v^2}{4}. \tag{10.61}$$

These considerations imply that for $M_H = 115$–200 GeV,

$$\Lambda_{SM} \leq 2\text{–}3 \text{ TeV}. \tag{10.62}$$

This is the conventional estimate of the scale at which, to avoid fine-tuning, new physics beyond the SM should appear.[2]

In any supersymmetric extension of the SM (provided SUSY is softly broken, see Section 9.2) such quadratic divergences disappear – and with them, this (acute) form of the fine-tuning problem; this was, of course, the primary motivation for the MSSM, as sketched in Section 1.1. Nevertheless, there may still be quite large (logarithmic) loop corrections to the tree-level parameters in the Higgs sector potential, which might imply the necessity for some residual fine-tuning, albeit of a much milder degree.

Within the context of the softly-broken MSSM, one such 'large logarithm' arises from the evolution of the parameter $m^2_{H_u}$, which is approximately given (to one-loop order) by equation (9.52) together with (9.55). Assuming that the dominant mass terms are those of the top squarks, we may further approximate (9.52) by

$$\frac{dm^2_{H_u}}{dt} \approx \frac{12y_t^2}{16\pi^2} m_S^2, \tag{10.63}$$

where m_S is, as before, the average of the two top squark squared masses. Thus, in running down from a high scale Λ_U to the weak scale, $m^2_{H_u}$ receives a contribution

$$\delta m^2_{H_u} \approx -\frac{3y_t^2}{4\pi^2} m_S^2 \ln\left(\frac{\Lambda_U}{m_S}\right). \tag{10.64}$$

If we take $\Lambda_U \sim 10^{16-18}$ (as in an mSUGRA theory, see Section 9.2), and $y_t \approx 1$, then the magnitude of (10.64) is of order $2 - 3m_S^2$.

Consider now equations (10.16) and (10.17) which express the minimization conditions on the Higgs potential at tree-level. Eliminating the parameter 'b' and using (10.21) and (10.26) we obtain

$$\frac{1}{2}m_Z^2 = -|\mu^2| + \frac{m^2_{H_d} - m^2_{H_u}\tan^2\beta}{\tan^2\beta - 1}. \tag{10.65}$$

For simplicity let's consider the case of large $\tan\beta$, so that (10.65) becomes

$$\frac{1}{2}m_Z^2 = -|\mu^2| - m^2_{H_u}. \tag{10.66}$$

In order to satisfy this minimization condition 'naturally' (no fine tuning) we may demand that terms on the right-hand side of (10.66) are no more than an order

[2] However, if, as noted by Veltman [17], M_H happens to lie close to the value that cancels $\delta_q(-\mu_H^2)$, namely $M_H \approx 316$ GeV, then Λ_{SM} could consistently be much higher than (10.62). The implications of such a value of M_H are discussed in [108], for example.

of magnitude larger than the left-hand side, analogously to (10.61). Applying this criterion to the shift (10.64) in $m^2_{\mathrm{H_u}}$, we find

$$m^2_{\mathrm{S}} < 10\,m^2_{\mathrm{Z}}/(\sim 5), \tag{10.67}$$

which implies

$$m_{\mathrm{S}} < 150\ \mathrm{GeV,\ say.} \tag{10.68}$$

Surprisingly, this is a substantially lower value than the one we arrived at for the 'scale of new physics' according to the previous SM argument, equation (10.62). Numerically, this is essentially because the large logarithm in (10.64), the strong coupling y_{t}, and the factor 12 combine to compensate the usual loop factor $1/(16\pi^2)$.[3]

Returning to (10.68), it is clear that such a relatively low value of m_{S} will prevent $m_{\mathrm{h^0}}$ from meeting the experimental bound (10.56). In fact, a significant increase in m_{S} above the value (10.68) is required, because this quantity enters only logarithmically in (10.57). On the other hand, m^2_{S} enters linearly in the fine-tuning argument involving $\delta m^2_{\mathrm{H_u}}$. In short, within the context of the MSSM, fine-tuning gets exponentially worse as $m_{\mathrm{h^0}}$ increases. If we take $m_{\mathrm{S}} \sim 500$ GeV as roughly the smallest value consistent with (10.56), then the factor of 10 in (10.67) is replaced by about 150, suggesting that the MSSM is already fine-tuned at the percentage level; and the tuning becomes rapidly more severe as m_{S} is increased.

The foregoing discussion is intended to illustrate in simple terms the nature of the perceived fine-tuning problem in the MSSM, and to give a rough idea of its quantitative extent. In fact, concerns about fine-tuning in models which require supersymmetry to be manifest not too far from the weak scale have been expressed for some time, and there are now extensive sub-literatures analysing the problem in detail, and proposing responses to it. Of course, there can be no absolute definition of the amount of fine-tuning that is 'acceptable' (1 part in 10? in 100? in 1000?), but, in the absence of new guidance from experiment, the *relative* amount of fine-tuning has been widely invoked as a useful criterion for guiding the search for physics beyond the SM, or for concentrating on certain regions of parameter space (cf. [106], for example). However, these developments lie beyond our scope.

10.4 Tree-level couplings of neutral Higgs bosons to SM particles

The phenomenolgy of the Higgs-sector particles depends, of course, not only on their masses but also on their couplings, which enter into production and decay

[3] Recall that it was this term that was responsible for the mechanism of radiative electroweak symmetry breaking via a significant and negative contribution to $m^2_{\mathrm{H_u}}$, as discussed in Section 9.3, following equation (9.56).

processes. In this section we shall derive some of the more important couplings of the neutral Higgs states h^0, H^0 and A^0, for illustrative purposes.

First, note that after transforming to the mass-diagonal basis, the relation (8.10) and similar ones for m_{dij} and m_{eij} become

$$m_{u,c,t} = v_u y_{u,c,t} \tag{10.69}$$

$$m_{d,s,b} = v_d y_{d,s,b} \tag{10.70}$$

$$m_{e,\mu,\tau} = v_d y_{e,\mu,\tau}. \tag{10.71}$$

In this basis, the Yukawa couplings in the superpotential are therefore (making use of (10.22))

$$y_{u,c,t} = \frac{m_{u,c,t}}{v_u} = \frac{g m_{u,c,t}}{\sqrt{2} m_W \sin\beta} \tag{10.72}$$

$$y_{d,s,b} = \frac{m_{d,s,b}}{v_d} = \frac{g m_{d,s,b}}{\sqrt{2} m_W \cos\beta} \tag{10.73}$$

$$y_{e,\mu,\tau} = \frac{m_{e,\mu,\tau}}{v_d} = \frac{g m_{e,\mu,\tau}}{\sqrt{2} m_W \cos\beta}. \tag{10.74}$$

Relations (10.72) and (10.73) suggest that very rough upper and lower bounds may be placed on $\tan\beta$ by requiring that neither y_t nor y_b is non-perturbatively large. For example, if $\tan\beta \geq 1$ then $y_t \leq 1.4$, and if $\tan\beta \leq 50$ then $y_b \leq 1.25$. Some GUT models can unify the running values of y_t, y_b and y_τ at the unification scale; this requires $\tan\beta \approx m_t/m_b \simeq 40$, as noted at the end of Section 9.3.

To find the couplings of the neutral MSSM Higgs bosons to fermions, we return to the Yukawa couplings (8.8) (together with the analogous ones for y_d^{ij} and y_e^{ij}), and expand H_u^0 and H_d^0 about their vacuum values. In order to get the result in terms of the physical fields h^0, H^0, however, we need to know how the latter are related to $\mathrm{Re}H_u^0$ and $\mathrm{Re}H_d^0$; that is, we require expressions for the eigenmodes of the (mass)^2 matrix (10.49) corresponding to the eigenvalues $m_{h^0}^2$ and $m_{H^0}^2$ of (10.50) and (10.51). We can write (10.49) as

$$\mathbf{M}_{h,H}^{sq} = \frac{1}{2}\begin{pmatrix} A + Bc & -As \\ -As & A - Bc \end{pmatrix}, \tag{10.75}$$

where $A = (m_{A^0}^2 + m_Z^2)$, $B = (m_{A^0}^2 - m_Z^2)$, $c = \cos 2\beta$, $s = \sin 2\beta$, and we have used (10.41). Expression (10.75) is calculated in the basis $(\sqrt{2}(\mathrm{Re}H_u^0 - v_u),$ $\sqrt{2}(\mathrm{Re}H_d^0 - v_d))$. Let us denote the normalized eigenvectors of (10.75) by u_h and u_H, where

$$u_h = \begin{pmatrix} \cos\alpha \\ -\sin\alpha \end{pmatrix}, \quad u_H = \begin{pmatrix} \sin\alpha \\ \cos\alpha \end{pmatrix}, \tag{10.76}$$

with eigenvalues $m^2_{\mathrm{h}^0}$ and $m^2_{\mathrm{H}^0}$, respectively, where

$$m^2_{\mathrm{h}^0} = \frac{1}{2}(A - C) \tag{10.77}$$

$$m^2_{\mathrm{H}^0} = \frac{1}{2}(A + C), \tag{10.78}$$

with $C = [A^2 - (A^2 - B^2)c^2]^{1/2}$. The equation determining u_{h} is then

$$\begin{pmatrix} A + Bc & -As \\ -As & A - Bc \end{pmatrix} \begin{pmatrix} \cos\alpha \\ -\sin\alpha \end{pmatrix} = (A - C) \begin{pmatrix} \cos\alpha \\ -\sin\alpha \end{pmatrix}, \tag{10.79}$$

which leads to

$$(C + Bc)\cos\alpha = -As\sin\alpha \tag{10.80}$$

$$(-C + Bc)\sin\alpha = As\cos\alpha. \tag{10.81}$$

It is conventional to rewrite (10.80) and (10.81) more conveniently, as follows. Multiplying (10.80) by $\sin\alpha$ and (10.81) by $\cos\alpha$ and then subtracting the results, we obtain

$$\sin 2\alpha = -\frac{As}{C} = -\frac{\left(m^2_{\mathrm{A}^0} + m^2_{\mathrm{Z}}\right)}{\left(m^2_{\mathrm{H}^0} - m^2_{\mathrm{h}^0}\right)}\sin 2\beta. \tag{10.82}$$

Again, multiplying (10.80) by $\cos\alpha$ and (10.81) by $\sin\alpha$ and adding the results gives

$$\cos 2\alpha = -\frac{Bc}{C} = -\frac{\left(m^2_{\mathrm{A}^0} - m^2_{\mathrm{Z}}\right)}{\left(m^2_{\mathrm{H}^0} - m^2_{\mathrm{h}^0}\right)}\cos 2\beta. \tag{10.83}$$

Equations (10.82) and (10.83) serve to define the correct quadrant for the mixing angle α, namely $-\pi/2 \le \alpha \le 0$. Note that in the limit $m^2_{\mathrm{A}^0} \gg m^2_{\mathrm{Z}}$ we have $\sin 2\alpha \approx -\sin 2\beta$ and $\cos 2\alpha \approx -\cos 2\beta$, and so

$$\alpha \approx \beta - \pi/2 \ \text{ for } m^2_{\mathrm{A}^0} \gg m^2_{\mathrm{Z}}. \tag{10.84}$$

The physical states are defined by

$$\begin{pmatrix} h^0 \\ H^0 \end{pmatrix} = \sqrt{2}\begin{pmatrix} \cos\alpha & -\sin\alpha \\ \sin\alpha & \cos\alpha \end{pmatrix} \begin{pmatrix} \mathrm{Re}H^0_{\mathrm{u}} - v_{\mathrm{u}} \\ \mathrm{Re}H^0_{\mathrm{d}} - v_{\mathrm{d}} \end{pmatrix}, \tag{10.85}$$

which we can write as

$$\mathrm{Re}H^0_{\mathrm{u}} = \left[v_{\mathrm{u}} + \frac{1}{\sqrt{2}}(\cos\alpha\, h^0 + \sin\alpha\, H^0)\right] \tag{10.86}$$

$$\mathrm{Re}H^0_{\mathrm{d}} = \left[v_{\mathrm{d}} + \frac{1}{\sqrt{2}}(-\sin\alpha\, h^0 + \cos\alpha\, H^0)\right]. \tag{10.87}$$

We also have, from (10.39) and (10.40),

$$\mathrm{Im}H_{\mathrm{u}}^{0} = \frac{1}{\sqrt{2}}(\sin\beta\, H_{\mathrm{Z}} + \cos\beta\, A^{0}) \tag{10.88}$$

$$\mathrm{Im}H_{\mathrm{d}}^{0} = \frac{1}{\sqrt{2}}(-\cos\beta\, H_{\mathrm{Z}} + \sin\beta\, A^{0}) \tag{10.89}$$

where H_{Z} is the massless field 'swallowed' by the Z^0.

We can now derive the couplings to fermions. For example, the Yukawa coupling (8.8) in the mass eigenstate basis, and for the third generation, is

$$-y_{\mathrm{t}}\big[\chi_{\bar{\mathrm{t}}\mathrm{L}}\cdot\chi_{\mathrm{tL}}(\mathrm{Re}H_{\mathrm{u}}^{0} + \mathrm{i}\,\mathrm{Im}\,H_{\mathrm{u}}^{0}) + \chi_{\mathrm{tL}}^{\dagger}\cdot\chi_{\bar{\mathrm{t}}\mathrm{L}}^{\dagger}(\mathrm{Re}H_{\mathrm{u}}^{0} - \mathrm{i}\,\mathrm{Im}\,H_{\mathrm{u}}^{0})\big]. \tag{10.90}$$

Substituting (10.86) for $\mathrm{Re}H_{\mathrm{u}}^{0}$, the '$v_{\mathrm{u}}$' part simply produces the Dirac mass m_{u} via (8.9), while the remaining part gives

$$\begin{aligned} &-\frac{m_{\mathrm{t}}}{\sqrt{2}v_{\mathrm{u}}}(\chi_{\bar{\mathrm{t}}\mathrm{L}}\cdot\chi_{\mathrm{tL}} + \chi_{\mathrm{tL}}^{\dagger}\cdot\chi_{\bar{\mathrm{t}}\mathrm{L}}^{\dagger})(\cos\alpha\, h^{0} + \sin\alpha\, H^{0}) \\ &= -\left(\frac{g m_{\mathrm{t}}}{2m_{\mathrm{W}}}\right)\bar{\Psi}_{\mathrm{t}}\Psi_{\mathrm{t}}\left(\frac{\cos\alpha}{\sin\beta}h^{0} + \frac{\sin\alpha}{\sin\beta}H^{0}\right), \end{aligned} \tag{10.91}$$

where '$\bar{\Psi}_{\mathrm{t}}\Psi_{\mathrm{t}}$' is the 4-component Dirac bilinear. The corresponding expression in the SM would be just

$$-\left(\frac{g m_{\mathrm{t}}}{2m_{\mathrm{W}}}\right)\bar{\Psi}_{\mathrm{t}}\Psi_{\mathrm{t}}\, H_{\mathrm{SM}}, \tag{10.92}$$

where H_{SM} is the SM Higgs boson. Equation (10.91) shows how the SM coupling is modified in the MSSM. Similarly, the coupling to the b quark is

$$-\left(\frac{g m_{\mathrm{b}}}{2m_{\mathrm{W}}}\right)\bar{\Psi}_{\mathrm{b}}\Psi_{\mathrm{b}}\left(-\frac{\sin\alpha}{\cos\beta}h^{0} + \frac{\cos\alpha}{\cos\beta}H^{0}\right), \tag{10.93}$$

which is to be compared with the SM coupling

$$-\left(\frac{g m_{\mathrm{b}}}{2m_{\mathrm{W}}}\right)\bar{\Psi}_{\mathrm{b}}\Psi_{\mathrm{b}}\, H_{\mathrm{SM}}. \tag{10.94}$$

The coupling to the τ lepton has the same form as (10.93), with the obvious replacement of m_{b} by m_{τ}. Finally the t–A^0 coupling is found by substituting (10.88) into (10.90), with the result

$$-\mathrm{i}\frac{m_{\mathrm{t}}}{v_{\mathrm{u}}}(\chi_{\bar{\mathrm{t}}\mathrm{L}}\cdot\chi_{\mathrm{tL}} - \chi_{\mathrm{tL}}^{\dagger}\cdot\chi_{\bar{\mathrm{t}}\mathrm{L}}^{\dagger})\frac{1}{\sqrt{2}}\cos\beta\, A^{0} = \mathrm{i}\left(\frac{g m_{\mathrm{t}}}{2m_{\mathrm{W}}}\right)\cot\beta\,\bar{\Psi}_{\mathrm{t}}\gamma_5\Psi_{\mathrm{t}}\, A^{0}, \tag{10.95}$$

where we have used (8.3); and similarly the b–A^0 coupling is

$$\mathrm{i}\left(\frac{g m_{\mathrm{b}}}{2m_{\mathrm{W}}}\right)\tan\beta\,\bar{\Psi}_{\mathrm{b}}\gamma_5\Psi_{\mathrm{b}}\, A^{0}. \tag{10.96}$$

Incidentally, the γ_5 coupling in (10.95) and (10.96) shows that the A^0 is a pseudoscalar boson (**CP** $= -1$), while the couplings (10.91) and (10.93) show that h^0 and H^0 are scalars (**CP** $= +1$). Once again, the τ–A^0 coupling is the same as (10.96), with m_b replaced by m_τ.

The limit of large m_{A^0} is interesting: in this case, α and β are related by (10.84), which implies

$$\sin\alpha \approx -\cos\beta \tag{10.97}$$

$$\cos\alpha \approx \sin\beta. \tag{10.98}$$

It then follows from (10.91) and (10.93) that in this limit the couplings of h^0 become those of the SM Higgs, while the couplings of H^0 are the same as those of the A^0. For small m_{A^0} and large $\tan\beta$ on the other hand, the h^0 couplings differ substantially from the SM couplings, b-states being relatively enhanced and t-states being relatively suppressed, while the H^0 couplings are independent of β.

The couplings of the neutral Higgs bosons to the gauge bosons are determined by the $SU(2)_L \times U(1)_y$ gauge invariance, that is by the terms (10.18) with D_μ given by (10.19). The terms involving W_μ^1, W_μ^2, $\mathrm{Re}H_u^0$ and $\mathrm{Re}H_d^0$ are easily found to be

$$\frac{g^2}{4}(W_\mu^1 W^{1\mu} + W_\mu^2 W^{2\mu})\left[(\mathrm{Re}H_u^0)^2 + (\mathrm{Re}H_d^0)^2\right]. \tag{10.99}$$

Substituting (10.86) and (10.87), the v_u^2 and v_d^2 parts generate the W-boson (mass)^2 term via (10.22), while trilinear W–W–(h^0,H^0) couplings are generated when one of the neutral Higgs fields is replaced by its vacuum value:

$$\begin{aligned}\frac{g^2}{4}&(W_\mu^1 W^{1\mu} + W_\mu^2 W^{2\mu})\sqrt{2}[v_u(\cos\alpha\, h^0 + \sin\alpha\, H^0) + v_d(-\sin\alpha\, h^0 + \cos\alpha\, H^0)]\\ &= \frac{g m_W}{2}(W_\mu^1 W^{1\mu} + W_\mu^2 W^{2\mu})[\sin(\beta-\alpha)\, h^0 + \cos(\beta-\alpha)\, H^0].\end{aligned} \tag{10.100}$$

Similarly, the trilinear Z–Z–(h^0,H^0) couplings are

$$\frac{g m_Z}{2\cos\theta_W} Z_\mu Z^\mu [\sin(\beta-\alpha)\, h^0 + \cos(\beta-\alpha)\, H^0]. \tag{10.101}$$

Again, these are the same as the couplings of the SM Higgs to W and Z, but modified by a factor $\sin(\beta-\alpha)$ for the h^0, and a factor $\cos(\beta-\alpha)$ for the H^0.[4] Once again, there is a simple large $m_{A^0}^2$ limit:

$$\sin(\beta-\alpha) \approx 1, \quad \cos(\beta-\alpha) \approx 0, \tag{10.102}$$

[4] This is essential for the viability of Drees's suggestion [105]: the excess of events near 98 GeV amounts to about 10% of the signal for a SM Higgs with that mass, and hence interpreting it as the h^0 requires that $\sin^2(\beta-\alpha) \approx 0.1$. It then follows that ZH^0 production at LEP would occur with nearly SM strength, if allowed kinematically. Hence the identification of the excess at around 115 GeV with the H^0.

showing that in this limit h^0 has SM couplings to gauge bosons, while the H^0 decouples from them entirely. At tree level, the A^0 has no coupling to pairs of gauge bosons.

There are also quadrilinear couplings which are generated when both neutral Higgs fields are varying:

$$\frac{1}{8}[g^2(W^1_\mu W^{1\mu} + W^2_\mu W^{2\mu}) + (g\cos\theta_W + g'\sin\theta_W)^2 Z_\mu Z^\mu](h^{02} + H^{02} + A^{02}). \tag{10.103}$$

Finally, there are trilinear couplings between the Z^0 and the neutral Higgs fields, which involve derivatives of the latter:

$$\frac{1}{2}(g\cos\theta_W + g'\sin\theta_W)[\cos(\alpha-\beta)(A^0\partial^\mu h^0 - h^0\partial^\mu A^0) - \sin(\alpha-\beta)(H^0\partial^\mu A^0 - A^0\partial^\mu H^0)]. \tag{10.104}$$

Note that couplings of Z^0 to h^0h^0, H^0H^0 and h^0H^0 pairs are absent due to the assumed **CP** invariance of the Higgs sector; **CP** allows Z^0 to couple to a scalar boson and a pseudoscalar boson. The complete set of Higgs sector couplings, including those to superpartners, is given in Section 8.4.3 of [49].

We have seen that, in the MSSM, the mass of the h^0 state is expected to be smaller than about 140 GeV. Consequently, the decays of the h^0 to $t\bar{t}$, Z^0Z^0 and W^+W^- are kinematically forbidden, as are (most probably) decays to superpartners. Since the strength of the h^0 interaction with any field is proportional to the mass of that field, the dominant decays will be to $b\bar{b}$ and $\tau\bar{\tau}$ pairs. The partial widths for these channels are easily calculated from (10.93), at tree level. Let the 4-momenta and spins of the final state b and $\bar{b}$ be p_1, s_1 and p_2, s_2. Then, borrowing formulae (12.7) and (12.10) from Chapter 12, we have (for three colours)

$$\Gamma(h^0 \to b\bar{b}) = \frac{3}{8\pi m^2_{h^0}} \frac{g^2 m_b^2 \sin^2\alpha}{4m_W^2\cos^2\beta}\, p \sum_{s_1 s_2} |\bar{u}(p_1, s_1)v(p_2, s_2)|^2, \tag{10.105}$$

where p is the magnitude of the 3-momentum of the final state particles in the rest frame of the h^0. The spinor factor is

$$\mathrm{Tr}[(\not{p}_2 + m_b)(\not{p}_1 - m_b)] = 2m^2_{h^0}(1 - 4m_b^2/m^2_{h^0}), \tag{10.106}$$

and

$$p = \frac{1}{2} m_{h^0}(1 - 4m_b^2/m^2_{h^0})^{1/2}. \tag{10.107}$$

Hence

$$\Gamma(h^0 \to b\bar{b}) = \frac{3g^2 m_b^2\sin^2\alpha}{32\pi m_W^2\cos^2\beta} m_{h^0}(1 - 4m_b^2/m^2_{h^0})^{3/2} \tag{10.108}$$

in agreement with (C.1b) of [49]. The partial width to $\tau\bar{\tau}$ is given by an analogous formula, without the factor of 3, so that the branching ratio of these two modes (at tree level) is

$$\frac{\Gamma(\mathrm{h}^0 \to \mathrm{b}\bar{\mathrm{b}})}{\Gamma(\mathrm{h}^0 \to \tau\bar{\tau})} \approx \frac{3m_{\mathrm{b}}^2}{m_\tau^2} \approx 20. \tag{10.109}$$

The partial width for $\mathrm{H}^0 \to \mathrm{b}\bar{\mathrm{b}}$ is given by (10.108) with $\sin\alpha$ replaced by $\cos\alpha$, and m_{h^0} by m_{H^0}.

Exercise 10.1 Show that

$$\Gamma(\mathrm{A}^0 \to \mathrm{b}\bar{\mathrm{b}}) = \frac{3g^2 m_{\mathrm{b}}^2 \tan^2\beta}{32\pi m_{\mathrm{W}}^2} m_{\mathrm{A}^0} \left(1 - 4m_{\mathrm{b}}^2/m_{\mathrm{A}^0}^2\right)^{1/2}. \tag{10.110}$$

The widths of the MSSM Higgs bosons depend sensitively on $\tan\beta$. The production rate at the LHC also depends on $\tan\beta$. The dominant production mechanism, as in the SM, is expected to be gluon fusion, proceeding via quark (or squark) loops. In the SM case, the top quark loop dominates; in the MSSM, if $\tan\beta$ is large and m_{A^0} not too large, the $\bar{\mathrm{b}}\mathrm{bh}$ coupling is relatively enhanced, as noted after equation (10.98), and the bottom quark loop becomes important. Searches for MSSM Higgs bosons are reviewed by Igo-Kemenes in [59].

11
Sparticle masses in the MSSM

In the two final chapters, we shall give an introduction to the physics of the various SUSY particle states – 'sparticles' – in the MSSM. The first step is to establish formulae for the masses of the sparticles, which we do in the present chapter. In the following one, we calculate some simple decay widths and production cross-sections at tree level, and also discuss very briefly some of the signatures that have been used in sparticle searches, together with some search results. We also mention the idea of 'benchmark sets' of SUSY parameters.

As in the scalar Higgs sector, the discussion of sparticle masses is complicated by mixing phenomena. In particular, after $SU(2)_L \times U(1)_y$ breaking, mixing will in general occur between any two (or more) fields which have the same colour, charge and spin. We shall begin with the simplest case, that of gluinos, for which no mixing occurs.

11.1 Gluinos

The gluino $\tilde{g}$ is the only colour octet fermion and so, since $SU(3)_c$ is unbroken, it cannot mix with any other MSSM particle, even if R-parity is violated. Its mass arises simply from the soft SUSY-breaking gluino mass term in (9.30):

$$-\frac{1}{2}M_3\tilde{g}^a \cdot \tilde{g}^a + \text{h.c.} \tag{11.1}$$

where the colour index a runs from 1 to 8. The expression (11.1) is written in 2-component notation, but is easily translated to Majorana form, as usual:

$$-\frac{1}{2}M_3\bar{\Psi}_{\rm M}^{\tilde{g}a}\Psi_{\rm M}^{\tilde{g}a}. \tag{11.2}$$

As noted after (9.30), even if we take M_3 to be real, it need not be positive – that is, of the sign to be conventionally associated with a mass term. The mass of the physical particle is $|M_3|$. Suppose that M_3 is negative, so that the mass term (11.1) is

$$+\frac{1}{2}|M_3|\tilde{g}^a \cdot \tilde{g}^a + \text{h.c.} \tag{11.3}$$

The sign is easily reversed by a redefinition of the spinor field $\tilde{g}^a$, for example

$$\tilde{g}^a = -\mathrm{i}\tilde{g}^{a\prime}, \tag{11.4}$$

which will leave the kinetic energy term unchanged. In the equivalent Majorana description, we have

$$\Psi_{\mathrm{M}}^{\tilde{g}a} = \begin{pmatrix} \mathrm{i}\sigma_2\tilde{g}^{a*} \\ \tilde{g}^a \end{pmatrix} = \begin{pmatrix} \mathrm{i}\sigma_2\mathrm{i}\tilde{g}^{a\prime *} \\ -\mathrm{i}\tilde{g}^{a\prime} \end{pmatrix} = \mathrm{i}\gamma_5\Psi_{\mathrm{M}}^{\tilde{g}a\prime}. \tag{11.5}$$

Hence, in the notation of Baer and Tata [49], we may generally allow for the possibility of negative Majorana mass parameters by redefining the relevant field for the sparticle $\tilde{\mathrm{p}}$ as

$$\Psi_{\mathrm{M}}^{\tilde{p}} \to (\mathrm{i}\gamma_5)^{\theta_{\tilde{\mathrm{p}}}}\Psi_{\mathrm{M}}^{\tilde{p}} \tag{11.6}$$

where $\theta_{\tilde{\mathrm{p}}} = 0$ if $m_{\tilde{\mathrm{p}}} > 0$, and $\theta_{\tilde{\mathrm{p}}} = 1$ if $m_{\tilde{\mathrm{p}}} < 0$. We have discussed the evolution of M_3 in Section 9.3, in mSUGRA-type models.

11.2 Neutralinos

We consider next the sector consisting of the neutral Higgsinos $\tilde{H}_{\mathrm{u}}^0$ and $\tilde{H}_{\mathrm{d}}^0$, and the neutral gauginos $\tilde{B}$ (bino) and $\tilde{W}^0$ (wino) (see Tables 8.1 and 8.2). These are all L-type spinor fields in our presentation (but they can equivalently be represented as Majorana fields, as explained in Section 2.3). In the absence of electroweak symmetry breaking, the $\tilde{B}$ and $\tilde{W}^0$ fields would have masses given by just the soft SUSY-breaking mass terms of (9.30):

$$-\frac{1}{2}M_1\tilde{B}\cdot\tilde{B} - \frac{1}{2}M_2\tilde{W}^0\cdot\tilde{W}^0 + \text{h.c.} \tag{11.7}$$

However, bilinear combinations of one of $(\tilde{B}, \tilde{W}^0)$ with one of $(\tilde{H}_{\mathrm{u}}^0, \tilde{H}_{\mathrm{d}}^0)$ are generated by the term '$-\sqrt{2}g[\ldots\ldots]$' in (7.72), when the neutral scalar Higgs fields acquire a vev. Such bilinear terms will, as in the Higgs sector, appear as non-zero off-diagonal entries in the 4×4 mass matrix for the four fields $\tilde{B}$, $\tilde{W}^0$, $\tilde{H}_{\mathrm{u}}^0$, and $\tilde{H}_{\mathrm{d}}^0$; that is, they will cause mixing. After the mass matrix is diagonalized, the resulting four neutral mass eigenstates are called neutralinos, which we shall denote by $\tilde{\chi}_i^0$ $(i = 1, 2, 3, 4)$, with the convention that the masses are ordered as $m_{\tilde{\chi}_1^0} < m_{\tilde{\chi}_2^0} < m_{\tilde{\chi}_3^0} < m_{\tilde{\chi}_4^0}$.

Consider for example the SU(2) contribution in (7.72) from the H_u supermultiplet, with $\alpha = 3$, $T^3 \equiv \tau^3/2$, $\lambda^3 \equiv \tilde{W}^0$, which is

$$-\sqrt{2}g\left(H_\mathrm{u}^{+\dagger}\ H_\mathrm{u}^{0\dagger}\right)\frac{\tau^3}{2}\begin{pmatrix}\tilde{H}_\mathrm{u}^+ \\ \tilde{H}_\mathrm{u}^0\end{pmatrix}\cdot \tilde{W}^0 + \text{h.c.} \tag{11.8}$$

When the field $H_\mathrm{u}^{0\dagger}$ acquires a vev v_u (which we have already chosen to be real), expression (11.8) contains the piece

$$+\frac{g}{\sqrt{2}}v_\mathrm{u}\tilde{H}_\mathrm{u}^0\cdot\tilde{W}^0 + \text{h.c.}, \tag{11.9}$$

which we shall re-write as

$$-\frac{1}{2}[-\sin\beta\cos\theta_\mathrm{W}m_\mathrm{Z}]\left(\tilde{H}_\mathrm{u}^0\cdot\tilde{W}^0 + \tilde{W}^0\cdot\tilde{H}_\mathrm{u}^0\right) + \text{h.c.}, \tag{11.10}$$

using (10.26) and (10.21), and the result of Exercise 2.3. In a gauge-eigenstate basis

$$\tilde{G}^0 = \begin{pmatrix}\tilde{B} \\ \tilde{W}^0 \\ \tilde{H}_\mathrm{d}^0 \\ \tilde{H}_\mathrm{u}^0\end{pmatrix}, \tag{11.11}$$

this will contribute a mixing between the (2,4) and (4,2) components. Similarly, the U(1) contribution from the H_u supermultiplet, after electroweak symmetry breaking, leads to the mixing term

$$-\frac{g'}{\sqrt{2}}v_\mathrm{u}\tilde{H}_\mathrm{u}^0\cdot\tilde{B} + \text{h.c.} \tag{11.12}$$

$$= -\frac{1}{2}[\sin\beta\sin\theta_\mathrm{W}m_\mathrm{Z}]\left(\tilde{H}_\mathrm{u}^0\cdot\tilde{B} + \tilde{B}\cdot\tilde{H}_\mathrm{u}^0\right) + \text{h.c.}, \tag{11.13}$$

which involves the (1,4) and (4,1) components. The SU(2) and U(1) contributions of the H_d supermultiplet to such bilinear terms can be evaluated similarly.

In addition to this mixing caused by electroweak symmetry breaking, mixing between $\tilde{H}_\mathrm{u}^0$ and $\tilde{H}_\mathrm{d}^0$ is induced by the SUSY-invariant 'μ term' in (8.14), namely

$$-\frac{1}{2}(-\mu)\left(\tilde{H}_\mathrm{u}^0\cdot\tilde{H}_\mathrm{d}^0 + \tilde{H}_\mathrm{d}^0\cdot\tilde{H}_\mathrm{u}^0\right) + \text{h.c.} \tag{11.14}$$

Putting all this together, mass terms involving the fields in $\tilde{G}^0$ can be written as

$$-\frac{1}{2}\tilde{G}^{0\mathrm{T}}\mathbf{M}_{\tilde{G}^0}\tilde{G}^0 + \text{h.c.} \tag{11.15}$$

where

$$\mathbf{M}_{\tilde{G}^0} = \begin{pmatrix} M_1 & 0 & -c_\beta s_W m_Z & s_\beta s_W m_Z \\ 0 & M_2 & c_\beta c_W m_Z & -s_\beta c_W m_Z \\ -c_\beta s_W m_Z & c_\beta c_W m_Z & 0 & -\mu \\ s_\beta s_W m_Z & -s_\beta c_W m_Z & -\mu & 0 \end{pmatrix}, \tag{11.16}$$

with $c_\beta \equiv \cos\beta$, $s_\beta \equiv \sin\beta$, $c_W \equiv \cos\theta_W$, and $s_W \equiv \sin\theta_W$.

In general, the parameters M_1, M_2 and μ can have arbitrary phases. Most analyses, however, assume the 'gaugino unification' condition (9.47) which implies (9.49) at the electroweak scale, so that one of M_1 and M_2 is fixed in terms of the other. A redefinition of the phases of $\tilde{B}$ and $\tilde{W}^0$ then allows us to make both M_1 and M_2 real and positive. The entries proportional to m_Z are real by virtue of the phase choices made for the Higgs fields in Section 10.1, which made v_u and v_d both real. It is usual to take μ to be real, but the sign of μ is unknown, and not fixed by Higgs-sector physics (see the comment following equation (10.26)). The neutralino sector is then determined by three real parameters, M_1 (or M_2), $\tan\beta$ and μ (as well as by m_Z and θ_W, of course).[1]

However, while the eigenvalues of $\mathbf{M}_{\tilde{G}^0}$ will now be real, there is no guarantee that they will be positive. As for the gluinos, we allow for this by redefining the Majorana fields for the neutralino mass eigenstates as

$$\Psi_M^{\tilde{\chi}_i^0} \rightarrow (i\gamma_5)^{\theta_{\tilde{\chi}_i^0}} \Psi_M^{\tilde{\chi}_i^0}, \tag{11.17}$$

where $\theta_{\tilde{\chi}_i^0} = 0$ if $m_{\tilde{\chi}_i^0} > 0$, and $\theta_{\tilde{\chi}_i^0} = 1$ if $m_{\tilde{\chi}_i^0} < 0$.

Clearly there is not a lot to be gained by pursuing the algebra of this 4×4 mixing problem, in general. A simple special case is that in which the m_Z-dependent terms in (11.16) are a relatively small perturbation on the other entries, which would imply that the neutralinos $\tilde{\chi}_1^0$ and $\tilde{\chi}_2^0$ are close to the weak eigenstates bino and wino, respectively, with masses approximately equal to M_1 and M_2, while the Higgsinos are mixed by the μ entries to form (approximately) the combinations

$$\tilde{H}_S^0 = \frac{1}{\sqrt{2}}(\tilde{H}_d^0 + \tilde{H}_u^0), \quad \text{and} \quad \tilde{H}_A^0 = \frac{1}{\sqrt{2}}(\tilde{H}_d^0 - \tilde{H}_u^0), \tag{11.18}$$

each having mass $\sim |\mu|$.

Assuming it is the LSP, the lightest neutralino, $\tilde{\chi}_1^0$, is an attractive candidate for non-baryonic dark matter [110].[2] Taking account of the restricted range of $\Omega_{CDM}h^2$ consistent with the WMAP data, calculations show [111–113] that $\tilde{\chi}_1^0$'s provide the

[1] In the general case of complex M_1, M_2 and μ, certain combinations of phases must be restricted to avoid unacceptably large **CP**-violating effects [109].

[2] Other possibilities exist. For example, in gauge-mediated SUSY breaking, the gravitino is naturally the LSP. For this and other dark matter candidates within a softly-broken SUSY framework, see [47] Section 6.

desired thermal relic density in certain quite well-defined regions in the space of the mSUGRA parameters ($m_{1/2}$, m_0, $\tan\beta$ and the sign of μ; A_0 was set to zero). Dark matter is reviewed by Drees and Gerbier in [59].

11.3 Charginos

The charged analogues of neutralinos are called 'charginos': there are two positively charged ones associated (before mixing) with $(\tilde{W}^+, \tilde{H}_{\rm u}^+)$, and two negatively charged ones associated with $(\tilde{W}^-, \tilde{H}_{\rm d}^-)$. Mixing between $\tilde{H}_{\rm u}^+$ and $\tilde{H}_{\rm d}^-$ occurs via the μ term in (8.14). Furthermore, as in the neutralino case, mixing between the charged gauginos and Higgsinos will occur via the '$-\sqrt{2}g[\dots]$' term in (7.72) after electroweak symmetry breaking. Consider for example the $H_{\rm u}$ supermultiplet terms in (7.72) involving $\tilde{W}^1$ and $\tilde{W}^2$, after the scalar Higgs $H_{\rm u}^0$ has acquired a vev $v_{\rm u}$. These terms are

$$-\frac{g}{\sqrt{2}}\left\{(0\ v_{\rm u})\left[\tau^1\begin{pmatrix}\tilde{H}_{\rm u}^+\\ \tilde{H}_{\rm u}^0\end{pmatrix}\cdot\tilde{W}^1+\tau^2\begin{pmatrix}\tilde{H}_{\rm u}^+\\ \tilde{H}_{\rm u}^0\end{pmatrix}\cdot\tilde{W}^2\right]\right\}+\text{h.c.} \tag{11.19}$$

$$=-\frac{g}{\sqrt{2}}v_{\rm u}\tilde{H}_{\rm u}^+\cdot(\tilde{W}^1+{\rm i}\tilde{W}^2)+\text{h.c.} \tag{11.20}$$

$$\equiv -gv_{\rm u}\tilde{H}_{\rm u}^+\cdot\tilde{W}^-+\text{h.c.} \tag{11.21}$$

$$=-\frac{1}{2}\sqrt{2}s_\beta m_{\rm W}(\tilde{H}_{\rm u}^+\cdot\tilde{W}^-+\tilde{W}^-\cdot\tilde{H}_{\rm u}^+)+\text{h.c.} \tag{11.22}$$

The corresponding terms from the $H_{\rm d}$ supermultiplet are

$$-gv_{\rm d}\tilde{H}_{\rm d}^-\cdot\tilde{W}^++\text{h.c.} \tag{11.23}$$

$$=-\frac{1}{2}\sqrt{2}c_\beta m_{\rm W}(\tilde{H}_{\rm d}^-\cdot\tilde{W}^++\tilde{W}^+\cdot\tilde{H}_{\rm d}^-)+\text{h.c.} \tag{11.24}$$

If we define a gauge-eigenstate basis

$$\tilde{g}^+=\begin{pmatrix}\tilde{W}^+\\ \tilde{H}_{\rm u}^+\end{pmatrix} \tag{11.25}$$

for the positively charged states, and similarly

$$\tilde{g}^-=\begin{pmatrix}\tilde{W}^-\\ \tilde{H}_{\rm d}^-\end{pmatrix} \tag{11.26}$$

for the negatively charged states, then the chargino mass terms can be written as

$$-\frac{1}{2}[\tilde{g}^{+{\rm T}}\mathbf{X}^{\rm T}\cdot\tilde{g}^-+\tilde{g}^{-{\rm T}}\mathbf{X}\cdot\tilde{g}^+]+\text{h.c.}, \tag{11.27}$$

where

$$\mathbf{X} = \begin{pmatrix} M_2 & \sqrt{2}s_\beta m_\mathrm{W} \\ \sqrt{2}c_\beta m_\mathrm{W} & \mu \end{pmatrix}. \tag{11.28}$$

Since $\mathbf{X}^\mathrm{T} \neq \mathbf{X}$ (unless $\tan\beta = 1$), two distinct 2×2 matrices are needed for the diagonalization. Let us define the mass-eigenstate bases by

$$\tilde{\chi}^+ = \mathbf{V}\tilde{g}^+, \quad \tilde{\chi}^+ = \begin{pmatrix} \tilde{\chi}_1^+ \\ \tilde{\chi}_2^+ \end{pmatrix} \tag{11.29}$$

$$\tilde{\chi}^- = \mathbf{U}\tilde{g}^-, \quad \tilde{\chi}^- = \begin{pmatrix} \tilde{\chi}_1^- \\ \tilde{\chi}_2^- \end{pmatrix}, \tag{11.30}$$

where $\mathbf{U}$ and $\mathbf{V}$ are unitary. Then the second term in (11.27) becomes

$$-\frac{1}{2}\tilde{\chi}^{-\mathrm{T}}\mathbf{U}^*\mathbf{X}\mathbf{V}^{-1} \cdot \tilde{\chi}^+, \tag{11.31}$$

and we require

$$\mathbf{U}^*\mathbf{X}\mathbf{V}^{-1} = \begin{pmatrix} m_{\tilde{\chi}_1^\pm} & 0 \\ 0 & m_{\tilde{\chi}_2^\pm} \end{pmatrix}. \tag{11.32}$$

What about the first term in (11.27)? It becomes

$$-\frac{1}{2}\tilde{\chi}^{+\mathrm{T}}\mathbf{V}^*\mathbf{X}^\mathrm{T}\mathbf{U}^\dagger \cdot \tilde{\chi}^-, \tag{11.33}$$

but since $\mathbf{V}^*\mathbf{X}^\mathrm{T}\mathbf{U}^\dagger = (\mathbf{U}^*\mathbf{X}\mathbf{V}^{-1})^\mathrm{T}$ it follows that the expression (11.33) is also diagonal, with the same eigenvalues $m_{\tilde{\chi}_1^\pm}$ and $m_{\tilde{\chi}_2^\pm}$.

Now note that the hermitian conjugate of (11.32) gives

$$\mathbf{V}\mathbf{X}^\dagger\mathbf{U}^\mathrm{T} = \begin{pmatrix} m^*_{\tilde{\chi}_1^\pm} & 0 \\ 0 & m^*_{\tilde{\chi}_2^\pm} \end{pmatrix}. \tag{11.34}$$

Hence

$$\mathbf{V}\mathbf{X}^\dagger\mathbf{X}\mathbf{V}^{-1} = \mathbf{V}\mathbf{X}^\dagger\mathbf{U}^\mathrm{T}\mathbf{U}^*\mathbf{X}\mathbf{V}^{-1} = \begin{pmatrix} |m_{\tilde{\chi}_1^\pm}|^2 & 0 \\ 0 & |m_{\tilde{\chi}_2^\pm}|^2 \end{pmatrix}, \tag{11.35}$$

and we see that the positively charged states $\tilde{\chi}^+$ diagonalize $\mathbf{X}^\dagger\mathbf{X}$. Similarly,

$$\mathbf{U}^*\mathbf{X}\mathbf{X}^\dagger\mathbf{U}^\mathrm{T} = \mathbf{U}^*\mathbf{X}\mathbf{V}^{-1}\mathbf{V}\mathbf{X}^\dagger\mathbf{U}^\mathrm{T} = \begin{pmatrix} |m_{\tilde{\chi}_1^\pm}|^2 & 0 \\ 0 & |m_{\tilde{\chi}_2^\pm}|^2 \end{pmatrix}, \tag{11.36}$$

and the negatively charged states $\tilde{\chi}^-$ diagonalize $\mathbf{X}\mathbf{X}^\dagger$. The eigenvalues of $\mathbf{X}^\dagger\mathbf{X}$ (or

$\mathbf{XX}^\dagger$) are easily found to be

$$\begin{pmatrix} |m_{\tilde{\chi}_1^\pm}|^2 \\ |m_{\tilde{\chi}_2^\pm}|^2 \end{pmatrix} = \frac{1}{2}\big[\big(M_2^2 + |\mu|^2 + 2m_{\mathrm{W}}^2\big) \mp \big\{\big(M_2^2 + |\mu|^2 + 2m_{\mathrm{W}}^2\big)^2 - 4|\mu M_2 - m_{\mathrm{W}}^2 \sin 2\beta|^2\big\}^{1/2}\big]. \quad (11.37)$$

It may be worth noting that, because $\mathbf{X}$ is diagonalized by the operation $\mathbf{U}^*\mathbf{X}\mathbf{V}^{-1}$, rather than by $\mathbf{V}\mathbf{X}\mathbf{V}^{-1}$ or $\mathbf{U}^*\mathbf{X}\mathbf{U}^{\mathrm{T}}$, these eigenvalues are not the squares of the eigenvalues of $\mathbf{X}$.

The expression (11.37) is not particularly enlightening, but as in the neutralino case it simplifies greatly if m_{W} can be regarded as a perturbation. Taking M_2 and μ to be real, the eigenvalues are then given approximately by $m_{\tilde{\chi}_1^\pm} \approx M_2$, and $m_{\tilde{\chi}_2^\pm} \approx |\mu|$ (the labelling assumes $M_2 < |\mu|$). In this limit, we have the approximate degeneracies $m_{\tilde{\chi}_1^\pm} \approx m_{\tilde{\chi}_2^0}$, and $m_{\tilde{\chi}_2^\pm} \approx m_{\tilde{H}_{\mathrm{S}}^0} \approx m_{\tilde{H}_{\mathrm{A}}^0}$. In general, the physics is sensitive to the ratio $M_2/|\mu|$.

11.4 Squarks and sleptons

The scalar partners of the SM fermions form the largest collection of new particles in the MSSM. Since separate partners are required for each chirality state of the massive fermions, there are altogether 21 new fields (the neutrinos are treated as massless here): four squark flavours and chiralities $\tilde{u}_{\mathrm{L}}, \tilde{u}_{\mathrm{R}}, \tilde{d}_{\mathrm{L}}, \tilde{d}_{\mathrm{R}}$ and three slepton flavours and chiralities $\tilde{\nu}_{\mathrm{eL}}, \tilde{e}_{\mathrm{L}}, \tilde{e}_{\mathrm{R}}$ in the first family, all repeated for the other two families.[3] These are all (complex) scalar fields, and so the 'L' and 'R' labels do not, of course, here signify chirality, but are just labels showing which SM fermion they are partnered with (and hence in particular what their $\mathrm{SU}(2) \times \mathrm{U}(1)$ quantum numbers are, see Table 8.1).

In principle, any scalars with the same electric charge, R-parity and colour quantum numbers can mix with each other, across families, via the soft SUSY-breaking parameters in (9.31), (9.33) and (9.37). This would lead to a 6×6 mixing problem for the u-type squark fields ($\tilde{u}_{\mathrm{L}}, \tilde{u}_{\mathrm{R}}, \tilde{c}_{\mathrm{L}}, \tilde{c}_{\mathrm{R}}, \tilde{t}_{\mathrm{L}}, \tilde{t}_{\mathrm{R}}$), and for the d-type squarks and the charged sleptons, and a 3×3 one for the sneutrinos. However, as we saw in Section 9.2, phenomenological constraints imply that interfamily mixing among the SUSY states must be very small. As before, therefore, we shall adopt the 'mSUGRA' form of the soft parameters as given in equations (9.40) and (9.42), which guarantees the suppression of unwanted interfamily mixing terms (although one must remember that other, and more general, parametrizations are not excluded). As in

[3] In the more general family-index notation of Section 9.2 (see equations (9.31), (9.33) and (9.37)), '$\tilde{Q}_1$' is the doublet ($\tilde{u}_{\mathrm{L}}, \tilde{d}_{\mathrm{L}}$), '$\tilde{Q}_2$' is ($\tilde{c}_{\mathrm{L}}, \tilde{s}_{\mathrm{L}}$), '$\tilde{Q}_3$' is ($\tilde{t}_{\mathrm{L}}, \tilde{b}_{\mathrm{L}}$), '$\tilde{\bar{u}}_1$' is $\tilde{u}_{\mathrm{R}}$, '$\tilde{\bar{d}}_1$' is $\tilde{d}_{\mathrm{R}}$ (and similarly for '$\tilde{\bar{u}}_{2,3}$' and '$\tilde{\bar{d}}_{2,3}$'), while '$\tilde{L}_1$' is ($\tilde{\nu}_{\mathrm{eL}}, \tilde{e}_{\mathrm{L}}$), '$\tilde{\bar{e}}_1$' is e_{R}, etc.

the cases considered previously in this section, we shall also have to include various effects due to electroweak symmetry breaking.

Consider first the soft SUSY-breaking (mass)^2 parameters of the sfermions (squarks and sleptons) of the first two families. In the model of (9.40) they are all degenerate at the high (Planck?) scale. The RGE evolution down to the electroweak scale is governed by equations of the same type as (9.53) and (9.54) but without the X_{t} terms, since the Yukawa couplings are negligible for the first two families. Thus the soft masses of the first and second families evolve by purely gauge interactions, which (see the comment following equation (9.56)) tend to increase the masses at low scales. Their evolution can be parametrized (following [46] equations (7.53)–(7.57)) by

$$m^2_{\tilde{\mathrm{u}}_{\mathrm{L}},\tilde{\mathrm{d}}_{\mathrm{L}}} = m^2_{\tilde{\mathrm{c}}_{\mathrm{L}},\tilde{\mathrm{s}}_{\mathrm{L}}} = m_0^2 + K_3 + K_2 + \frac{1}{9}K_1 \tag{11.38}$$

$$m^2_{\tilde{\mathrm{u}}_{\mathrm{R}}} = m^2_{\tilde{\mathrm{c}}_{\mathrm{R}}} = m_0^2 + K_3 \qquad + \frac{16}{9}K_1 \tag{11.39}$$

$$m^2_{\tilde{\mathrm{d}}_{\mathrm{R}}} = m^2_{\tilde{\mathrm{s}}_{\mathrm{R}}} = m_0^2 + K_3 \qquad + \frac{4}{9}K_1 \tag{11.40}$$

$$m^2_{\tilde{\nu}_{\mathrm{eL}},\tilde{\mathrm{e}}_{\mathrm{L}}} = m^2_{\tilde{\nu}_{\mu\mathrm{L}},\tilde{\mu}_{\mathrm{L}}} = m_0^2 \qquad + K_2 + K_1 \tag{11.41}$$

$$m^2_{\tilde{\mathrm{e}}_{\mathrm{R}}} = m^2_{\tilde{\mu}_{\mathrm{R}}} = m_0^2 \qquad + 4K_1. \tag{11.42}$$

Here K_3, K_2 and K_1 are the RGE contributions from SU(3), SU(2) and U(1) gauginos respectively: all the chiral supermultiplets couple to the gauginos with the same ('universal') gauge couplings. The different numerical coefficients in front of the K_1 terms are the squares of the y-values of each field (see Table 8.1), which enter into the relevant loops (our y is twice that of [46]). All the K's are positive, and are roughly of the same order of magnitude as the gaugino (mass)^2 parameter $m^2_{1/2}$, but with K_3 significantly greater than K_2, which in turn is greater than K_1 (this is because of the relative sizes of the different gauge couplings at the weak scale: $g_3^2 \sim 1.5$, $g_2^2 \sim 0.4$, $g_1^2 \sim 0.2$, see Section 8.3). The large 'K_3' contribution is likely to be quite model-independent, and it is therefore reasonable to expect that squark (mass)^2 values will be greater than slepton ones.

Equations (11.38)–(11.42) give the soft (mass)^2 parameters for the fourteen states involved, in the first two families (we defer consideration of the third family for the moment). In addition to these contributions, however, there are further terms to be included which arise as a result of electroweak symmetry breaking. For the first two families, the most important of such contributions are those coming from SUSY-invariant D-terms (see (7.74)) of the form (squark)^2(Higgs)^2 and (slepton)^2(Higgs)^2, after the scalar Higgs fields H_{u}^0 and H_{d}^0 have acquired vevs.

Returning to equation (7.73), the SU(2) contribution to D^α is

$$D^\alpha = g\left\{(\tilde{u}_{\rm L}^\dagger \tilde{d}_{\rm L}^\dagger)\frac{\tau^\alpha}{2}\begin{pmatrix}\tilde{u}_{\rm L}\\ \tilde{d}_{\rm L}\end{pmatrix} + (\tilde{\nu}_{\rm eL}^\dagger \tilde{e}_{\rm L}^\dagger)\frac{\tau^\alpha}{2}\begin{pmatrix}\tilde{\nu}_{\rm eL}\\ \tilde{e}_{\rm L}\end{pmatrix}\right.$$

$$\left. + (H_{\rm u}^{+\dagger} H_{\rm u}^{0\dagger})\frac{\tau^\alpha}{2}\begin{pmatrix}H_{\rm u}^+\\ H_{\rm u}^0\end{pmatrix} + (H_{\rm d}^{0\dagger} H_{\rm d}^{-\dagger})\frac{\tau^\alpha}{2}\begin{pmatrix}H_{\rm d}^0\\ H_{\rm d}^-\end{pmatrix}\right\} \tag{11.43}$$

$$\to g\left\{(\tilde{u}_{\rm L}^\dagger \tilde{d}_{\rm L}^\dagger)\frac{\tau^\alpha}{2}\begin{pmatrix}\tilde{u}_{\rm L}\\ \tilde{d}_{\rm L}\end{pmatrix} + (\tilde{\nu}_{\rm eL}^\dagger \tilde{e}_{\rm L}^\dagger)\frac{\tau^\alpha}{2}\begin{pmatrix}\tilde{\nu}_{\rm eL}\\ \tilde{e}_{\rm L}\end{pmatrix} - \frac{1}{2}v_{\rm u}^2\delta_{\alpha 3} + \frac{1}{2}v_{\rm d}^2\delta_{\alpha 3}\right\}, \tag{11.44}$$

after symmetry breaking. When this is inserted into the Lagrangian term $-\frac{1}{2}D^\alpha D^\alpha$, pieces which are quadratic in the scalar fields – and are therefore (mass)^2 terms – will come from cross terms between the '$\tau^\alpha/2$' and '$\delta_{\alpha 3}$' terms. These cross terms are proportional to $\tau^3/2$, and therefore split apart the $T^3 = +1/2$ weak isospin components from the $T^3 = -1/2$ components, but they are diagonal in the weak eigenstate basis. Their contribution to the sfermion (mass)^2 matrix is therefore

$$+\frac{1}{2}g^2\,2\,\frac{1}{2}(v_{\rm d}^2 - v_{\rm u}^2)T^3, \tag{11.45}$$

where $T^3 = \tau^3/2$. Similarly, the U(1) contribution to 'D' is

$$D_y = g'\left\{\sum_{\tilde{f}}\frac{1}{2}\tilde{f}^\dagger y_{\tilde{f}}\tilde{f} - \frac{1}{2}(v_{\rm d}^2 - v_{\rm u}^2)\right\}, \tag{11.46}$$

after symmetry breaking, where the sum is over all sfermions (squarks and sleptons). Expression (11.46) leads to the sfermion (mass)^2 term

$$+\frac{1}{2}g'^2 2\left(-\frac{1}{2}y\right)\frac{1}{2}(v_{\rm d}^2 - v_{\rm u}^2). \tag{11.47}$$

Since $y/2 = Q - T^3$, where Q is the electromagnetic charge, we can combine (11.45) and (11.47) to give a total (mass)^2 contribution for each sfermion:

$$\begin{aligned}\Delta_{\tilde{f}} &= \frac{1}{2}(v_{\rm d}^2 - v_{\rm u}^2)[(g^2 + g'^2)T^3 - g'^2 Q]\\ &= m_{\rm Z}^2\cos 2\beta[T^3 - \sin^2\theta_{\rm W} Q],\end{aligned} \tag{11.48}$$

using (10.21). As remarked earlier, $\Delta_{\tilde{f}}$ is diagonal in the weak eigenstate basis, and the appropriate contributions simply have to be added to the right-hand side of equations (11.38)–(11.42). It is interesting to note that the splitting between the doublet states is predicted to be

$$-m_{\tilde{\rm u}_{\rm L}}^2 + m_{\tilde{\rm d}_{\rm L}}^2 = -m_{\tilde{\nu}_{\rm eL}}^2 + m_{\tilde{\rm e}_{\rm L}}^2 = -\cos 2\beta m_{\rm W}^2, \tag{11.49}$$

and similarly for the second family. On the assumption that $\tan\beta$ is most probably greater than 1 (see the comments following equation (10.74)), the 'down' states are heavier.

Sfermion (mass)^2 terms are also generated by SUSY-invariant F-terms, after symmetry breaking; that is, terms in the Lagrangian of the form

$$-\left|\frac{\partial W}{\partial \phi_i}\right|^2 \tag{11.50}$$

for every scalar field ϕ_i (see equations (5.19) and (5.22)); for these purposes we regard W of (8.4) as being written in terms of the scalar fields, as in Section 5.1. Remembering that the Yukawa couplings are proportional to the associated fermion masses (see (8.10) and (10.69)–(10.71)), we see that on the scale expected for the masses of the sfermions, only terms involving the Yukawas of the third family can contribute significantly. Thus to a very good approximation we can write

$$\begin{aligned} W \approx\ & y_{\mathrm{t}}\tilde{t}_{\mathrm{R}}^{\dagger}(\tilde{t}_{\mathrm{L}}H_{\mathrm{u}}^{0} - \tilde{b}_{\mathrm{L}}H_{\mathrm{u}}^{+}) - y_{\mathrm{b}}\tilde{b}_{\mathrm{R}}^{\dagger}(\tilde{t}_{\mathrm{L}}H_{\mathrm{d}}^{-} - \tilde{b}_{\mathrm{L}}H_{\mathrm{d}}^{0}) - y_{\tau}\tilde{\tau}_{\mathrm{R}}^{\dagger}(\tilde{\nu}_{\tau\mathrm{L}}H_{\mathrm{d}}^{-} - \tilde{\tau}_{\mathrm{L}}H_{\mathrm{d}}^{0}) \\ & + \mu(H_{\mathrm{u}}^{+}H_{\mathrm{d}}^{-} - H_{\mathrm{u}}^{0}H_{\mathrm{d}}^{0}) \end{aligned} \tag{11.51}$$

as in (8.12) (with $\tilde{t}_{\mathrm{L}}$ replaced by $\tilde{t}_{\mathrm{R}}^{\dagger}$ etc.) and (8.13). Then we have, for example,

$$-\left|\frac{\partial W}{\partial \tilde{t}_{\mathrm{R}}^{\dagger}}\right|^2 = -y_{\mathrm{t}}^2\tilde{t}_{\mathrm{L}}^{\dagger}\tilde{t}_{\mathrm{L}}|H_{\mathrm{u}}^0|^2 \rightarrow -y_{\mathrm{t}}^2 v_{\mathrm{u}}^2\tilde{t}_{\mathrm{L}}^{\dagger}\tilde{t}_{\mathrm{L}} = -m_{\mathrm{t}}^2\tilde{t}_{\mathrm{L}}^{\dagger}\tilde{t}_{\mathrm{L}}, \tag{11.52}$$

after H_{u}^0 acquires the vev v_{u}. The L-type top squark ('stop') therefore gets a (mass)^2 term equal to the top quark (mass)^2. There will be an identical term for the R-type stop squark, coming from $-|\partial W/\partial \tilde{t}_{\mathrm{L}}|^2$. Similarly, there will be (mass)^2 terms m_{b}^2 for $\tilde{b}_{\mathrm{L}}$ and $\tilde{b}_{\mathrm{R}}$, and m_{τ}^2 for $\tilde{\tau}_{\mathrm{L}}$ and $\tilde{\tau}_{\mathrm{R}}$, although these are probably negligible in this context.

We also need to consider derivatives of W with respect to the Higgs fields. For example, we have

$$-\left|\frac{\partial W}{\partial H_{\mathrm{u}}^0}\right|^2 = -|y_{\mathrm{t}}\tilde{t}_{\mathrm{R}}^{\dagger}\tilde{t}_{\mathrm{L}} - \mu H_{\mathrm{d}}^0|^2 \rightarrow -|y_{\mathrm{t}}\tilde{t}_{\mathrm{R}}^{\dagger}\tilde{t}_{\mathrm{L}} - \mu v_{\mathrm{d}}|^2, \tag{11.53}$$

after symmetry breaking. The expression (11.53) contains the off-diagonal bilinear term

$$\mu v_{\mathrm{d}} y_{\mathrm{t}}(\tilde{t}_{\mathrm{R}}^{\dagger}\tilde{t}_{\mathrm{L}} + \tilde{t}_{\mathrm{L}}^{\dagger}\tilde{t}_{\mathrm{R}}) = \mu m_{\mathrm{t}}\cot\beta(\tilde{t}_{\mathrm{R}}^{\dagger}\tilde{t}_{\mathrm{L}} + \tilde{t}_{\mathrm{L}}^{\dagger}\tilde{t}_{\mathrm{R}}), \tag{11.54}$$

which mixes the R and L fields. Similarly, $-|\partial W/\partial H_{\mathrm{d}}^0|^2$ contains the mixing terms

$$\mu m_{\mathrm{b}}\tan\beta(\tilde{b}_{\mathrm{R}}^{\dagger}\tilde{b}_{\mathrm{L}} + \tilde{b}_{\mathrm{L}}^{\dagger}\tilde{b}_{\mathrm{R}}) \tag{11.55}$$

and

$$\mu m_\tau \tan\beta(\tilde{\tau}_{\rm R}^\dagger \tilde{\tau}_{\rm L} + \tilde{\tau}_{\rm L}^\dagger \tilde{\tau}_{\rm R}). \tag{11.56}$$

Finally, bilinear terms can also arise directly from the soft triple scalar couplings (9.37), after the scalar Higgs fields acquire vevs. Assuming the conditions (9.42), and retaining only the third family contribution as before, the relevant terms from (9.37) are

$$-A_0 y_{\rm t} v_{\rm u}(\tilde{t}_{\rm R}^\dagger \tilde{t}_{\rm L} + \tilde{t}_{\rm L}^\dagger \tilde{t}_{\rm R}) = -A_0 m_{\rm t}(\tilde{t}_{\rm R}^\dagger \tilde{t}_{\rm L} + \tilde{t}_{\rm L}^\dagger \tilde{t}_{\rm R}), \tag{11.57}$$

together with similar $\tilde{b}_{\rm R} - \tilde{b}_{\rm L}$ and $\tilde{\tau}_{\rm R} - \tilde{\tau}_{\rm L}$ mixing terms.

Putting all this together, then, the (mass)^2 values for the squarks and sleptons of the first two families are given by the expressions (11.38)–(11.42), together with the relevant contribution $\Delta_{\tilde{f}}$ of (11.48). For the third family, we discuss the $\tilde{\rm t}$, $\tilde{\rm b}$ and $\tilde{\tau}$ sectors separately. The (mass)^2 term for the top squarks is

$$-(\tilde{t}_{\rm L}^\dagger \tilde{t}_{\rm R}^\dagger)\, \mathbf{M}_{\tilde{\rm t}}^2 \begin{pmatrix} \tilde{t}_{\rm L} \\ \tilde{t}_{\rm R} \end{pmatrix}, \tag{11.58}$$

where

$$\mathbf{M}_{\tilde{\rm t}}^2 = \begin{pmatrix} m_{\tilde{\rm t}_{\rm L},\tilde{\rm b}_{\rm L}}^2 + m_{\rm t}^2 + \Delta_{\tilde{\rm u}_{\rm L}} & m_{\rm t}(A_0 - \mu\cot\beta) \\ m_{\rm t}(A_0 - \mu\cot\beta) & m_{\tilde{\rm t}_{\rm R}}^2 + m_{\rm t}^2 + \Delta_{\tilde{\rm u}_{\rm R}} \end{pmatrix}, \tag{11.59}$$

with

$$\Delta_{\tilde{\rm u}_{\rm L}} = \left(\frac{1}{2} - \frac{2}{3}\sin^2\theta_{\rm W}\right) m_{\rm Z}^2 \cos 2\beta \tag{11.60}$$

and

$$\Delta_{\tilde{\rm u}_{\rm R}} = -\frac{2}{3}\sin^2\theta_{\rm W} m_{\rm Z}^2 \cos 2\beta. \tag{11.61}$$

Here $m_{\tilde{\rm t}_{\rm L},\tilde{\rm b}_{\rm L}}^2$ and $m_{\tilde{\rm t}_{\rm R}}^2$ are given approximately by (9.53) and (9.54) respectively. In contrast to the corresponding equations for the first two families, the $X_{\rm t}$ term is now present, and will tend to reduce the running masses of $\tilde{\rm t}_{\rm L}$ and $\tilde{\rm t}_{\rm R}$ at low scales (the second more than the first), relative to those of the corresponding states in the first two families; on the other hand, the $m_{\rm t}^2$ term tends to work in the other direction.

The real symmetric matrix $\mathbf{M}_{\tilde{\rm t}}^2$ can be diagonalized by the orthogonal transformation

$$\begin{pmatrix} \tilde{t}_1 \\ \tilde{t}_2 \end{pmatrix} = \begin{pmatrix} \cos\theta_{\rm t} & -\sin\theta_{\rm t} \\ \sin\theta_{\rm t} & \cos\theta_{\rm t} \end{pmatrix} \begin{pmatrix} \tilde{t}_{\rm L} \\ \tilde{t}_{\rm R} \end{pmatrix}; \tag{11.62}$$

the eigenvalues are denoted by $m_{\tilde{\rm t}_1}^2$ and $m_{\tilde{\rm t}_2}^2$, with $m_{\tilde{\rm t}_1}^2 < m_{\tilde{\rm t}_2}^2$. Because of the large value of $m_{\rm t}$ in the off-diagonal positions in (11.59), mixing effects in the stop sector are likely to be substantial, and will probably result in the mass of the lighter stop, $m_{\tilde{t}_1}$, being significantly smaller than the mass of any other squark. Of course, the mixing effect must not become too large, or else $m^2_{\tilde{t}_1}$ is driven to negative values, which would imply (as in the electroweak Higgs case) a spontaneous breaking of colour symmetry. This requirement places a bound on the magnitude of the unknown parameter A_0, which cannot be much greater than $m_{\tilde{t}_L,\tilde{b}_L}$.

Turning now to the $\tilde{b}$ sector, the (mass)^2 matrix is

$$\mathbf{M}^2_{\tilde{b}} = \begin{pmatrix} m^2_{\tilde{t}_L,\tilde{b}_L} + m^2_{b} + \Delta_{\tilde{d}_L} & m_b(A_0 - \mu \tan\beta) \\ m_b(A_0 - \mu\tan\beta) & m^2_{\tilde{b}_R} + m^2_{b} + \Delta_{\tilde{d}_R} \end{pmatrix}, \tag{11.63}$$

with

$$\Delta_{\tilde{d}_L} = \left(-\frac{1}{2} + \frac{1}{3}\sin^2\theta_W\right) m_Z^2 \cos 2\beta \tag{11.64}$$

and

$$\Delta_{\tilde{d}_R} = \frac{1}{3}\sin^2\theta_W m_Z^2 \cos 2\beta. \tag{11.65}$$

Here, since X_t enters into the evolution of the mass of $\tilde{b}_L$ but not of $\tilde{b}_R$, we expect that the running mass of $\tilde{b}_R$ will be much the same as those of $\tilde{d}_R$ and $\tilde{s}_R$, but that $m_{\tilde{b}_L}$ may be less than $m_{\tilde{d}_L}$ and $m_{\tilde{s}_L}$. Similarly, the (mass)^2 matrix in the $\tilde{\tau}$ sector is

$$\mathbf{M}^2_{\tilde{\tau}} = \begin{pmatrix} m^2_{\tilde{\nu}_{\tau L},\tilde{\tau}_L} + m^2_{\tau} + \Delta_{\tilde{e}_L} & m_\tau(A_0 - \mu \tan\beta) \\ m_\tau(A_0 - \mu\tan\beta) & m^2_{\tilde{\tau}_R} + m^2_{\tau} + \Delta_{\tilde{e}_R} \end{pmatrix}, \tag{11.66}$$

with

$$\Delta_{\tilde{e}_L} = \left(-\frac{1}{2} + \sin^2\theta_W\right) m_Z^2 \cos 2\beta \tag{11.67}$$

and

$$\Delta_{\tilde{e}_R} = \frac{1}{3}\sin^2\theta_W m_Z^2 \cos 2\beta. \tag{11.68}$$

Mixing effects in the $\tilde{b}$ and $\tilde{\tau}$ sectors depend on how large $\tan\beta$ is (see the off-diagonal terms in (11.63) and (11.66)). It seems that for $\tan\beta$ less than about 5(?), mixing effects will not be large, so that the masses of $\tilde{b}_R$, $\tilde{\tau}_R$ and $\tilde{\tau}_L$ will all be approximately degenerate with the corresponding states in the first two families, while $\tilde{b}_L$ will be lighter than $\tilde{d}_L$ and $\tilde{s}_L$. For larger values of $\tan\beta$, strong mixing may take place, as in the stop sector. In this case, $\tilde{b}_1$ and $\tilde{\tau}_1$ may be significantly lighter than their analogues in the first two families (also, $\tilde{\nu}_{\tau L}$ may be lighter than $\tilde{\nu}_{eL}$ and $\tilde{\nu}_{\mu L}$). Neutralinos and charginos will then decay predominantly to taus and staus.

12

Some simple tree-level calculations in the MSSM

To complete our introduction to the physics of sparticles in the MSSM, we now present some calculations of sparticle decay widths and production cross sections. We work at tree-level only, with the choice of unitary gauge in the gauge sectors, where only physical fields appear (see, for example, [7] Sections 19.5 and 19.6). We shall see how the interactions written down in Chapters 7 and 8 in rather abstract and compressed notation translate into more physical expressions, and there will be further opportunities to practise using Majorana spinors. However, since we shall only be considering a limited number of particular processes, we shall not derive general Feynman rules for Majorana particles (they can be found in [45, 47, 114, 115], for example); instead, the matrix elements which arise will be directly evaluated by the elementary 'reduction' procedure, as described in Section 6.3.1 of [15], for example. Our results will be compared with those quoted in the book by Baer and Tata [49], which conveniently contains a compendium of tree-level formulae for sparticle decay widths and production cross sections. Representative calculations of cross sections for sparticle production at hadron colliders may be found in [116]. Experimental methods for measuring superparticle masses and cross sections at the LHC are summarized in [117].

12.1 Sparticle decays

12.1.1 The gluino decays $\tilde{g} \to u\,\bar{\tilde{u}}_L$ *and* $\tilde{g} \to t\,\bar{\tilde{t}}_1$

We consider first (Figure 12.1) the decay of a gluino $\tilde{g}$ of mass $m_3(= m_{\tilde{g}})$, 4-momentum k_3, spin s_3 and colour label c_3, to a quark u of mass $m_1(= m_u)$, 4-momentum k_1, spin s_1 and colour label c_1, and an anti-squark $\bar{\tilde{u}}_L$ of mass $m_2(= m_{\tilde{u}_L})$, 4-momentum k_2 and colour label c_2. We assume that the decay is kinematically allowed. Squark mixing may be neglected for this first-generation final state; we shall include it for $\tilde{g} \to t\,\bar{\tilde{t}}_1$.

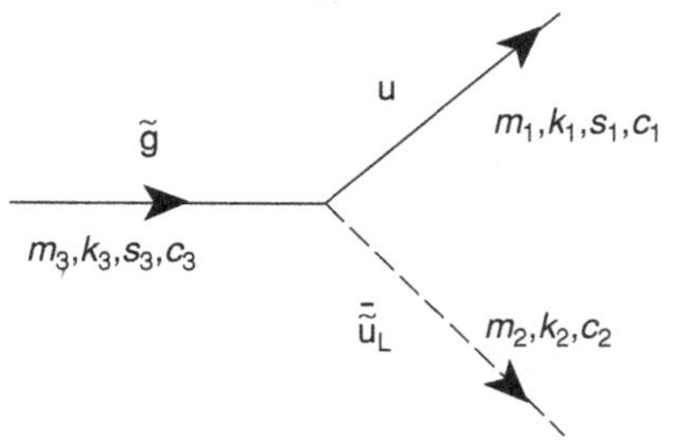

Figure 12.1 Lowest-order diagram for the decay $\tilde{g} \to u\bar{\tilde{u}}_L$.

The gluino, quark and squark fields are denoted by the L-spinors $\tilde{g}$, χ_u and the complex scalar $\tilde{u}_L$, respectively. The relevant interaction is contained in (7.72), namely

$$-\sqrt{2}g_s\tilde{g}^{a\dagger}\cdot\chi^{\dagger}_{u\alpha}\frac{1}{2}(\lambda^a)_{\alpha\beta}\tilde{u}_{L\beta}, \tag{12.1}$$

where the colour indices are such that a runs from 1 to 8 and α, β run from 1 to 3. We note that the strength of the interaction is determined by the QCD coupling constant g_s. In calculating the decay rate it is convenient to make use of the trace techniques for spin sums which are familiar from SM physics. We therefore begin by converting (12.1) to 4-component form – Dirac (Ψ) for the quark field, Majorana (Ψ_M) for the gluino. We have

$$\begin{aligned}\tilde{g}^{a\dagger}\cdot\chi^{\dagger}_{u\alpha} = \chi^{\dagger}_{u\alpha}\cdot\tilde{g}^{a\dagger} &= \bar{\Psi}^{\chi_u}_{M\alpha}P_R\Psi^{\tilde{g}a}_M \quad \text{from (2.116)}\\ &= \left(P_L\Psi^{\chi_u}_{M\alpha}\right)^{\dagger}\beta\Psi^{\tilde{g}a}_M\\ &= (P_L\Psi_{u\alpha})^{\dagger}\beta\Psi^{\tilde{g}a}_M \quad \text{from (8.27)}\\ &= \bar{\Psi}_{u\alpha}P_R\Psi^{\tilde{g}a}_M. \end{aligned}\tag{12.2}$$

We can allow for the possibility that the gluino mass parameter M_3 of (11.1) is negative by replacing $\Psi^{\tilde{g}}_M$ by $(i\gamma_5)^{\theta_{\tilde{g}}}\Psi^{\tilde{g}}_M$, as discussed in Section 11.1.1. Then (12.2) becomes

$$(i)^{\theta_{\tilde{g}}}\bar{\Psi}_{u\alpha}P_R\Psi^{\tilde{g}a}_M, \tag{12.3}$$

using $P_R\gamma_5 = P_R$. This refinement will only be relevant when we include squark mixing.

To lowest order in g_s, the decay amplitude is then

$$-i\sqrt{2}g_s(i)^{\theta_{\tilde{g}}}\langle u, k_1, s_1, c_1; \bar{\tilde{u}}_L, k_2, c_2 \,|\int d^4x\;\bar{\Psi}_{u\alpha}(x)P_R\Psi^{\tilde{g}a}_M(x)\frac{1}{2}\lambda^a_{\alpha\beta}\tilde{u}_{L\beta}(x)|\tilde{g}, k_3, s_3, c_3\rangle. \tag{12.4}$$

The matrix element may be evaluated by 'reducing' the particles in the initial and final states. For example,

$$\begin{aligned}
&\langle \mathrm{u}, k_1, s_1, c_1|\bar{\Psi}_{\mathrm{u}\alpha}(x)\\
&= \langle 0|c_{\mathrm{u},s_1c_1}(k_1)\sqrt{2E_{k_1}}\int \frac{\mathrm{d}^3\boldsymbol{k}'}{(2\pi)^3\sqrt{2E_{k'}}}\sum_{s'c'}[c^\dagger_{\mathrm{u},s'c'}(k')\bar{u}(k',s')\omega^*_\alpha(c')\mathrm{e}^{\mathrm{i}k'\cdot x}\\
&\quad + d_{\mathrm{u},s'c'}(k')\bar{v}(k',s')\omega^*_\alpha(c')\mathrm{e}^{-\mathrm{i}k'\cdot x}]\\
&= \langle 0|\bar{u}(k_1,s_1)\omega^*_\alpha(c_1)\mathrm{e}^{\mathrm{i}k_1\cdot x} \text{ using (2.129);} \qquad (12.5)
\end{aligned}$$

here $\omega(c)$ is the 3-component colour wavefunction for a colour triplet with colour label 'c'. Proceeding in the same way for the other two fields, (12.4) reduces to

$$\begin{aligned}
&-\mathrm{i}\sqrt{2}g_{\mathrm{s}}(\mathrm{i})^{\theta_{\mathrm{g}}}\bar{u}(k_1,s_1)P_{\mathrm{R}}u(k_3,s_3)\Omega_a(c_3)\left(\omega^\dagger(c_1)\frac{1}{2}\lambda^a\omega(c_2)\right)(2\pi)^4\delta^4(k_1+k_2-k_3)\\
&\quad\equiv (2\pi)^4\delta^4(k_1+k_2-k_3)\mathrm{i}\mathcal{M}, \qquad (12.6)
\end{aligned}$$

where $\Omega_a(c_3)$ $(a = 1, 2, \ldots 8)$ is the colour wavefunction for the gluino, and $\mathrm{i}\mathcal{M}$ is the invariant amplitude for the process.

The decay rate (partial width) is given by (see equation (6.59) of [15])

$$\Gamma = \frac{1}{2E_3}(2\pi)^4\int \delta^4(k_1+k_2-k_3)\overline{|\mathcal{M}|^2}\frac{\mathrm{d}^3\boldsymbol{k}_1}{(2\pi)^3 2E_{k_1}}\frac{\mathrm{d}^3\boldsymbol{k}_2}{(2\pi)^3 2E_{k_2}} \qquad (12.7)$$

where $\overline{|\mathcal{M}|^2}$ is the result of averaging over initial spins and colours, and summing over final spins and colours:

$$\overline{|\mathcal{M}|^2} = \frac{1}{8}\sum_{c_1,c_2,c_3}\frac{1}{2}\sum_{s_1,s_2}|\mathcal{M}|^2. \qquad (12.8)$$

The colour factor is evaluated in problem 14.4 of [7], and is equal to 1/2. The spinor part is

$$\begin{aligned}
&\frac{1}{2}\mathrm{Tr}\left[\left(\frac{1+\gamma_5}{2}\right)(\not{k}_3+m_3)\left(\frac{1-\gamma_5}{2}\right)(\not{k}_1+m_1)\right]\\
&= \frac{1}{2}\mathrm{Tr}\left[\left(\frac{1+\gamma_5}{2}\right)\not{k}_3\left(\frac{1-\gamma_5}{2}\right)\not{k}_1\right]\\
&= \frac{1}{2}\mathrm{Tr}\left[\left(\frac{1+\gamma_5}{2}\right)\not{k}_3\not{k}_1\right]\\
&= \frac{1}{4}\mathrm{Tr}[\not{k}_3\not{k}_1] = k_3\cdot k_1 = \frac{1}{2}\left(m_3^2+m_1^2-m_2^2\right). \qquad (12.9)
\end{aligned}$$

Finally, the phase space integral is (see equation (6.64) of [15])

$$\frac{1}{2E_3}(2\pi)^4 \int \delta^4(k_1 + k_2 - k_3)\frac{\mathrm{d}^3\boldsymbol{k}_1}{(2\pi)^3 2E_{k_1}}\frac{\mathrm{d}^3\boldsymbol{k}_2}{(2\pi)^3 2E_{k_2}} = \frac{1}{8\pi m_3^2}k(m_1, m_2, m_3), \tag{12.10}$$

where k is the magnitude of the 3-momentum of the final state particles 1,2 in the rest frame of the decaying particle 3:

$$k(m_1, m_2, m_3) = \left[m_1^4 + m_2^4 + m_3^4 - 2m_1^2 m_2^2 - 2m_2^2 m_3^2 - 2m_3^2 m_1^2\right]/2m_3. \tag{12.11}$$

In the present case, $m_1 = m_{\mathrm{u}}, m_2 = m_{\tilde{\bar{\mathrm{u}}}_{\mathrm{L}}}$ and $m_3 = m_{\tilde{\mathrm{g}}}$. So we find

$$\Gamma(\tilde{\mathrm{g}} \to \mathrm{u}\tilde{\bar{\mathrm{u}}}_{\mathrm{L}}) = \frac{\alpha_{\mathrm{s}}}{4}\left(1 + \frac{m_{\mathrm{u}}^2}{m_{\tilde{\mathrm{g}}}^2} - \frac{m_{\tilde{\bar{\mathrm{u}}}_{\mathrm{L}}}^2}{m_{\tilde{\mathrm{g}}}^2}\right) k(m_{\mathrm{u}}, m_{\tilde{\bar{\mathrm{u}}}_{\mathrm{L}}}, m_{\tilde{\mathrm{g}}}), \tag{12.12}$$

in agreement with formula (B.1a) of [49]. If, for the sake of illustration, we take $k \approx 100$ GeV, $\alpha_{\mathrm{s}} \approx 0.1$, then the partial width for this mode is $\Gamma \sim$ few GeV, with a corresponding lifetime of order 10^{-25}s.

We turn now to the decay $\tilde{\mathrm{g}} \to \mathrm{t}\tilde{\bar{\mathrm{t}}}_1$. We recall that the fields $\tilde{t}_{1,2}$ which correspond to the mass eigenstates are given in terms of the unmixed fields $\tilde{t}_{\mathrm{R,L}}$ by (11.62). In addition to the amplitude for

$$\tilde{\mathrm{g}} \to \mathrm{t}\tilde{\bar{\mathrm{t}}}_{\mathrm{L}}, \tag{12.13}$$

we therefore also need the amplitude for

$$\tilde{\mathrm{g}} \to \mathrm{t}\tilde{\bar{\mathrm{t}}}_{\mathrm{R}}. \tag{12.14}$$

The interaction responsible for (12.13) is simply (12.1) with 'u' replaced by 't':

$$-\sqrt{2}g_{\mathrm{s}}\tilde{g}^{a\dagger}\cdot\chi_{\mathrm{t}\alpha}^{\dagger}\frac{1}{2}(\lambda^a)_{\alpha\beta}\tilde{t}_{\mathrm{L}\beta} \to -\sqrt{2}g_{\mathrm{s}}(\mathrm{i})^{\theta_{\tilde{\mathrm{g}}}}\bar{\Psi}_{\mathrm{t}\alpha}P_{\mathrm{R}}\Psi_{\mathrm{M}}^{\tilde{g}a}\frac{1}{2}(\lambda^a)_{\alpha\beta}\tilde{t}_{\mathrm{L}\beta}, \tag{12.15}$$

and the component for producing a $\tilde{\bar{\mathrm{t}}}_1$ is

$$-\sqrt{2}g_{\mathrm{s}}(\mathrm{i})^{\theta_{\tilde{\mathrm{g}}}}\bar{\Psi}_{\mathrm{t}\alpha}P_{\mathrm{R}}\Psi_{\mathrm{M}}^{\tilde{g}a}\frac{1}{2}(\lambda^a)_{\alpha\beta}\cos\theta_{\mathrm{t}}\,\tilde{t}_{1\beta}. \tag{12.16}$$

For (12.14), we note that the field $\tilde{t}_{\mathrm{R}}^{\dagger}$ creates the scalar partner of the weak singlet quark and destroys the scalar partner of the weak singlet antiquark. So, in the notation of Sections 8.1 and 8.2, $\tilde{t}_{\mathrm{R}}^{\dagger}$ and $\chi_{\bar{\mathrm{t}}}$ form a chiral multiplet, which belongs to the $\bar{\mathbf{3}}$ representation of $\mathrm{SU(3)_c}$. The term in (7.72) responsible for the decay (12.14) is then

$$-\sqrt{2}g_{\mathrm{s}}\tilde{t}_{\mathrm{R}\alpha}\frac{1}{2}(-\lambda^{a*})_{\alpha\beta}\chi_{\bar{\mathrm{t}}\beta}\cdot\tilde{g}^a. \tag{12.17}$$

We now convert the spinors to 4-component form. We have

$$\chi_{\bar{\mathrm{t}}\beta}\cdot\tilde{g}^a = \bar{\Psi}_{\mathrm{M}\beta}^{\chi_{\bar{\mathrm{t}}}}P_{\mathrm{L}}\Psi_{\mathrm{M}}^{\tilde{g}a}, \tag{12.18}$$

as usual, where we recall from (8.19) that

$$\chi_{\bar{\rm t}} = -{\rm i}\sigma_2 \psi_{\rm t}^*, \tag{12.19}$$

so that

$$\Psi_{\rm M}^{\chi_{\bar{\rm t}}} = \begin{pmatrix} {\rm i}\sigma_2 \chi_{\bar{\rm t}}^* = {\rm i}\sigma_2(-{\rm i}\sigma_2)\psi_{\rm t} = \psi_{\rm t} \\ \chi_{\bar{\rm t}} = -{\rm i}\sigma_2 \psi_{\rm t}^* \end{pmatrix} = \Psi_{\rm M}^{\psi_{\rm t}}, \tag{12.20}$$

using (8.26). Hence

$$\bar{\Psi}_{\rm M}^{\chi_{\bar{\rm t}}} P_{\rm L} = \left(P_{\rm R}\Psi_{\rm M}^{\psi_{\rm t}}\right)^\dagger \beta = (P_{\rm R}\Psi_{\rm t})^\dagger \beta = \bar{\Psi}_{\rm t} P_{\rm L}, \tag{12.21}$$

where we have used (8.27). The interaction (12.17) can then be written as

$$-\sqrt{2} g_{\rm s} \tilde{t}_{{\rm R}\alpha} \frac{1}{2}(-\lambda^{a*})_{\alpha\beta} (-{\rm i})^{\theta_{\rm g}} \bar{\Psi}_{{\rm t}\beta} P_{\rm L} \Psi_{\rm M}^{\tilde{g}a}, \tag{12.22}$$

where we have included the phase factor to allow for negative M_3, and used $P_{\rm L}\gamma_5 = -P_{\rm L}$. The component for producing a $\bar{\tilde{\rm t}}_1$ is

$$-\sqrt{2} g_{\rm s} (-\sin\theta_{\rm t}\, \tilde{t}_{1\alpha}) \frac{1}{2}(-\lambda^{a*})_{\alpha\beta} (-{\rm i})^{\theta_{\rm g}} \bar{\Psi}_{{\rm t}\beta} P_{\rm L} \Psi_{\rm M}^{\tilde{g}a}. \tag{12.23}$$

The matrix element of (12.23) can be evaluated as before, in (12.4)–(12.6). Consider in particular the colour part, which is

$$\omega_\alpha(c_2)(-\lambda^{a*})_{\alpha\beta}\omega_\beta^*(c_1), \tag{12.24}$$

where c_1, c_2 are the colour labels of the quark and anti-squark. Since the ω's are not operators, (12.24) can equally be written as

$$\omega_\beta^*(c_1)(-\lambda^{a\dagger})_{\beta\alpha}\omega_\alpha(c_2) = -\omega^\dagger(c_1)\lambda^a\omega(c_2), \tag{12.25}$$

where we have used the hermiticity of the λ's. We see that this is now the same as the colour factor for (12.6), and hence for (12.16), but with a minus sign.

Putting all this together, we find that the amplitude for the decay $\tilde{\rm g} \to {\rm t}\bar{\tilde{\rm t}}_1$ takes the same form as the left-hand side of (12.6), but with the replacement

$$\begin{aligned} \bar{u}(k_1, s_1)({\rm i})^{\theta_{\rm g}} P_{\rm R} u(k_3, s_3) &\to \bar{u}(k_1, s_1)[({\rm i})^{\theta_{\rm g}} P_{\rm R}\cos\theta_{\rm t} + (-{\rm i})^{\theta_{\rm g}} P_{\rm L}\sin\theta_{\rm t}]u(k_3, s_3) \\ &\equiv \bar{u}(k_1, s_1)[A + B\gamma_5]u(k_3, s_3), \end{aligned} \tag{12.26}$$

where

$$\begin{aligned} A &= \frac{1}{2}({\rm i})^{\theta_{\rm g}}\cos\theta_{\rm t} + \frac{1}{2}(-{\rm i})^{\theta_{\rm g}}\sin\theta_{\rm t}, \\ B &= \frac{1}{2}({\rm i})^{\theta_{\rm g}}\cos\theta_{\rm t} - \frac{1}{2}(-{\rm i})^{\theta_{\rm g}}\sin\theta_{\rm t}. \end{aligned} \tag{12.27}$$

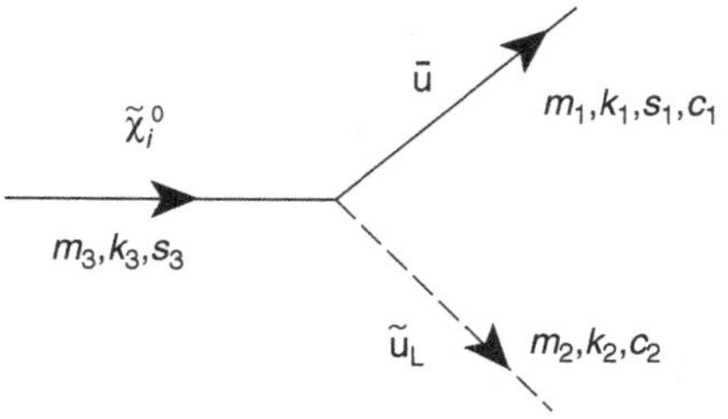

Figure 12.2 Lowest-order diagram for the decay $\tilde{\chi}_i^0 \to \bar{u}\tilde{u}_L$.

The spinor trace is then

$$\frac{1}{2}\mathrm{Tr}[(A + B\gamma_5)(\not{k}_3 + m_3)(-B^*\gamma_5 + A^*)(\not{k}_1 + m_1)]$$
$$= (|A|^2 + |B|^2)\left(m_3^2 + m_1^2 - m_2^2\right) + (|A|^2 - |B|^2)2m_1m_3, \quad (12.28)$$

and

$$|A|^2 + |B|^2 = \frac{1}{2}, \quad |A|^2 - |B|^2 = (-)^{\theta_{\tilde{g}}}\frac{1}{2}\sin 2\theta_t. \quad (12.29)$$

The partial width $\Gamma(\tilde{g} \to t\bar{\tilde{t}}_1)$ is then given by (12.12), but with the replacement

$$\left(1 + \frac{m_u^2}{m_{\tilde{g}}^2} - \frac{m_{\bar{\tilde{u}}_L}^2}{m_{\tilde{g}}^2}\right) \to \left(1 + \frac{m_u^2}{m_{\tilde{g}}^2} - \frac{m_{\bar{\tilde{u}}_L}^2}{m_{\tilde{g}}^2}\right) + 2(-)^{\theta_{\tilde{g}}}\sin 2\theta_t\frac{m_t}{m_{\tilde{g}}}, \quad (12.30)$$

in agreement with formula (B.1b) of [49].

There are of course many such two-body modes: these channels may be repeated for all the other flavours. If all such two-body decays to squarks are kinematically forbidden, the dominant gluino decay would be via a virtual squark, which then decays weakly to charginos and neutralinos (we saw earlier, in Section 9.3, that most models assume that the gluino mass is significantly greater than that of the neutralinos and charginos).

12.1.2 The neutralino decays $\tilde{\chi}_i^0 \to \bar{u}\tilde{u}_L$ and $\tilde{\chi}_i^0 \to \bar{t}\tilde{t}_1$

We consider the decay (Figure 12.2) of a neutralino $\tilde{\chi}_i^0$ of mass $m_3(= m_{\tilde{\chi}_i^0})$, 4-momentum k_3 and spin s_3, to an anti-quark ū of mass $m_1(= m_u)$, 4-momentum k_1, spin s_1 and colour c_1, and a squark $\tilde{u}_L$ of mass $m_2(= m_{\tilde{u}_L})$, 4-momentum k_2, and colour c_2. The process is similar to the first gluino decay considered in the previous subsection, and as in that case we shall neglect squark mixing in this first generation process. By the same token, we shall only consider the $\tilde{W}^0$ and $\tilde{B}$ components of the neutralinos, neglecting the coupling to their Higgsino components which arises from the first generation Yukawa coupling in the superpotential.

The relevant interaction is contained in the electroweak part of (7.72), namely

$$-\frac{1}{\sqrt{2}}\tilde{u}_{\mathrm{L}}^{\dagger}\left(g\tilde{W}^0+\frac{g'}{3}\tilde{B}\right)\cdot\chi_{\mathrm{u}}, \tag{12.31}$$

where a sum over the colour indices of the quark and squark fields is understood. We re-write the gauge-eigenstate fields $\tilde{G}^0$ of (11.11) in terms of the mass-eigenstate fields $\tilde{\chi}^0$ by

$$\tilde{G}^0=\mathbf{V}\tilde{\chi}^0 \tag{12.32}$$

where $\mathbf{V}$ is an orthogonal matrix, so that

$$\tilde{B}=\sum_i \mathrm{V}_{\tilde{B}i}\tilde{\chi}_i^0, \quad \text{and} \quad \tilde{W}^0=\sum_i \mathrm{V}_{\tilde{W}^0 i}\tilde{\chi}_i^0. \tag{12.33}$$

Then (12.31) becomes

$$\begin{aligned}&-\frac{1}{\sqrt{2}}\tilde{u}_{\mathrm{L}}^{\dagger}\sum_i\left(g\mathrm{V}_{\tilde{W}^0 i}+\frac{g'}{3}\mathrm{V}_{\tilde{B}i}\right)\tilde{\chi}_i^0\cdot\chi_{\mathrm{u}}\\ &=-\frac{1}{\sqrt{2}}\tilde{u}_{\mathrm{L}\alpha}^{\dagger}\sum_i\left(g\mathrm{V}_{\tilde{W}^0 i}+\frac{g'}{3}\mathrm{V}_{\tilde{B}i}\right)(-\mathrm{i})^{\theta_{\tilde{\chi}_i^0}}\bar{\Psi}_{\mathrm{M}}^{\tilde{\chi}_i^0}P_{\mathrm{L}}\Psi_{\mathrm{u}\alpha},\end{aligned} \tag{12.34}$$

where in the second line we have re-instated the colour indices, which are summed over $\alpha=1$ to 3, and included the phase factor (11.17) to take care of negative mass eigenvalues, using $\gamma_5 P_{\mathrm{L}}=-P_{\mathrm{L}}$.

The amplitude for the decay of the i-th neutralino state is then obtained as in (12.4)–(12.6), and we find the result

$$(2\pi)^4\delta^4(k_1+k_2-k_3)\mathrm{i}A_{\mathrm{u}}^{\tilde{\chi}_i^0}\bar{v}(k_3,s_3)P_{\mathrm{L}}v(k_1,s_1)\omega^{\dagger}(c_2)_{\alpha}\omega(c_1)_{\alpha}, \tag{12.35}$$

where

$$A_{\mathrm{u}}^{\tilde{\chi}_i^0}=\frac{-1}{\sqrt{2}}(-\mathrm{i})^{\theta_{\tilde{\chi}_i^0}}\left(g\mathrm{V}_{\tilde{W}i}+\frac{g'}{3}\mathrm{V}_{\tilde{B}i}\right). \tag{12.36}$$

For the decay rate, the spinor trace is very similar to (12.9), and yields the same answer:

$$\frac{1}{2}\sum_{s_1,s_3}|\bar{v}(k_3,s_3)P_{\mathrm{L}}v(k_1,s_1)|^2=\frac{1}{2}\left(m_3^2+m_1^2-m_2^2\right). \tag{12.37}$$

The colour factor is

$$\sum_{c_1,c_2}|\omega^{\dagger}(c_2)\omega(c_1)|^2=3, \tag{12.38}$$

since $\omega^{\dagger}(c_2)\omega(c_1)=\delta_{c_2c_1}$. The phase space factor is as in (12.10), and the partial rate is obtained as

$$\Gamma(\tilde{\chi}_i^0 \to \mathrm{u}\bar{\tilde{\mathrm{u}}}_\mathrm{L}) = \frac{3}{16\pi}|A_\mathrm{u}^{\tilde{\chi}_i^0}|^2 \left(1 + \frac{m_\mathrm{u}^2}{m_{\tilde{\chi}_i^0}^2} - \frac{m_{\tilde{\mathrm{u}}_\mathrm{L}}^2}{m_{\tilde{\chi}_i^0}^2}\right) k(m_\mathrm{u}, m_{\tilde{\mathrm{u}}_\mathrm{L}}, m_{\tilde{\chi}_i^0}), \tag{12.39}$$

in agreement with formula (B.66) of [49].

The calculation of the partial width for $\tilde{\chi}_i^0 \to \bar{\mathrm{t}}\tilde{\mathrm{t}}_1$ is complicated both by squark mixing, as discussed for $\tilde{\mathrm{g}} \to \mathrm{t}\bar{\tilde{\mathrm{t}}}_1$, and by the inclusion of Higgsino components.

To include squark mixing, we require the amplitude for both

$$\tilde{\chi}_i^0 \to \bar{\mathrm{t}}\tilde{\mathrm{t}}_\mathrm{L} \tag{12.40}$$

and

$$\tilde{\chi}_i^0 \to \bar{\mathrm{t}}\tilde{\mathrm{t}}_\mathrm{R}. \tag{12.41}$$

The $\tilde{W}^0 - \tilde{B}$ part of the interaction responsible for (12.40) is of course the same as (12.31), with 'u' replaced by 't'. For (12.41), only $\tilde{B}$ contributes, and the relevant interaction is

$$\frac{1}{\sqrt{2}}\frac{4}{3}g'\tilde{t}_\mathrm{R}^\dagger \bar{\Psi}_\mathrm{M}^{\tilde{B}} P_\mathrm{R}\Psi_\mathrm{t}. \tag{12.42}$$

For the i-th neutralino mass-eigenstate field, the required interaction, so far, is then

$$\tilde{t}_{\mathrm{L}\alpha}^\dagger A_\mathrm{u}^{\tilde{\chi}_i^0} \bar{\Psi}_\mathrm{M}^{\tilde{\chi}_i^0} P_\mathrm{L}\Psi_{\mathrm{t}\alpha} + \tilde{\mathrm{t}}_{\mathrm{R}\alpha}^\dagger B_\mathrm{u}^{\tilde{\chi}_i^0} \bar{\Psi}_\mathrm{M}^{\tilde{\chi}_i^0} P_\mathrm{R}\Psi_{\mathrm{t}\alpha}, \tag{12.43}$$

where

$$B_\mathrm{u}^{\tilde{\chi}_i^0} = \frac{1}{\sqrt{2}}\frac{4}{3}g'(\mathrm{i})^{\theta_{\tilde{\chi}_i^0}} \mathrm{V}_{\tilde{B}i}. \tag{12.44}$$

The relevant part of the superpotential is

$$W = y_\mathrm{t}\tilde{t}_\mathrm{R}^\dagger \tilde{t}_\mathrm{L} H_\mathrm{u}^0 \ldots, \tag{12.45}$$

where $\tilde{t}_\mathrm{R}^\dagger$ could alternatively be written as $\bar{\tilde{t}}_\mathrm{L}$. The resultant Yukawa couplings to the Higgsino fields are

$$-y_\mathrm{t}\tilde{t}_\mathrm{R}^\dagger \tilde{H}_\mathrm{u}^0 \cdot \chi_\mathrm{t} \to -y_\mathrm{t}\tilde{t}_\mathrm{R}^\dagger \mathrm{V}_{\tilde{H}_\mathrm{u}^0 i}\, \tilde{\chi}_i^0 \cdot \chi_\mathrm{t} = -y_\mathrm{t}\tilde{t}_{\mathrm{R}\alpha}^\dagger \mathrm{V}_{\tilde{H}_\mathrm{u}^0 i}\, (-\mathrm{i})^{\theta_{\tilde{\chi}_i^0}} \bar{\Psi}_\mathrm{M}^{\tilde{\chi}_i^0} P_\mathrm{L}\Psi_{\mathrm{t}\alpha}, \tag{12.46}$$

and

$$-y_\mathrm{t}\tilde{t}_\mathrm{L}^\dagger \bar{\tilde{H}}_\mathrm{u}^0 \cdot \bar{\chi}_\mathrm{t} \to -y_\mathrm{t}\tilde{t}_\mathrm{L}^\dagger \mathrm{V}_{\tilde{H}_\mathrm{u}^0 i}\, \bar{\tilde{\chi}}_i^0 \cdot \bar{\chi}_\mathrm{t} = -y_\mathrm{t}\tilde{t}_{\mathrm{L}\alpha}^\dagger \mathrm{V}_{\tilde{H}_\mathrm{u}^0 i}\, (\mathrm{i})^{\theta_{\tilde{\chi}_i^0}} \bar{\Psi}_\mathrm{M}^{\tilde{\chi}_i^0} P_\mathrm{R}\Psi_{\mathrm{t}\alpha}, \tag{12.47}$$

using manipulations similar to those in (12.17)–(12.22). Combining (12.46) and (12.47) with (12.43), and retaining the $\tilde{t}_1$ component only, we arrive at the interaction

$$\tilde{t}_{1\alpha}^\dagger \bar{\Psi}_\mathrm{M}^{\tilde{\chi}_i^0} [a + b\gamma_5]\Psi_{\mathrm{t}\alpha} \tag{12.48}$$

where

$$a = \frac{1}{2}\left[\cos\theta_t\left(A_u^{\tilde{\chi}_i^0} - y_t(i)^{\theta_{\tilde{\chi}_i^0}} V_{\tilde{H}_u^0 i}\right) + \sin\theta_t\left(-B_u^{\tilde{\chi}_i^0} + y_t(-i)^{\theta_{\tilde{\chi}_i^0}} V_{\tilde{H}_u^0 i}\right)\right] \quad (12.49)$$

and

$$b = \frac{1}{2}\left[-\sin\theta_t\left(B_u^{\tilde{\chi}_i^0} + y_t(-i)^{\theta_{\tilde{\chi}_i^0}} V_{\tilde{H}_u^0 i}\right) - \cos\theta_t\left(A_u^{\tilde{\chi}_i^0} + y_t(i)^{\theta_{\tilde{\chi}_i^0}} V_{\tilde{H}_u^0 i}\right)\right]. \quad (12.50)$$

The decay amplitude is then the same as (12.35), but with

$$\bar{v}(k_3, s_3) A_u^{\tilde{\chi}_i^0} P_L v(k_1, s_1) \quad (12.51)$$

replaced by

$$\bar{v}(k_3, s_3)(a + b\gamma_5) v(k_1, s_1). \quad (12.52)$$

The spinor trace calculation is similar to that in (12.28), and the partial width is found to be

$$\Gamma(\tilde{\chi}_i^0 \to \bar{t}\tilde{t}_1) = \frac{3}{8\pi m_{\tilde{\chi}_i^0}^2}\{|a|^2[(m_t + m_{\tilde{\chi}_i^0})^2 - m_{\tilde{t}_1}^2] + |b|^2[(m_t - m_{\tilde{\chi}_i^0})^2 - m_{\tilde{t}_1}^2]\}$$
$$\times k(m_t, m_{\tilde{t}_1}, m_{\tilde{\chi}_i^0}) \quad (12.53)$$

in agreement with formula (B.65) of [49].

Exercise 12.1 The squark decay $\tilde{t}_1 \to t\tilde{\chi}_i^0$

The interaction responsible for this decay is closely related to that for $\tilde{\chi}_i^0 \to \bar{t}\tilde{t}_1$ – in fact, it is the hermitian conjugate of (12.48), namely

$$\tilde{t}_{1\alpha}\bar{\Psi}_{t\alpha}[a^* - b^*\gamma_5]\Psi_M^{\tilde{\chi}_i^0}. \quad (12.54)$$

Assuming that the decay is kinematically allowed, $m_{\tilde{t}_1} > m_t + m_{\tilde{\chi}_i^0}$, show that

$$\Gamma(\tilde{t}_1 \to t\tilde{\chi}_i^0) = \frac{1}{4\pi m_{\tilde{t}_1}^2}\{|a|^2[m_{\tilde{t}_1}^2 - (m_t + m_{\tilde{\chi}_i^0})^2] + |b|^2[m_{\tilde{t}_1}^2 - (m_t - m_{\tilde{\chi}_i^0}^2)^2]\}$$
$$\times k(m_t, m_{\tilde{\chi}_i^0}, m_{\tilde{t}_1}), \quad (12.55)$$

in agreement with formula (B.39) of [49].

12.1.3 The neutralino decay $\tilde{\chi}_i^0 \to \tilde{\chi}_j^0 + Z^0$

We recall that the Z^0 field is given by the linear combination (10.20). Since B^μ has weak hypercharge equal to zero, it does not couple to the corresponding gaugino field $\tilde{B}$. On the other hand, the coupling of the $SU(2)_L$ gauge fields $\boldsymbol{W}^\mu$ to the gaugino triplet $\tilde{\boldsymbol{W}}$ is given by (7.28). Because of the antisymmetry of the ϵ symbol,

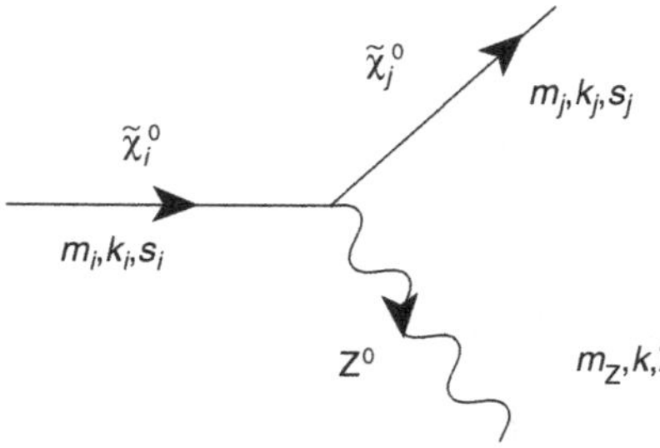

Figure 12.3 Lowest-order diagram for the decay $\tilde{\chi}_i^0 \to \tilde{\chi}_j^0 + Z^0$.

it is clear that W_3^μ couples only to $\tilde{W}^1$ and $\tilde{W}^2$, not to $\tilde{W}^0$. Hence the couplings of Z^0 to neutralinos arise only via their Higgsino components.

The $SU(2)_L \times U(1)_y$ gauge interactions of the two Higgsino doublets are given by the terms (in 2-component notation)

$$\begin{aligned} &\mathrm{i}(\tilde{H}_\mathrm{u}^{+\dagger}\tilde{H}_\mathrm{u}^{0\dagger})\bar{\sigma}^\mu\left(\mathrm{i}\frac{g}{2}\boldsymbol{\tau}\cdot\boldsymbol{W}_\mu + \mathrm{i}\frac{g'}{2}B_\mu\right)\begin{pmatrix}\tilde{H}_\mathrm{u}^+\\ \tilde{H}_\mathrm{u}^0\end{pmatrix} \\ &\quad + \mathrm{i}(\tilde{H}_\mathrm{d}^{0\dagger}\tilde{H}_\mathrm{d}^{-\dagger})\bar{\sigma}^\mu\left(\mathrm{i}\frac{g}{2}\boldsymbol{\tau}\cdot\boldsymbol{W}_\mu - \mathrm{i}\frac{g'}{2}B_\mu\right)\begin{pmatrix}\tilde{H}_\mathrm{d}^0\\ \tilde{H}_\mathrm{d}^-\end{pmatrix}. \end{aligned} \tag{12.56}$$

In converting to Majorana form via (2.120), we must remember that while the L-parts of the doublets

$$\begin{pmatrix}\Psi_\mathrm{M}^{\tilde{H}_\mathrm{u}^+}\\ \Psi_\mathrm{M}^{\tilde{H}_\mathrm{u}^0}\end{pmatrix}, \quad \begin{pmatrix}\Psi_\mathrm{M}^{\tilde{H}_\mathrm{d}^0}\\ \Psi_\mathrm{M}^{\tilde{H}_\mathrm{d}^-}\end{pmatrix} \tag{12.57}$$

transform as a **2**-dimensional representation of $SU(2)_L$, the R-parts transform as a $\bar{\mathbf{2}}$ (see for example (8.25)–(8.29)). The parts of (12.56) involving the neutral Higgsinos then become

$$-\frac{1}{4}(g^2 + g'^2)^{1/2}\left[\bar{\Psi}_\mathrm{M}^{\tilde{H}_\mathrm{u}^0}\gamma^\mu\gamma_5\Psi_\mathrm{M}^{\tilde{H}_\mathrm{u}^0} - \bar{\Psi}_\mathrm{M}^{\tilde{H}_\mathrm{d}^0}\gamma^\mu\gamma_5\Psi_\mathrm{M}^{\tilde{H}_\mathrm{d}^0}\right]Z_\mu. \tag{12.58}$$

Finally, converting to the neutralino mass-eigenstate fields and including the phase factors of (11.17), we obtain the interaction for $\tilde{\chi}_i^0 \to \tilde{\chi}_j^0 + Z^0$ as

$$W_{ij}\bar{\Psi}_\mathrm{M}^{\tilde{\chi}_j^0}\gamma^\mu(\gamma_5)^{\theta_{\tilde{\chi}_i^0}+\theta_{\tilde{\chi}_j^0}+1}\Psi_\mathrm{M}^{\tilde{\chi}_i^0}Z_\mu \tag{12.59}$$

where

$$W_{ij} = \frac{1}{4}(g^2 + g'^2)^{1/2}(-\mathrm{i})^{\theta_{\tilde{\chi}_j^0}}(\mathrm{i})^{\theta_{\tilde{\chi}_i^0}}\left(\mathrm{V}_{\tilde{H}_\mathrm{d}^0 i}\mathrm{V}_{\tilde{H}_\mathrm{d}^0 j} - \mathrm{V}_{\tilde{H}_\mathrm{u}^0 i}\mathrm{V}_{\tilde{H}_\mathrm{u}^0 j}\right). \tag{12.60}$$

We denote (see Figure 12.3) the mass, 4-momentum and spin of the decaying $\tilde{\chi}_i^0$ by $m_i(= m_{\tilde{\chi}_i^0})$, k_i and s_i, of the final $\tilde{\chi}_j^0$ by $m_j(= m_{\tilde{\chi}_j^0})$, k_j and s_j, and the mass,

4-momentum and polarization of the Z^0 by m_Z, k and λ. Evaluating the appropriate matrix element of (12.59), we obtain the decay amplitude

$$i(2\pi)^4\delta^4(k_i - k_j - k)W_{ij}\bar{u}(k_j, s_j)\gamma^\mu(\gamma_5)^{\theta_{\tilde{\chi}_i^0}+\theta_{\tilde{\chi}_j^0}+1}u(k_i, s_i)\epsilon_\mu(k, \lambda). \quad (12.61)$$

For the decay rate we need to evaluate the contraction

$$N \equiv N^{\mu\nu}P_{\mu\nu} \quad (12.62)$$

where (see, for example, equation (19.19) of [7])

$$P_{\mu\nu} = \sum_\lambda \epsilon_\mu(k, \lambda)\epsilon_\nu^*(k, \lambda) = -g_{\mu\nu} + k_\mu k_\nu/m_Z^2 \quad (12.63)$$

and

$$\begin{aligned} N^{\mu\nu} &= \frac{1}{2}\sum_{s_i,s_j}\left[\bar{u}_j\gamma^\mu(\gamma_5)^{\theta_{\tilde{\chi}_i^0}+\theta_{\tilde{\chi}_j^0}+1}u_i\bar{u}_i(-\gamma_5)^{\theta_{\tilde{\chi}_i^0}+\theta_{\tilde{\chi}_j^0}+1}\gamma^\nu u_j\right] \\ &= -\frac{1}{2}\mathrm{Tr}\left[(\not{k}_j + m_j)\gamma^\mu(-\not{k}_i + (-)^{\theta_{\tilde{\chi}_i^0}+\theta_{\tilde{\chi}_j^0}}m_i)\gamma^\nu\right] \\ &= 2\left(k_i^\mu k_j^\nu + k_i^\nu k_j^\mu - g^{\mu\nu}k_i \cdot k_j\right) - (-)^{\theta_{\tilde{\chi}_i^0}+\theta_{\tilde{\chi}_j^0}}2g^{\mu\nu}m_i m_j. \end{aligned} \quad (12.64)$$

Performing the contraction (12.62) yields the result

$$N = m_{\tilde{\chi}_i^0}^2 + m_{\tilde{\chi}_j^0}^2 - m_Z^2 + \frac{\left(m_{\tilde{\chi}_i^0}^2 - m_{\tilde{\chi}_j^0}^2\right)^2 - m_Z^4}{m_Z^2} + 6(-)^{\theta_{\tilde{\chi}_i^0}+\theta_{\tilde{\chi}_j^0}}m_{\tilde{\chi}_i^0}m_{\tilde{\chi}_j^0}. \quad (12.65)$$

The decay rate is then

$$\Gamma\left(\tilde{\chi}_i^0 \to \tilde{\chi}_j^0 + Z^0\right) = \frac{1}{8\pi m_{\tilde{\chi}_i^0}^2}|W_{ij}|^2 N k\left(m_{\tilde{\chi}_j^0}, m_Z, m_{\tilde{\chi}_i^0}\right). \quad (12.66)$$

This differs from formula (B.61b) of [49] by a factor of 4.

12.2 Sparticle production processes

12.2.1 Squark pair production in qq collisions

We begin by considering the process

$$q_1 q_2 \to \tilde{q}_{1L}\tilde{q}_{2R}, \quad (12.67)$$

where q_1 and q_2 are non-identical quarks, belonging in practice to the first or second generation. The relevant Feynman diagram is shown in Figure 12.4. The momenta, spins and colour labels of the quarks are p_1, s_1, c_1 and p_2, s_2, c_2, and the momenta and colour labels of the squarks are k_1, c_1' and k_2, c_2'. The interaction which produces

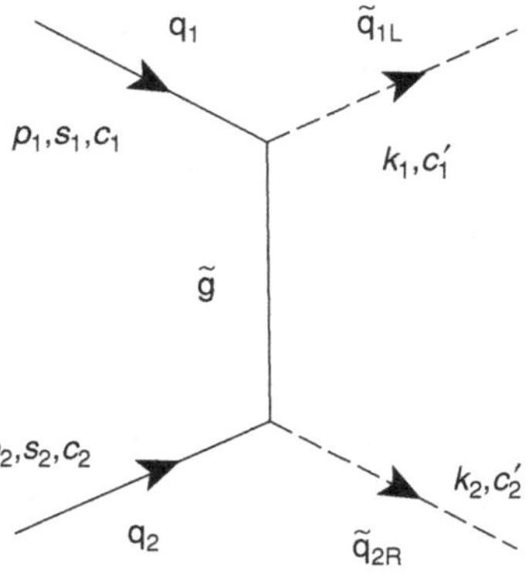

Figure 12.4 Lowest-order diagram for the process $q_1q_2 \to \tilde{q}_{1L}\tilde{q}_{2R}$.

the $\tilde{q}_{1L}$ is the hermitian conjugate of (12.15), namely

$$\mathcal{L}_{1L} = -\sqrt{2}g_s(-i)^{\theta_g}\tilde{q}_L^\dagger \bar{\Psi}_M^{\tilde{g}a} P_L \frac{1}{2}\lambda^a \Psi_q. \tag{12.68}$$

Similarly, $\tilde{q}_{2R}$ is produced by the hermitian conjugate of (12.22), namely

$$\mathcal{L}_{2R} = +\sqrt{2}g_s(i)^{\theta_g}\tilde{q}_R^\dagger \bar{\Psi}_M^{\tilde{g}a} P_R \frac{1}{2}\lambda^a \Psi_q. \tag{12.69}$$

The amplitude for Figure 12.4 is then

$$\left\langle \tilde{q}_{1L}, k_1, c_1'; \tilde{q}_{2R}, k_2, c_2' \right| \int d^4x d^4y T[i\mathcal{L}_{1L}(x) i\mathcal{L}_{2R}(y)] \left| q_1, p_1, s_1, c_1; q_2, p_2, s_2, c_2 \right\rangle. \tag{12.70}$$

After reducing the particles in the initial and final states, this becomes

$$\begin{aligned} 2g_s^2 \int d^4x d^4y\, e^{i(k_1\cdot x + k_2 \cdot y - p_1 \cdot x - p_2 \cdot y)} \omega^\dagger(c_1')\frac{\lambda^a}{2}\omega(c_1)\omega^\dagger(c_2')\frac{\lambda^b}{2}\omega(c_2) \\ \times \left\langle 0 \right| T\left[\bar{\Psi}_{M\alpha}^{\tilde{g}a}(x)\bar{\Psi}_{M\gamma}^{\tilde{g}b}(y)\right]\left| 0 \right\rangle P_{L\alpha\beta}u_\beta(p_1, s_1) P_{R\gamma\delta}u_\delta(p_2, s_2), \end{aligned} \tag{12.71}$$

where we have indicated the spinor indices explicitly. We use (2.144) to write the spinor part as

$$\begin{aligned} &\delta^{ab} C_{\alpha\epsilon}^{T} S_{F\epsilon\gamma}(x-y) P_{L\alpha\beta}u_\beta(p_1, s_1) P_{R\gamma\delta}u_\delta(p_2, s_2) \\ &\quad = \delta^{ab}\left(C P_L u(p_1, s_1)\right)^T S_F(x-y) P_R u(p_2, s_2). \end{aligned} \tag{12.72}$$

Now

$$(C P_L u)^T = u^T P_L C^T = \bar{v} C^T P_L C^T = \bar{v} P_L (C^T)^2 = -\bar{v} P_L, \tag{12.73}$$

where we have used $C\gamma_5 = \gamma_5 C$, and equations (2.133) and (2.142). Writing $S_F(x-y)$ in terms of its Fourier transform (2.138) allows us to perform the integrals over x and y, and we obtain the amplitude

$$\mathrm{i}(2\pi)^4\delta^4(p_1+p_2-k_1-k_2)\mathcal{M}_{\mathrm{q}\tilde{\mathrm{q}}}, \tag{12.74}$$

where

$$\mathcal{M}_{\mathrm{q}\tilde{\mathrm{q}}} = -2g_s^2\,\omega_{1'}^\dagger\frac{\lambda^a}{2}\omega_1\,\omega_{2'}^\dagger\frac{\lambda^a}{2}\omega_2\,\bar{v}(p_1,s_1)P_\mathrm{L}\frac{1}{\not{k}_1-\not{p}_1-m_{\tilde{\mathrm{g}}}}P_\mathrm{R}u(p_2,s_2). \tag{12.75}$$

For the cross section, we require the modulus squared of the amplitude, summed over final state colours and averaged over intial state colours and spins, which we denote by $\overline{|\mathcal{M}_{\mathrm{q}\tilde{\mathrm{q}}}|^2}$. For the colour part, we note that

$$\sum_{c_1,c_1'}\omega^\dagger(c_1')\frac{\lambda^a}{2}\omega(c_1)\omega^\dagger(c_1)\frac{\lambda^b}{2}\omega(c_1') = \frac{1}{4}\mathrm{Tr}\lambda^a\lambda^b = \frac{1}{2}\delta^{ab}, \tag{12.76}$$

and hence the colour factor is

$$\frac{1}{4}\frac{1}{9}\sum_{ab}\delta^{ab}\delta^{ab} = \frac{2}{9}. \tag{12.77}$$

The spinor factor is

$$\begin{aligned}&\frac{1}{4}\frac{1}{\left(\hat{t}-m_{\tilde{\mathrm{g}}}^2\right)^2}\mathrm{Tr}[\not{p}_1P_\mathrm{L}(\not{k}_1-\not{p}_1+m_{\tilde{\mathrm{g}}})P_\mathrm{R}\not{p}_2P_\mathrm{L}(\not{k}_1-\not{p}_1+m_{\tilde{\mathrm{g}}})P_\mathrm{R}]\\ &= \frac{1}{\left(\hat{t}-m_{\tilde{\mathrm{g}}}^2\right)^2}\frac{1}{4}\left[-\left(\hat{t}-m_{\tilde{\mathrm{q}}_\mathrm{L}}^2\right)\left(\hat{t}-m_{\tilde{\mathrm{q}}_\mathrm{R}}^2\right)-\hat{s}\hat{t}\right],\end{aligned} \tag{12.78}$$

where we are using the 'hatted' Mandelstam variables (conventional in parton kinematics) defined by

$$\begin{aligned}&\hat{s} = (p_1+p_2)^2 = (k_1+k_2)^2, \quad \hat{t} = (p_1-k_1)^2 = (p_2-k_2)^2,\\ &\hat{u} = (p_1-k_2)^2 = (p_2-k_1)^2.\end{aligned} \tag{12.79}$$

The differential cross section in the centre of mass system is given in terms of $\overline{|\mathcal{M}_{\mathrm{q}\tilde{\mathrm{q}}}|^2}$ by (see, for example, [15] Section 6.3.4)

$$\frac{\mathrm{d}\sigma}{\mathrm{d}\Omega} = \frac{1}{4p\hat{W}}\frac{p'}{16\pi^2\hat{W}^2}\overline{|\mathcal{M}_{\mathrm{q}\tilde{\mathrm{q}}}|^2}, \tag{12.80}$$

where $\hat{W} = \sqrt{\hat{s}}$, and p, p' are the magnitudes of the initial and final state momenta, in the CM system. The kinematics is simplified if we neglect the quark masses, and

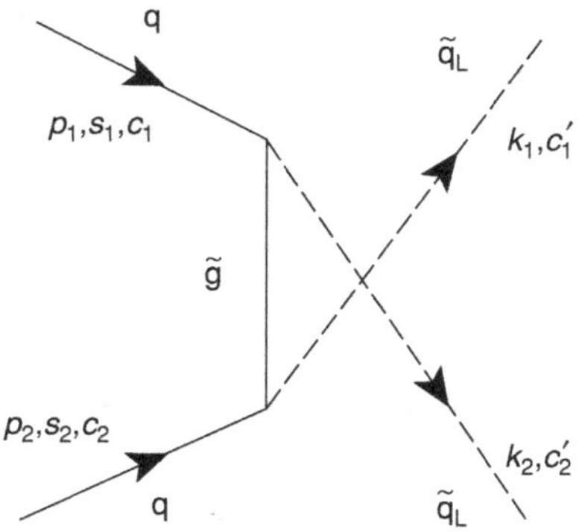

Figure 12.5 Lowest-order exchange diagram for the process $qq \to \tilde{q}_L\tilde{q}_L$.

assume the final squarks have equal mass. Then

$$\mathrm{d}\hat{t} = \hat{W}p'\mathrm{d}\cos\theta = \hat{W}p'\mathrm{d}\Omega/2\pi, \quad p = \hat{W}/2. \tag{12.81}$$

Putting all this together, we arrive at

$$\frac{\mathrm{d}\sigma}{\mathrm{d}\hat{t}}(q_1q_2 \to \tilde{q}_{1L}\tilde{q}_{2R}) = \frac{2\pi\alpha_s^2}{9\hat{s}^2}\frac{\left[-\left(\hat{t}-m_{\tilde{q}_L}^2\right)\left(\hat{t}-m_{\tilde{q}_R}^2\right)-\hat{s}\hat{t}\right]}{\left(\hat{t}-m_{\tilde{g}}^2\right)^2} \tag{12.82}$$

in agreement with formula (A.7d) of [49].

Exercise 12.2 Show that the cross section for $q_1q_2 \to \tilde{q}_{1L}\tilde{q}_{2L}$ is

$$\frac{\mathrm{d}\sigma}{\mathrm{d}\hat{t}}(q_1q_2 \to \tilde{q}_{1L}\tilde{q}_{2L}) = \frac{2\pi\alpha_s^2}{9}\frac{m_{\tilde{g}}^2\hat{s}}{\left(\hat{t}-m_{\tilde{g}}^2\right)^2} \tag{12.83}$$

in agreement with formula (A.7e) of [49].

A new complication arises when we consider the analogous calculations for the case in which the initial state quarks are identical, $q_1 = q_2$, for example

$$qq \to \tilde{q}_L\tilde{q}_L. \tag{12.84}$$

There is now a 'direct' amplitude of the form (12.75), corresponding to Figure 12.4, which is obtained straightforwardly as

$$\mathcal{M}_{q\tilde{q}}^{(d)} = 2g_s^2\,\omega_{1'}^\dagger\frac{\lambda^a}{2}\omega_1\,\omega_{2'}^\dagger\frac{\lambda^a}{2}\omega_2\,\bar{v}(p_1,s_1)P_L\frac{1}{\not{k}_1-\not{p}_1-m_{\tilde{g}}}P_L u(p_2,s_2)(-i)^{2\theta_{\tilde{g}}}. \tag{12.85}$$

In addition, to ensure antisymmetry for identical fermions, we must subtract from this the 'exchange' amplitude, corresponding to the diagram of Figure 12.5, given by the interchanges $p_1 \leftrightarrow p_2$, $s_1 \leftrightarrow s_2$, $c_1 \leftrightarrow c_2$, so that the total amplitude is $\mathcal{M}_{q\tilde{q}}^{(d)} + \mathcal{M}_{q\tilde{q}}^{(e)}$ where

$$\mathcal{M}_{q\tilde{q}}^{(e)} = -2g_s^2\, \omega_{1'}^{\dagger}\frac{\lambda^a}{2}\omega_2\, \omega_{2'}^{\dagger}\frac{\lambda^a}{2}\omega_1\, \bar{v}(p_2,s_2)P_{\mathrm{L}}\frac{1}{\not k_1 - \not p_2 - m_{\tilde{g}}}P_{\mathrm{L}}u(p_1,s_1)(-\mathrm{i})^{2\theta_{\tilde{g}}}. \tag{12.86}$$

This result can, of course, also be obtained by evaluating the matrix element of the relevant interaction. The cross section will therefore involve the sum of three terms: the square of the direct amplitude

$$|\mathcal{M}_{q\tilde{q}}^{(d)}|^2, \tag{12.87}$$

the square of the exchange amplitude

$$|\mathcal{M}_{q\tilde{q}}^{(e)}|^2, \tag{12.88}$$

and the interference term

$$2\,\mathrm{Re}\big[\mathcal{M}_{q\tilde{q}}^{(d)}\mathcal{M}_{q\tilde{q}}^{(e)*}\big]. \tag{12.89}$$

The part of the cross section arising from (12.87) is given by (12.83) as before, and the part from (12.88) is the same but with $\hat{t}$ replaced by $\hat{u}$; we must also remember to include an overall factor of $1/2$ due to identical particles in the final state.

The evaluation of the interference contribution is more involved. Consider first the colour factor, which (apart from the averaging factor $1/9$) is

$$\begin{aligned}&\sum_{c_1,c_2,c_1',c_2',a,b} \omega_{1'}^{\dagger}\frac{\lambda^a}{2}\omega_1\omega_1^{\dagger}\frac{\lambda^b}{2}\omega_{2'}\omega_{2'}^{\dagger}\frac{\lambda^a}{2}\omega_2\omega_2^{\dagger}\frac{\lambda^b}{2}\omega_{1'}\\ &= \frac{1}{16}\sum_{c_1',a,b}\omega_{1'}^{\dagger}\lambda^a\lambda^b\lambda^a\lambda^b\omega_{1'} = \frac{1}{16}\sum_{a,b}\mathrm{Tr}(\lambda^a\lambda^b\lambda^a\lambda^b).\end{aligned} \tag{12.90}$$

Now we have

$$\lambda^b\lambda^a = \lambda^a\lambda^b - 2\mathrm{i}f_{abc}\lambda^c, \tag{12.91}$$

where the coefficients f_{abc} are as usual the structure constants of SU(3). The part of (12.90) not involving the fs is then

$$\mathrm{Tr}\sum_a[(\lambda^a/2)(\lambda^a/2)]\sum_b[(\lambda^b/2)(\lambda^b/2)] = \mathrm{Tr}\left[\left(\frac{4}{3}\frac{4}{3}\right)I_3\right] = 16/3 \tag{12.92}$$

where I_3 is the unit 3×3 matrix. In (12.92) we have used the fact that $\sum_a[(\lambda^a/2)(\lambda^a/2)]$ is the Casimir operator C_2 of SU(3) (see [7], Section M.5), having the value $4/3\, I_3$ in the representation **3**. The remaining part of (12.90) is

$$\frac{2\mathrm{i}}{16}\sum_{a,b,c} f_{bac}\mathrm{Tr}(\lambda^b\lambda^a\lambda^c) = \frac{\mathrm{i}}{16}\sum_{a,b,c} f_{bac}\mathrm{Tr}([\lambda^b,\lambda^a]\lambda^c)$$
$$= -\frac{1}{8}\sum_{a,b,c,d} f_{bac}f_{bad}\mathrm{Tr}(\lambda^d\lambda^c) = -\frac{1}{4}\sum_{a,b,c} f_{bac}f_{bac}. \qquad (12.93)$$

To evaluate the product of fs, we note that the generators of SU(3) in the representation **8** are given by (see [7], equation (12.84))

$$\left(G_a^{(8)}\right)_{bc} = -\mathrm{i}f_{abc}, \qquad (12.94)$$

and that the value of C_2 in this representation is $3I_8$, where I_8 is the unit 8×8 matrix, so that

$$\sum_{a,b,c}\left(G_a^{(8)}\right)_{bc}\left(G_a^{(8)}\right)_{cb} = 24. \qquad (12.95)$$

Hence

$$\sum_{a,b,c} f_{abc}f_{abc} = 24, \qquad (12.96)$$

and expression (12.93) is equal to -6. Combining this result with (12.92), we find that (12.90) equals $-2/3$.

The spinor part of the interference term (12.89) is

$$-\bar{v}(p_1,s_1)P_{\mathrm{L}}\frac{1}{\not{k}_1-\not{p}_1-m_{\tilde{g}}}P_{\mathrm{L}}u(p_2,s_2)\bar{u}(p_1,s_1)P_{\mathrm{R}}\frac{1}{\not{k}_1-\not{p}_2-m_{\tilde{g}}}P_{\mathrm{R}}v(p_2,s_2), \qquad (12.97)$$

which has to be summed over s_1 and s_2 (with a factor of $1/4$ for the spin average). As it stands, (12.97) is not in a suitable form for using the standard Trace techniques: first of all, spinors referring to the same spin and momenta variables need to be adjacent to each other, and secondly we need expressions of the form $u\bar{u}$ and $v\bar{v}$, not $\bar{v}\bar{u}$ and uv. To deal with the first difficulty, we write out (12.97) including all the spinor labels explicitly, and rearrange it as

$$-\bar{v}_\alpha(p_1,s_1)\bar{u}_\lambda(p_1,s_1)(P_{\mathrm{R}})_{\lambda\mu}\left(\frac{1}{\not{k}_1-\not{p}_2-m_{\tilde{g}}}\right)_{\mu\nu}(P_{\mathrm{R}})_{\nu\tau}v_\tau(p_2,s_2)$$
$$\times\, u_\delta(p_2,s_2)\left(P_{\mathrm{L}}^{\mathrm{T}}\right)_{\delta\gamma}\left[\left(\frac{1}{\not{k}_1-\not{p}_1-m_{\tilde{g}}}\right)^{\mathrm{T}}\right]_{\gamma\beta}\left(P_{\mathrm{L}}^{\mathrm{T}}\right)_{\beta\alpha}, \qquad (12.98)$$

where ‘T’ denotes the transpose. We now use

$$\bar{v}_\alpha(p_1,s_1) = C^{\mathrm{T}}_{\alpha\sigma}u_\sigma(p_1,s_1), \quad \text{and} \quad v_\tau(p_2,s_2) = C_{\tau\eta}\bar{u}_\eta(p_2,s_2) \qquad (12.99)$$

from (2.133), which enables the spin-averaged expression to be written as

$$-\frac{1}{4}\mathrm{Tr}\left\{C^{\mathrm{T}}\not{p}_1 P_{\mathrm{R}}\frac{1}{\not{k}_1-\not{p}_2-m_{\tilde{\mathrm{g}}}}P_{\mathrm{R}}C\not{p}_2{}^{\mathrm{T}}P_{\mathrm{L}}\frac{1}{(\not{k}_1-\not{p}_1-m_{\tilde{\mathrm{g}}})^{\mathrm{T}}}P_{\mathrm{L}}\right\} \tag{12.100}$$

where we have used $\gamma_5^{\mathrm{T}}=\gamma_5$, and taken the quarks to be massless. We now note that

$$C\gamma_\mu^{\mathrm{T}}=-\gamma_\mu C \quad \text{and} \quad CC^{\mathrm{T}}=1, \tag{12.101}$$

so that (12.100) becomes

$$\begin{aligned}&\frac{-1}{4(\hat{u}-m_{\tilde{\mathrm{g}}}^2)(\hat{t}-m_{\tilde{\mathrm{g}}}^2)}\mathrm{Tr}[\not{p}_1 P_{\mathrm{R}}(\not{k}_1-\not{p}_2+m_{\tilde{\mathrm{g}}})P_{\mathrm{R}}\not{p}_2 P_{\mathrm{L}}(\not{k}_1-\not{p}_1-m_{\tilde{\mathrm{g}}})P_{\mathrm{L}}]\\ &=\frac{m_{\tilde{\mathrm{g}}}^2}{4(\hat{u}-m_{\tilde{\mathrm{g}}}^2)(\hat{t}-m_{\tilde{\mathrm{g}}}^2)}\mathrm{Tr}[\not{p}_1\not{p}_2 P_{\mathrm{L}}]=\frac{m_{\tilde{\mathrm{g}}}^2\hat{s}}{4(\hat{u}-m_{\tilde{\mathrm{g}}}^2)(\hat{t}-m_{\tilde{\mathrm{g}}}^2)}.\end{aligned} \tag{12.102}$$

Remembering now the factor of 2 in (12.89), the result (12.83) (and the corresponding one with $\hat{t}$ replaced by $\hat{u}$), and the overall factor of 1/2, we find that the cross section for $\mathrm{qq}\to\tilde{\mathrm{q}}_{\mathrm{L}}\tilde{\mathrm{q}}_{\mathrm{L}}$ is

$$\frac{\mathrm{d}\sigma}{\mathrm{d}\hat{t}}(\mathrm{qq}\to\tilde{\mathrm{q}}_{\mathrm{L}}\tilde{\mathrm{q}}_{\mathrm{L}})=\frac{\pi\alpha_{\mathrm{s}}^2 m_{\tilde{\mathrm{g}}}^2}{9\hat{s}}\left[\frac{1}{(\hat{t}-m_{\tilde{\mathrm{g}}}^2)^2}+\frac{1}{(\hat{u}-m_{\tilde{\mathrm{g}}}^2)^2}-\frac{2/3}{(\hat{t}-m_{\tilde{\mathrm{g}}}^2)(\hat{u}-m_{\tilde{\mathrm{g}}}^2)}\right] \tag{12.103}$$

in agreement with formula (A.7i) of [49]. Similar manipulations are presented in Appendix E of [45].

Exercise 12.3 Show that the interference term vanishes (in the limit of vanishing quark mass) for the case $\mathrm{qq}\to\tilde{\mathrm{q}}_{\mathrm{L}}\tilde{\mathrm{q}}_{\mathrm{R}}$, and hence that the cross section is

$$\begin{aligned}&\frac{\mathrm{d}\sigma}{\mathrm{d}\hat{t}}(\mathrm{qq}\to\tilde{\mathrm{q}}_{\mathrm{L}}\tilde{\mathrm{q}}_{\mathrm{R}})\\ &=\frac{2\pi\alpha_{\mathrm{s}}^2}{9\hat{s}^2}\left[\frac{[-(\hat{t}-m_{\tilde{\mathrm{q}}_{\mathrm{L}}}^2)(\hat{t}-m_{\tilde{\mathrm{q}}_{\mathrm{R}}}^2)-\hat{s}\hat{t}]}{(\hat{t}-m_{\tilde{\mathrm{g}}}^2)^2}+\frac{[-(\hat{u}-m_{\tilde{\mathrm{q}}_{\mathrm{L}}}^2)(\hat{u}-m_{\tilde{\mathrm{q}}_{\mathrm{R}}}^2)-\hat{s}\hat{u}]}{(\hat{u}-m_{\tilde{\mathrm{g}}}^2)^2}\right]\end{aligned} \tag{12.104}$$

in agreement with formula (A.7j) of [49].

The expressions for the cross sections of squark (or gluino) production in qq collisions are very similar to those obtained from standard QCD tree graphs (see, for example, [7] Section 14.3); the main qualitative difference is that the propagator factor $\hat{t}^{-2}$ for the massless gluon is replaced by $(\hat{t}-m_{\tilde{\mathrm{g}}}^2)^{-2}$ for the massive gluino (and similarly for the $\hat{u}$-channel terms). For an order of magnitude estimate, we

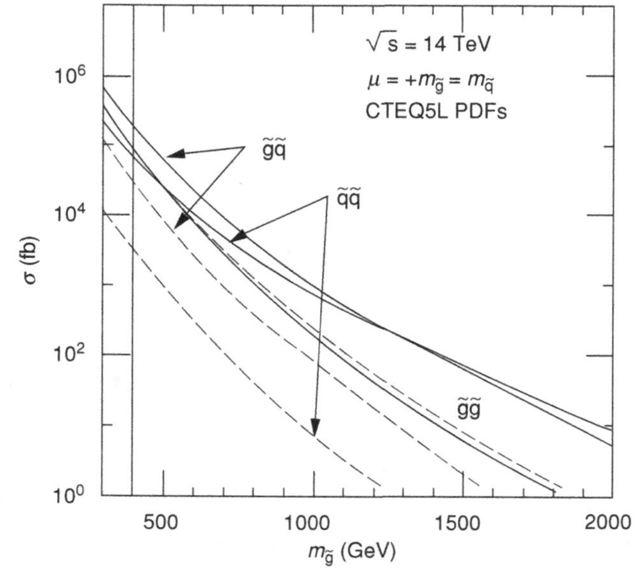

Figure 12.6 Cross sections for squark and gluino production at the CERN LHC pp collider for $m_{\tilde{q}} = m_{\tilde{g}}$ (solid) and for $m_{\tilde{q}} = 2m_{\tilde{g}}$ (dashed). [Figure reprinted with permission from *Weak Scale Supersymmetry* by H. Baer and X. Tata (Cambridge: Cambridge University Press, 2006), p. 318.]

may set

$$\sigma \sim \frac{2\pi\alpha_s^2}{9\hat{s}} \sim 250\,\text{fb} \tag{12.105}$$

for $\alpha_s \sim 0.15$ and $\sqrt{\hat{s}} = 5$ TeV. The initial state quarks are, of course, constituents of hadrons, and so these parton-level cross sections must be convoluted with appropriate parton distribution functions to obtain the cross sections for physical production processes in hadron–hadron collisions; see, for example, [118]. As an illustration of the predictions, we show in Figure 12.6 (taken from [49]) the cross sections for squark and gluino production at the CERN LHC pp collider. For $m_{\tilde{g}}$ and $m_{\tilde{q}}$ less than about 1 TeV, and an integrated luminosity of 10 fb^{-1}, one expects some thousands of $\tilde{g}$, $\tilde{q}$ events at the LHC.

We now turn to sparticle production via electroweak interactions.

12.2.2 Slepton and sneutrino pair production in $q\bar{q}$ *collisions*

We consider first the production of a charged slepton in association with its sneutrino partner

$$d\bar{u} \to \tilde{l}_L \bar{\tilde{\nu}}_L, \tag{12.106}$$

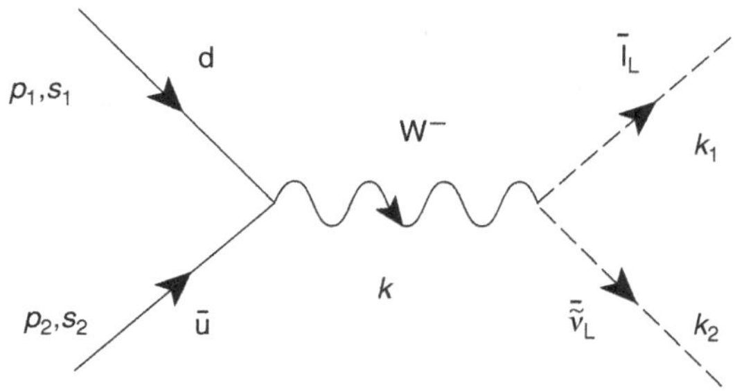

Figure 12.7 Lowest-order diagram for the process $\mathrm{d}\bar{\mathrm{u}} \to \tilde{\mathrm{l}}_{\mathrm{L}}\bar{\tilde{\nu}}_{\mathrm{L}}$.

proceeding via W^--exchange in the $\hat{s}$-channel, as shown in Figure 12.7. The 4-momenta and spins of the d and ū quarks are p_1, s_1 and p_2, s_2, and the 4-momenta of the slepton and sneutrino are k_1 and k_2; colour labels are suppressed. The SM interaction at the first vertex is contained in the quark analogue of (8.34), and is

$$\mathcal{L}_{\mathrm{q\,W}} = -\frac{g}{\sqrt{2}} V_{\mathrm{ud}} \bar{\Psi}_{\mathrm{u\,L}} \gamma^{\mu} \Psi_{\mathrm{d\,L}} W_{\mu}, \tag{12.107}$$

where (see, for example, [7] equation (22.25))

$$W_{\mu} = (W_{1\mu} - \mathrm{i}W_{2\mu})/\sqrt{2} \tag{12.108}$$

is the field which destroys the W^+ or creates the W^-, and V_{ud} is the appropriate element of the CKM matrix. The interaction at the sparticle vertex is contained in the SU(2) gauge invariant kinetic term (see (7.67))

$$(D_{\mu}\phi_{\tilde{\mathrm{l}}_{\mathrm{L}}})^{\dagger}(D^{\mu}\phi_{\tilde{\mathrm{l}}_{\mathrm{L}}}) \tag{12.109}$$

where

$$\phi_{\tilde{\mathrm{l}}_{\mathrm{L}}} = \begin{pmatrix} \tilde{\nu}_{\mathrm{L}} \\ \tilde{l}_{\mathrm{L}} \end{pmatrix}, \quad \text{and} \quad D_{\mu} = \partial_{\mu} + \mathrm{i}g\frac{\boldsymbol{\tau}}{2}\cdot \boldsymbol{W}_{\mu}. \tag{12.110}$$

The relevant term is

$$\mathcal{L}_{\mathrm{W\,sl}} = -\frac{g}{\sqrt{2}}\mathrm{i}(\tilde{l}_{\mathrm{L}}^{\dagger}\partial^{\nu}\tilde{\nu}_{\mathrm{L}} - (\partial^{\nu}\tilde{l}_{\mathrm{L}}^{\dagger})\tilde{\nu}_{\mathrm{L}})W_{\nu}^{\dagger}. \tag{12.111}$$

Note that (12.111) is the same as the hermitian conjugate of (12.107) with the fermionic current replaced by the corresponding bosonic one, and with $V_{\mathrm{ud}} \to 1$.

The amplitude for Figure 12.7 is then

$$\langle \tilde{\mathrm{l}}_{\mathrm{L}}, k_1; \bar{\tilde{\nu}}_{\mathrm{L}}, k_2 | \int \mathrm{d}^4x\mathrm{d}^4y T\left[\mathrm{i}\mathcal{L}_{\mathrm{q\,W}}(x)\mathrm{i}\mathcal{L}_{\mathrm{W\,sl}}(y)\right] | \mathrm{d}, p_1, s_1; \bar{\mathrm{u}}, p_2, s_2\rangle. \tag{12.112}$$

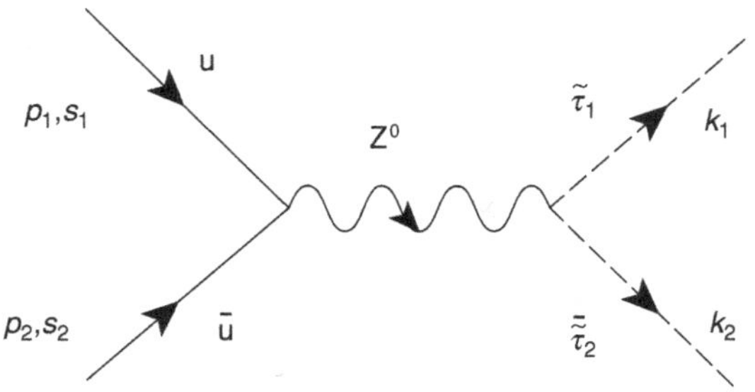

Figure 12.8 Z^0 exchange diagram for the process $u\bar{u} \to \tilde{\tau}_1 \bar{\tilde{\tau}}_2$.

The reduction of the final state sleptons leads to a factor $i(k_2^\nu - k_1^\nu)$ from the derivatives in $\mathcal{L}_{W\,sl}$, and (12.112) becomes

$$(2\pi)^4 \delta^4(p_1 + p_2 - k_1 - k_2) i\mathcal{M}_{ch\,sl}, \qquad (12.113)$$

where

$$\mathcal{M}_{ch\,sl} = \frac{g^2}{2} V_{ud} \bar{v}(p_2, s_2)\gamma^\mu P_L u(p_1, s_1) \left(\frac{-g_{\mu\nu} + k_\mu k_\nu / m_W^2}{\hat{s} - m_W^2 + i m_W \Gamma_W} \right) (k_2^\nu - k_1^\nu) \qquad (12.114)$$

and $k = k_1 + k_2 = p_1 + p_2$. The term

$$\bar{v}(p_2, s_2)\gamma^\mu P_L u(p_1, s_1) k_\mu = \bar{v}(p_2, s_2)(\not{p}_1 + \not{p}_2) P_L u(p_1, s_1) \qquad (12.115)$$

vanishes in the massless quark limit. The spinor factor in the cross section is

$$\begin{aligned} \frac{1}{12}\mathrm{Tr}[\not{p}_2(\not{k}_2 - \not{k}_1)P_L \not{p}_1 P_R(\not{k}_2 - \not{k}_1)] &= \frac{1}{24}\mathrm{Tr}[\not{p}_2(\not{k}_2 - \not{k}_1)\not{p}_1(\not{k}_2 - \not{k}_1)] \\ &= \frac{1}{3}\left(\hat{t}\hat{u} - m_{\tilde{l}}^2 m_{\tilde{\nu}}^2\right], \end{aligned} \qquad (12.116)$$

where $\hat{t} = (p_1 - k_1)^2$, $\hat{u} = (p_1 - k_2)^2$, and the factor of $1/3$ comes from the colour average. The cross section is then

$$\frac{d\sigma}{d\hat{t}}(d\bar{u} \to \tilde{l}_L \bar{\tilde{\nu}}_L) = \frac{g^4 |V_{ud}|^2}{192\pi \hat{s}^2} \frac{1}{\left(\hat{s} - m_W^2\right)^2 + m_W^2 \Gamma_W^2} \left(\hat{t}\hat{u} - m_{\tilde{l}_L}^2 m_{\tilde{\nu}_L}^2\right) \qquad (12.117)$$

in agreement with equation (A.14) of [49], setting V_{ud} to unity.

An analogous neutral current process is

$$u\bar{u} \to \tilde{\tau}_1 \bar{\tilde{\tau}}_2 \qquad (12.118)$$

proceeding via Z^0 exchange in the $\hat{s}$ channel, as shown in Figure 12.8. We take the 4-momenta and spins of the u and ū to be p_1, s_1 and p_2, s_2, and the 4-momenta of the $\tilde{\tau}_1$ and $\bar{\tilde{\tau}}_2$ to be k_1, k_2. The SM interaction at the quark vertex is[1]

$$\mathcal{L}_{\mathrm{uZ}} = -\bar{\Psi}_{\mathrm{u}}[\alpha_{\mathrm{u}}\gamma^\mu + \beta_{\mathrm{u}}\gamma^\mu\gamma_5]\Psi_{\mathrm{u}}Z_\mu \tag{12.119}$$

where (see, for example, [7] equations (22.53) and (22.54))

$$\alpha_{\mathrm{u}} = \frac{g}{12\cos\theta_{\mathrm{W}}}(3 - 8\sin^2\theta_{\mathrm{W}}), \quad \beta_{\mathrm{u}} = -\frac{g}{4\cos\theta_{\mathrm{W}}}. \tag{12.120}$$

At the stau vertex, we need the SU(2) × U(1) gauge invariant kinetic term for the L-doublet:

$$\phi_{\tilde{\tau}} = \begin{pmatrix} \tilde{\nu}_{\tau\mathrm{L}} \\ \tilde{\tau}_{\mathrm{L}} \end{pmatrix} \quad \text{with} \quad D_\mu = \partial_\mu + \mathrm{i}g\frac{\boldsymbol{\tau}}{2}\cdot\boldsymbol{W}_\mu - \frac{1}{2}g'B_\mu, \tag{12.121}$$

and also for the R-singlet $\tilde{\tau}_{\mathrm{R}}$ with

$$D_\mu = \partial_\mu - \mathrm{i}g'B_\mu. \tag{12.122}$$

The $\tilde{\tau}_{\mathrm{L}}$ interaction comes out to be

$$-j^\mu_{\mathrm{NC}}(\tilde{\tau}_{\mathrm{L}})Z_\mu, \tag{12.123}$$

where

$$j^\mu_{\mathrm{NC}}(\tilde{\tau}_{\mathrm{L}}) = g^\tau_{\mathrm{L}}(\mathrm{i}\tilde{\tau}^\dagger_{\mathrm{L}}\partial^\mu\tilde{\tau}_{\mathrm{L}} + \text{h.c.}) \tag{12.124}$$

and

$$g^\tau_{\mathrm{L}} = \frac{g}{\cos\theta_{\mathrm{W}}}(-1/2 + \sin^2\theta_{\mathrm{W}}). \tag{12.125}$$

Note that the coupling strength (12.125) is exactly the same as the one for τs in the SM (see, for example, equations (22.38) and (22.39) of [7]); once again, the bosonic current here replaces the fermionic one. Similarly, the $\tilde{\tau}_{\mathrm{R}}$ interaction is

$$-j^\mu_{\mathrm{NC}}(\tilde{\tau}_{\mathrm{R}})Z_\mu \tag{12.126}$$

where

$$j^\mu_{\mathrm{NC}}(\tilde{\tau}_{\mathrm{R}}) = g^\tau_{\mathrm{R}}(\mathrm{i}\tilde{\tau}^\dagger_{\mathrm{R}}\partial^\mu\tilde{\tau}_{\mathrm{R}} + \text{h.c.}) \tag{12.127}$$

and (cf. equation (22.40) of [7])

$$g^\tau_{\mathrm{R}} = \frac{g}{\cos\theta_{\mathrm{W}}}\sin^2\theta_{\mathrm{W}}. \tag{12.128}$$

[1] Our notation here is slightly different from that of [49]: they write $e\alpha_{\mathrm{u}}$ and $e\beta_{\mathrm{u}}$ for our α_{u} and β_{u}, and our Z_μ field is the negative of theirs.

We now convert (12.123) and (12.126) to the mass-basis fields $\tilde{\tau}_1$ and $\tilde{\tau}_2$ using the analogue of (11.62) with $\theta_{\rm t}$ replaced by θ_τ. The required interaction is then

$$\mathcal{L}_{Z\tilde{\tau}} = {\rm i}\beta_\tau \sin 2\theta_\tau (\tilde{\tau}_2^\dagger \partial^\nu \tilde{\tau}_1 + \tilde{\tau}_1^\dagger \partial^\nu \tilde{\tau}_2) Z_\nu + {\rm h.c.}, \tag{12.129}$$

where

$$\beta_\tau = \frac{g}{4\cos\theta_{\rm W}}. \tag{12.130}$$

Proceeding as before, the amplitude for the process (12.118) is

$$(2\pi)^4 \delta^4(p_1 + p_2 - k_1 - k_2) {\rm i}\mathcal{M}_{{\rm u}Z\tilde{\tau}}, \tag{12.131}$$

where

$$\mathcal{M}_{{\rm u}Z\tilde{\tau}} = \beta_\tau \sin 2\theta_\tau \bar{v}(p_2, s_2)\gamma^\mu(\alpha_{\rm u} + \beta_{\rm u}\gamma_5)u(p_1, s_1)\frac{Ak_{1\mu} + Bk_{2\mu}}{\hat{s} - m_Z^2 + {\rm i}m_Z\Gamma_Z} \tag{12.132}$$

and

$$A = \left(m_{\tilde{\tau}_1}^2 - m_{\tilde{\tau}_2}^2 - m_Z^2\right)/m_Z^2, \quad B = \left(m_Z^2 + m_{\tilde{\tau}_1}^2 - m_{\tilde{\tau}_2}^2\right)/m_Z^2. \tag{12.133}$$

In the limit of massless quarks, we find

$$\bar{v}(p_2, s_2)\gamma^\mu(\alpha_{\rm u}+\beta_{\rm u}\gamma_5)u(p_1, s_1)(Ak_{1\mu}+Bk_{2\mu}) = 2\bar{v}(p_2, s_2)\not{k}_2(\alpha_{\rm u}+\beta_{\rm u}\gamma_5)u(p_1, s_1). \tag{12.134}$$

The resulting spinor factor in the cross section is

$$\frac{1}{12}{\rm Tr}[\not{p}_2\not{k}_2(\alpha_{\rm u} + \beta_{\rm u}\gamma_5)\not{p}_1(\alpha_{\rm u} - \beta_{\rm u}\gamma_5)\not{k}_2] = \frac{1}{6}\left(\alpha_{\rm u}^2 + \beta_{\rm u}^2\right)\left(\hat{t}\hat{u} - m_{\tilde{\tau}_1}^2 m_{\tilde{\tau}_2}^2\right), \tag{12.135}$$

and we obtain

$$\frac{{\rm d}\sigma}{{\rm d}\hat{t}}({\rm u}\bar{\rm u} \to \tilde{\tau}_1\bar{\tilde{\tau}}_2) = \frac{1}{24\pi\hat{s}^2}\beta_\tau^2 \sin^2 2\theta_\tau \left(\alpha_{\rm u}^2 + \beta_{\rm u}^2\right)\frac{1}{\left(\hat{s} - m_Z^2\right)^2 + m_Z^2\Gamma_Z^2}\left(\hat{t}\hat{u} - m_{\tilde{\tau}_1}^2 m_{\tilde{\tau}_2}^2\right) \tag{12.136}$$

in agreement with (A.15b) of [49].

12.2.3 *Stop and stau pair production in* e^+e^- *collisions*

Hadron colliders are suitable for broad searches for new physics, by virtue of their high beam energy and relatively large sparticle production cross-sections. However, if sparticles are found at the LHC (for example), precision studies of their properties will be best undertaken with a TeV scale e^+e^- collider, operating with polarizable

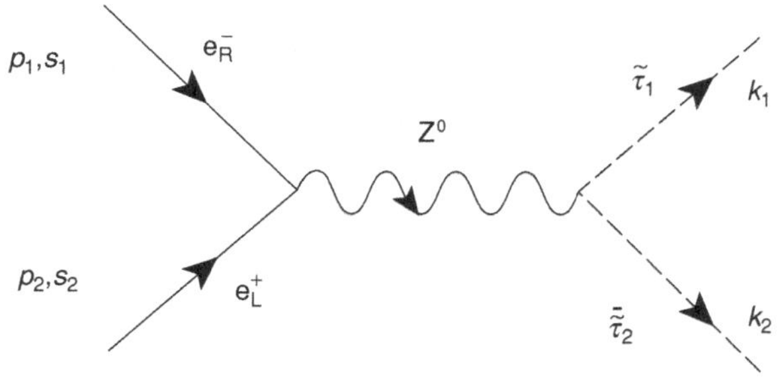

Figure 12.9 Z^0 exchange diagram for the process $e_L^+ e_R^- \to \tilde{\tau}_1 \bar{\tilde{\tau}}_2$.

beams. Here we shall consider just two simple processes:

$$e_L^+ e_R^- \to \tilde{\tau}_1 \bar{\tilde{\tau}}_2 \tag{12.137}$$

and

$$e_L^+ e_R^- \to \tilde{t}_1 \bar{\tilde{t}}_2. \tag{12.138}$$

Figure 12.9 shows the Feynman diagram for (12.137). We take the 4-momentum and spin of the e^- to be p_1, s_1, and those of the e^+ to be p_2, s_2; the 4-momenta of the $\tilde{\tau}_1$ and $\bar{\tilde{\tau}}_2$ are k_1 and k_2. The SM interaction at the electron vertex is

$$\mathcal{L}_{eZ} = -\bar{\Psi}_e[\alpha_e \gamma^\mu + \beta_e \gamma^\mu \gamma_5]\Psi_e Z_\mu \tag{12.139}$$

where (see, for example, (22.41) and (22.42) of [7])

$$\alpha_e = \frac{g}{4\cos\theta_W}(4\sin^2\theta_W - 1), \quad \beta_e = \frac{g}{4\cos\theta_W}. \tag{12.140}$$

The stau vertex has been given in (12.129). The amplitude is

$$(2\pi)^4 \delta^4(p_1 + p_2 - k_1 - k_2) \mathrm{i}\mathcal{M}_{eZ\tilde{\tau}}, \tag{12.141}$$

where

$$\mathcal{M}_{eZ\tilde{\tau}} = \beta_\tau \sin 2\theta_\tau \bar{v}(p_2, s_2)\gamma^\mu(\alpha_e + \beta_e\gamma_5)P_R u(p_1, s_1)\frac{Ak_{1\mu} + Bk_{2\mu}}{s - m_Z^2 + \mathrm{i}m_Z\Gamma_Z}, \tag{12.142}$$

where A and B are given in (12.133), and $s = (p_1 + p_2)^2$. The projection operator P_R has been inserted before the initial state electron spinor; this selects out the R polarization state in the limit in which the electron mass is neglected. In this limit, and using $\gamma_5 P_R = P_R$, we find (as in (12.134))

$$\mathcal{M}_{eZ\tilde{\tau}} = 2(\alpha_e + \beta_e)\beta_\tau \sin 2\theta_\tau \bar{v}(p_2, s_2)\not{k}_2 P_R u(p_1, s_1)\frac{1}{s - m_Z^2 + \mathrm{i}m_Z\Gamma_Z}. \tag{12.143}$$

For the cross section, we now formally sum over initial spins rather than average them: the operator $P_{\rm R}$ eliminates the unwanted states. The Trace factor is then

$$\mathrm{Tr}[\not p_2 \not k_2 P_{\rm R} \not p_1 P_{\rm L} \not k_2] = \left(tu - m^2_{\tilde\tau_1} m^2_{\tilde\tau_2}\right) \tag{12.144}$$

where $t = (p_1 - k_1)^2$ and $u = (p_1 - k_2)^2$, and the cross section is

$$\frac{\mathrm{d}\sigma}{\mathrm{d}t}(\mathrm{e}^+_{\rm L}\mathrm{e}^-_{\rm R} \to \tilde\tau_1 \bar{\tilde\tau}_2) = \frac{\pi\alpha^2}{4\hat s^2} \frac{\sin^2 2\theta_\tau}{\cos^4\theta_{\rm W}} \frac{1}{\left(s - m^2_{\rm Z}\right)^2 + m^2_{\rm Z}\Gamma^2_{\rm Z}} \left(tu - m^2_{\tilde\tau_1} m^2_{\tilde\tau_2}\right). \tag{12.145}$$

In this case, the order of magnitude of the cross section is

$$\sigma \sim \alpha^2/s \sim 20\,\mathrm{fb} \tag{12.146}$$

for $\sqrt{s} = 1$ TeV.

Exercise 12.4 Stop pair production in $\mathrm{e}^+\mathrm{e}^-$ collisions

Show that the Z-stop interaction is

$$\mathcal{L}_{\rm Z\tilde t} = \mathrm{i}\beta_{\rm t} \sin 2\theta_{\rm t} (\tilde{\rm t}_1^\dagger \partial^\nu \tilde{\rm t}_2 + \tilde{\rm t}_2^\dagger \partial^\nu \tilde{\rm t}_1) Z_\nu + \mathrm{h.c.} \tag{12.147}$$

where

$$\beta_{\rm t} = -\frac{g}{4\cos\theta_{\rm W}}, \tag{12.148}$$

and hence that the cross section is

$$\frac{\mathrm{d}\sigma}{\mathrm{d}t}(\mathrm{e}^+_{\rm L}\mathrm{e}^-_{\rm R} \to \tilde{\rm t}_1 \bar{\tilde{\rm t}}_2) = \frac{3\pi\alpha^2}{4s^2} \frac{\sin^2 2\theta_{\rm t}}{\cos^2\theta_{\rm W}} \frac{1}{\left(s - m^2_{\rm Z}\right)^2 + m^2_{\rm Z}\Gamma^2_{\rm Z}} \left(tu - m^2_{\tilde\tau_1} m^2_{\tilde\tau_2}\right). \tag{12.149}$$

This agrees with (A.21c) of [49] after transforming the variable t to $z = \cos\theta$, where θ is the angle between the initial e^- and the final $\tilde{\rm t}_1$ in the centre of mass system (see (12.81)).

12.3 Signatures and searches

Our introductory treatment would be incomplete without even a brief discussion of the extensive experimental searches for sparticles which have been made, and of some characteristic signatures by which sparticle production events might be distinguished from backgrounds due to SM processes.

(i) **Gluinos**

A useful signature for gluino pair ($\tilde g \tilde g$) production is the *like-sign dilepton signal* [119–121]. This arises if the gluino decays with a significant branching ratio to hadrons plus a chargino, which then decays to lepton $+ \nu + \tilde\chi^0_1$. Since the gluino is indifferent to electric charge, the single lepton from each $\tilde g$ decay will carry either charge with equal probability. Hence many events should contain two like-sign leptons (plus jets plus $\not{E}_T$). This has a low SM background, because in the SM isolated lepton pairs come from W^+W^-, Drell–Yan or $t\bar{t}$ production, all of which give opposite sign dileptons. Like-sign dilepton events can also arise from $\tilde{g}\tilde{q}$ and $\tilde{q}\bar{\tilde{q}}$ production.

Collider Detector at Fermilab (CDF) [122] reported no candidate events for like-sign dilepton pairs. Other searches based simply on dileptons (not required to be like-sign) plus two jets plus $\not{E}_T$ [123, 124] reported no sign of any excess events. Results were expressed in terms of exclusion contours for mSUGRA parameters.

(ii) **Neutralinos and charginos**

As an illustration of possible signatures for neutralino and chargino production (at hadron colliders, for example), we mention the *trilepton signal* [125–130], which arises from the production

$$p\bar{p} \text{ (or pp)} \to \tilde{\chi}_1^{\pm}\tilde{\chi}_2^0 + X \tag{12.150}$$

followed by the decays

$$\tilde{\chi}_1^{\pm} \to l'^{\pm}\nu\tilde{\chi}_1^0 \tag{12.151}$$

$$\tilde{\chi}_2^0 \to l\bar{l}\tilde{\chi}_1^0. \tag{12.152}$$

Here the two LSPs in the final state carry away $2m_{\tilde{\chi}_1^0}$ of missing energy, which is observed as missing transverse energy, $\not{E}_T$ (see Section 8.4). In addition, there are three energetic, isolated leptons, and little jet activity. The expected SM background is small. Using the data sample collected from the 1992–3 run of the Fermilab Tevatron, D0 [131] and CDF [132] reported no candidate trilepton events after applying all selection criteria; the expected background was roughly 2 ± 1 events. Upper limits on the product of the cross section times the branching ratio (single trilepton mode) were set, for various regions in the space of MSSM parameters. Later searches using the data sample from the 1994–5 run [133, 134] were similarly negative.

(iii) **Squarks and sleptons**

At e^+e^- colliders the $\tilde{t}_1$ pair production cross section depends on the mixing angle θ_t; for example, the contribution from Z exchange actually vanishes when $\cos^2\theta_{\tilde{t}} = \frac{4}{3}\sin^2\theta_W$ [135]. In contrast, $\tilde{t}_1$'s are pair-produced in hadron colliders with no mixing-angle dependence. Which decay modes of the $\tilde{t}_1$ dominate depends on the masses of charginos and sleptons. For example, if $m_{\tilde{t}_1}$ lies below all chargino and slepton masses, then the dominant decay is

$$\tilde{t}_1 \to c + \tilde{\chi}_1^0, \tag{12.153}$$

which proceeds through loops (a FCNC transition). If $m_{\tilde{t}_1} > m_{\tilde{\chi}^{\pm}}$,

$$\tilde{t}_1 \to b + \tilde{\chi}^{\pm} \tag{12.154}$$

is the main mode, with $\tilde{\chi}^{\pm}$ then decaying to $l\nu\tilde{\chi}_1^0$. D0 reported on a search for such light stops [136]; their signal was two acollinear jets plus $\not{E}_T$ (they did not attempt to identify flavour). Improved bounds on the mass of the lighter stop were obtained by CDF [137] using a vertex detector to tag c- and b-quark jets. More recent searches are reported in [138] and [139]. The bounds depend sensitively on the (assumed) mass of the neutralino $\tilde{\chi}_1^0$; data is presented in the form of excluded regions in a $m_{\tilde{\chi}_1^0} - m_{\tilde{t}_1}$ plot.

The search for a light $\tilde{b}_1$ decaying to $b + \tilde{\chi}_1^0$ is similar to that for $\tilde{t}_1 \to c + \tilde{\chi}_1^0$. D0 [140] tagged b-jets through semi-leptonic decays to muons. They observed five candidate events consistent with the final state $b\bar{b} + \not{E}_T$, as compared to an estimated background of 6.0 ± 1.3 events from $t\bar{t}$ and W and Z production; results were presented in the form of an excluded region in the $(m_{\tilde{\chi}_1^0}, m_{\tilde{b}_1})$ plane. Improved bounds were obtained in the CDF experiment [137].

Searches for SUSY particles are reviewed by Schmitt in [59], including in particular searches at LEP, which we have not discussed. Chapter 15 of [48] and chapter 15 of [49] also provide substantial reviews. In rough terms, the present status is that there is 'little room for SUSY particles lighter than m_Z.' (Schmitt, in [59].) With all LEP data analysed, and if there is still no signal from the Tevatron collaborations, it will be left to the LHC to provide definitive tests.

12.4 Benchmarks for SUSY searches

Assuming degeneracy between the first two families of sfermions, there are 25 distinct masses for the undiscovered states of the MSSM: seven squarks and sleptons in the first two families, seven in the third family, four Higgs states, four neutralinos, two charginos and one gluino. Many details of the phenomenology to be expected (production cross sections, decay branching ratios) will obviously depend on the precise ordering of these masses. These in turn depend, in the general MSSM, on a very large number (over 100) of parameters characterizing the soft SUSY-breaking terms, as noted in Section 9.2. Any kind of representative sampling of such a vast parameter space is clearly out of the question. On the other hand, in order (for example) to use simulations to assess the prospects for detecting and measuring these new particles at different accelerators, some consistent model must be adopted [141]. This is because, very often, a promising SUSY signal in one channel, which has a small SM background, actually turns out to have a large background from other SUSY production and decay processes. Faced with this situation, it seems necessary to reduce drastically the size of the parameter space, by adopting one of the more restricted models for SUSY breaking, such as the mSUGRA one. Such models typically have only three or four parameters; for instance, in mSUGRA they are, as we have seen, m_0, $m_{1/2}$, A_0, $\tan\beta$, and the sign of μ.

But a complete sampling of even a three- or four-dimensional parameter space, in order (say) to simulate experimental signatures within a detector, is beyond present capabilities. This is why such studies are performed only for certain specific points in parameter space, or in some cases along certain lines. Such parameter sets are called 'benchmark sets'.

Various choices of benchmark have been proposed. To a certain extent, which one is likely to be useful depends on what is being investigated. For example, the '$m_{\rm h}^{\rm max}$-scenario' [102] referred to in Section 10.2 is suitable for setting conservative bounds on $\tan\beta$ and $m_{\rm A^0}$, on the basis of the non-observation of the lightest Higgs state. Another approach is to require that the benchmark points used for studying collider phenomenology should be compatible with various experimental constraints – for example [142] the LEP searches for SUSY particles and for the Higgs boson, the precisely measured value of the anomalous magnetic moment of the muon, the decay $\rm b \to s\gamma$, and (on the assumption that $\tilde{\chi}_1^0$ is the LSP) the relic density $\Omega_{\tilde{\chi}_1^0} h^2$. The authors of [142] worked within the mSUGRA model, taking $A_0 = 0$ and considering 13 benchmark points (subject to these constraints) in the space of the remaining parameters (m_0, $m_{1/2}$, $\tan\beta$, sign μ). A more recent study [143] updates the analysis in the light of the more precise dark matter bounds provided by the WMAP data.

One possible drawback with this approach is that minor modifications to the SUSY-breaking model might significantly alter the cosmological bounds, or the rate for $\rm b \to s\gamma$, while having little effect on the collider phenomenology; thus important regions of parameter space might be excluded prematurely. In any case, it is clearly desirable to formulate benchmarks for other possibilities for SUSY-breaking, in particular. The 'Snowmass Points and Slopes' (SPS) [144] are a set of benchmark points and lines in parameter space, which include seven mSUGRA-type scenarios, two gauge-mediated symmetry-breaking scenarios (it should be noted that here the LSP is the gravitino), and one anomaly-mediated symmetry-breaking scenario. Another study [145] concentrates on models which imply that at least some superpartners are light enough to be detectable at the Tevatron (for 2 $\rm fb^{-1}$ integrated luminosity); such models are apparently common among effective field theories derived from the weakly coupled heterotic string.

The last two references conveniently provide diagrams or tables showing the SUSY particle spectrum (i.e. the 25 masses) for each of the benchmark points. They are, in fact, significantly different. For example, Figure 12.10 (taken from [144]) shows two sparticle spectra corresponding to the parameter values

$$\text{SPS 1a}: m_0 = 100\ \text{GeV}, m_{1/2} = 250\ \text{GeV}, A_0 = -100\ \text{GeV}, \tan\beta = 10, \mu > 0 \tag{12.155}$$

and

$$\text{SPS 2}: m_0 = 1450\ \text{GeV}, m_{1/2} = 300\ \text{GeV}, A_0 = 0, \tan\beta = 10, \mu > 0. \tag{12.156}$$

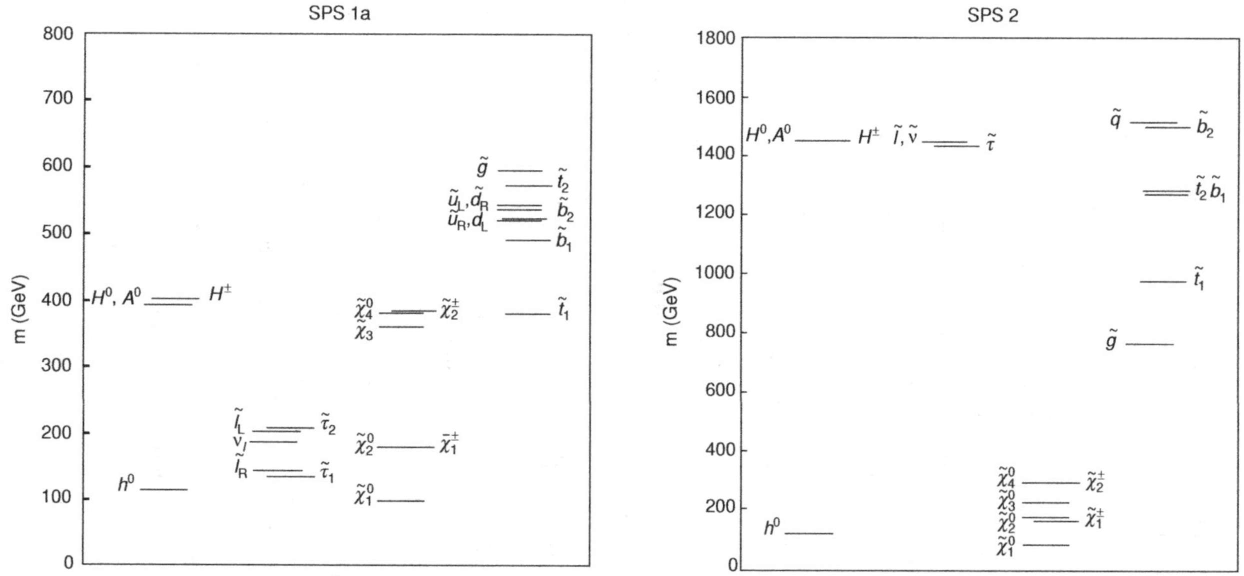

Figure 12.10 The SUSY particle spectra for the benchmark points corresponding to the parameter values SPS 1a (equation (12.155)) and SPS 2 (equation (12.156)). [From B. C. Allanach, *et al.*, *Eur. Phys. J.* **C25** Figure 1, p. 118 (2002), reprinted with the permission of Springer Science and Business Media.]

Such spectra may themselves be regarded as the benchmarks, rather than the values of the high-scale parameters which led to them. If and when sparticles are discovered, their masses and other properties may provide a window into the physics of SUSY breaking. However, as emphasized in Section 9.4 of [47], there are in principle not enough observables at hadron colliders to determine all the parameters of the soft SUSY-breaking Lagrangian; for this, data from a future e^+e^- collider will be required.

Ultimately, if supersymmetry is realized near the weak scale in nature, high precision data in the sparticle sector will become available. A correspondingly precise theoretical analysis will require the inclusion of higher-order corrections, for which a well defined theoretical framework is needed; one – the Supersymmetry Parameter Analysis Convention (SPA) – has already been proposed [146]. Such efforts have the ambitious aim of reconstructing the fundamental supersymmetric theory, and its breaking mechanism, from the data – as and when that may be forthcoming.

References

[1] Heisenberg, W. (1939), *Z. Phys.* **A113**, 61–86.
[2] Glashow, S. L. (1961), *Nucl. Phys.* **22**, 579–88.
[3] Weinberg, S. (1967), *Phys. Rev. Lett.* **19**, 1264–6.
[4] Salam, A. (1968), in *Elementary Particle Theory: Relativistic Groups and Analyticity* (Nobel Symposium No. 8) ed. Svartholm, N. (Stockholm: Almqvist and Wiksell) p. 367.
[5] 't Hooft, G. (1971), *Nucl. Phys.* **B35**, 167–88.
[6] Higgs, P. W. (1964), *Phys. Rev. Lett.* **13**, 508–9; (1966), *Phys. Rev.* **145**, 1156–63.
[7] Aitchison, I. J. R. and Hey, A. J. G. (2004), *Gauge Theories in Particle Physics. Volume II: QCD and the Electroweak Theory* (Bristol and Philadelphia: IoP Publishing).
[8] Susskind, L. (1979), *Phys. Rev.* **D20**, 2619–25.
[9] 't Hooft, G. (1980), in *Recent Developments in Gauge Theories, Proc. 1979 NATO Advanced Study Institute Cargèse, France*, Nato Advanced Study Institute Series, Series B, Physics, v.59, eds. 't Hooft, G., *et al.* (New York: Plenum), pp. 135–157; reprinted in *Dynamical Gauge Symmetry Breaking: A Collection of Reprints* 1983, pp. 345–67 eds. Farhi, E. and Jackiw, R. W. (Singapore: World Scientific).
[10] Weinberg, S. (1976), in *Gauge Theories and Modern Field Theory*, eds. Arnowitt, R. and Nath, P. (Cambridge MA: MIT Press) pp. 1–26; see also Weinberg, S. (1974), in *Proc. XVII Int. Conf. on High Energy Physics*, ed. Smith, J. R. (Chilton, Didcot, Oxon: The Science Research Council, Rutherford Laboratory) pp. III-59–III-65.
[11] Weinberg, S. (1975), *Phys. Rev.* **D13**, 974–96; (1979), *ibid* **D19**, 1277–80.
[12] Fahri, E. and Susskind, L. (1981), *Phys. Rep.* **74**, 277–321.
[13] Lane, K. D. (2002), *Two Lectures on Technicolor* hep-ph/0202255.
[14] Hill, C. T. and Simmons, E. H. (2003), *Phys. Rep.* **381**, 235–402.
[15] Aitchison, I. J. R. and Hey, A. J. G. (2003), *Gauge Theories in Particle Physics. Volume I: From Relativistic Quantum Mechanics to QED* (Bristol and Philadelphia: IoP Publishing).
[16] Witten, E. (1981), *Nucl. Phys.* **B188**, 513–54.
[17] Veltman, M. (1981), *Act. Phys. Polon.* **B12**, 437–57.
[18] Kaul, R. (1982), *Phys. Lett.* **109B**, 19–24.
[19] Wess, J. and Zumino, B. (1974), *Phys. Lett.* **49B**, 52–4.
[20] Dimopoulos, S. and Georgi, H. (1981), *Nucl. Phys.* **B193**, 150–62.
[21] Sakai, N. (1981), *Z. Phys.* **C11**, 153–7.
[22] Ramond, P. (1971), *Phys. Rev.* **D3**, 2415–18.

[23] Neveu, A. and Schwarz, J. H. (1971), *Nucl. Phys.* **B31**, 86–112.
[24] Gervais, J.-L. and Sakita, B. (1971), *Nucl. Phys.* **B34**, 632–9.
[25] Gol'fand, Y. A. and Likhtman, E. P. (1971), *JETP Lett.* **13**, 323–6.
[26] Haag, R., Łopusański, J. T. and Sohnius, M. (1975), *Nucl. Phys.* **B88**, 257–74.
[27] Volkov, D. V. and Akulov, V. P. (1973), *Phys. Lett.* **46B**, 109–10.
[28] Wess, J. and Zumino, B. (1974), *Nucl. Phys.* **B70**, 39–50; *ibid.* **B78**, 1–13.
[29] Salam, A. and Strathdee, J. (1974), *Nucl. Phys.* **B76**, 477–82.
[30] Fayet, P. (1975), *Nucl. Phys.* **B90**, 104–27.
[31] Fayet, P. (1976), *Phys. Lett.* **64B**, 159–62.
[32] Fayet, P. (1977), *Phys. Lett.* **69B**, 489–94.
[33] Fayet, P. (1978), in *New Frontiers in High-Energy Physics, Proc. Orbis Scientiae*, Coral Gables, FL, USA, eds. Perlmutter, A. and Scott, L. F. (New York: Plenum) p. 413.
[34] Farrar, G. R. and Fayet, P. (1978), *Phys. Lett.* **76B**, 575–9.
[35] Farrar, G. R. and Fayet, P. (1978), *Phys. Lett.* **79B**, 442–6.
[36] Farrar, G. R. and Fayet, P. (1980), *Phys. Lett.* **89B**, 191–4.
[37] Antoniadis, I. (1990), *Phys. Lett.* **246B**, 377–84; Arkani-Hamed, N., *et al.* (1998), *Phys. Lett.* **429B**, 263–72; Antoniadis, I., *et al.* (1998), *Phys. Lett.* **436B**, 257–63; Arkani-Hamed, N., *et al.* (1999), *Phys. Rev.* **D59**, 086004-1-21.
[38] Lykken, J. D. (2005), *Czech. J. Phys.* **55**, B577–B598.
[39] Ellis, J. (2002), in *2001 European School of High Energy Physics* CERN 2002–002 eds. Ellis, N. and March-Russell, J., p. 166.
[40] Coleman, S. R. and Mandula, J. (1967), *Phys. Rev.* **159**, 1251–6.
[41] Shifman, M. A. (1999), *ITEP Lectures on Particle Physics and Field Theory, Volume 1; World Scientific Lecture Notes in Physics – Vol. 62* (Singapore: World Scientific), Chapter IV.
[42] Bailin, D. and Love, A. (1994), *Supersymmetric Gauge Field Theory and String Theory* (Bristol and Philadelphia: IoP Publishing), Equation (1.38).
[43] Fayet, P. and Ferrara, S. (1977), *Phys. Rep.* **32**, 249–334.
[44] Nilles, H. P. (1984), *Phys. Rep.* **110**, 1–162.
[45] Haber, H. E. and Kane, G. L. (1985), *Phys. Rep.* **117**, 75–263.
[46] Martin, S. P. (1997), *A Supersymmetry Primer* hep-ph/9709356 version 4, June 2006.
[47] Chung, D., *et al.* (2005), *Phys. Rep.* **407**, 1–203.
[48] Drees, M., Godbole, R. M. and Roy, P. (2004), *Theory and Phenomenology of Sparticles* (Singapore: World Scientific).
[49] Baer, H. and Tata, X. (2006), *Weak Scale Supersymmetry* (Cambridge: Cambridge University Press).
[50] Peskin, M. E. and Schroeder, D. V. (1995), *An Introduction to Quantum Field Theory* (Reading, MA: Addison-Wesley).
[51] Iliopoulos, J. and Zumino, B. (1974), *Nucl. Phys.* **B76**, 310–32.
[52] Grisaru, M., Siegel, W. and Rocek, M. (1979), *Nucl. Phys.* **B159**, 429–50.
[53] Kim, J. E. and Nilles, H. P. (1984), *Phys. Lett.* **138B**, 150–4.
[54] Polonsky, N. (2000), in *Supersymmetry, Supergravity and Superstring*, in *Proc. KIAS-CTP International Symposium 1999*, Seoul, Korea, eds. Kim, J. E. and Lee, C. (Singapore: World Scientific), pp. 100–24.
[55] Georgi, H., Quinn, H. R. and Weinberg, S. (1974), *Phys. Rev. Lett.* **33**, 451–4.
[56] Langacker, P. (1990), in *Proc. PASCOS90 Symposium* eds. Nath, P. and Reucroft, S. (Singapore: World Scientific); Ellis, J., Kelley, S. and Nanopoulos, D. (1991), *Phys. Lett.* **260B**, 131–7; Amaldi, U., de Boer, W. and Furstenau, H. (1991), *Phys. Lett.*

260B, 447–55; Langacker, P. and Luo, M. (1991), *Phys. Rev.* **D44**, 817–22; Giunti, C., Kim, C. W. and Lee, U. W. (1991), *Mod. Phys. Lett.* **A6**, 1745–56; Ross, G. G. and Roberts, R. G. (1992), *Nucl. Phys.* **B377**, 571–92.
[57] Peskin, M. (1997), in *1996 European School of High Energy Physics*, CERN-97-03, eds. Ellis, N. and Neubert, M., pp. 49–142.
[58] Pokorski, S. (1999), *Act. Phys. Pol.* **B30**, 1759–74.
[59] Yao, W.-M., *et al.* (2006), *The Review of Particle Physics*: *J. Phys.* **G33**, 1–1232.
[60] Dimopoulos, S., Raby, S. and Wilczek, F. (1981), *Phys. Rev.* **D24**, 1681–3.
[61] Ibáñez, L. E. and Ross, G. G. (1981), *Phys. Lett.* **105B**, 439–42.
[62] Einhorn, M. B. and Jones, D. R. T. (1982), *Nucl. Phys.* **B196**, 475–88.
[63] O'Raifeartaigh, L. (1975), *Nucl. Phys.* **B96**, 331–52.
[64] Ferrara, S., Girardello, L. and Palumbo, F. (1979), *Phys. Rev.* **D20**, 403–8.
[65] Fayet, P. and Iliopoulos, J. (1974), *Phys. Lett.* **51B**, 461–4.
[66] Girardello, L. and Grisaru, M. T. (1982), *Nucl. Phys.* **B194**, 65–76.
[67] Dimopoulos, S. and Sutter, D. (1995), *Nucl. Phys.* **B452**, 496–512.
[68] Nilles, H. P. (1982), *Phys. Lett.* **115B**, 193–6; (1983), *Nucl. Phys.* **B217**, 366–80.
[69] Chamseddine, A. H., Arnowitt, R. and Nath, P. (1982), *Phys. Rev. Lett.* **49**, 970–4.
[70] Barbieri, R., Ferrara, S. and Savoy, C. A. (1982), *Phys. Lett.* **119B**, 343–7.
[71] Dine, M. and Nelson, A. E. (1993), *Phys. Rev.* **D48**, 1277–87; Dine, M., Nelson, A. E. and Shirman, Y. (1995), *Phys. Rev.* **D51**, 1362–70; Dine, M., Nelson, A. E., Nir, Y. and Shirman, Y. (1996), *Phys. Rev.* **D53**, 2658–69.
[72] Kaplan, D. E., Kribs, G. D. and Schmaltz, M. (2000), *Phys. Rev.* **D62**, 035010-1–10; Chacko, Z., Luty, M. A., Nelson, A. E. and Ponton, E. (2000), *JHEP* **01**, 003.
[73] Randall, L. and Sundrum, R. (1999), *Nucl. Phys.* **B557**, 79–118; Giudice, G. F., Luty, M. A., Murayama, H. and Rattazi, R. (1998), *JHEP* **12**, 027.
[74] Inoue, K., Kakuto, A., Komatsu, H. and Takeshita, H. (1982), *Prog. Theor. Phys.* **68**, 927–46 [Erratum: (1982), *ibid.* **70**, 330]; and (1984), *ibid.* **71**, 413–6.
[75] Ibáñez, L. E. and Ross, G. G. (1982), *Phys. Lett.* **110B**, 215–20.
[76] Ibáñez, L. E. (1982), *Phys. Lett.* **118B**, 73–8.
[77] Ellis, J., Nanopoulos, D. V. and Tamvakis, K. (1983), *Phys. Lett.* **121B**, 123–9.
[78] Ellis, J., Hagelin, J., Nanopoulos, D. V. and Tamvakis, K. (1983), *Phys. Lett.* **125B**, 275–81.
[79] Alvarez-Gaumé, L., Polchinski, J. and Wise, M. B. (1983), *Nucl. Phys.* **B221**, 495–523.
[80] Kane, G. L., Kolda, C., Roszkowski, L. and Wells, J. D. (1994), *Phys. Rev.* **D49**, 6173–210.
[81] Arason, H., *et al.* (1991), *Phys. Rev. Lett.* **67**, 2933–6.
[82] Barger, V., Berger, M. S. and Ohmann, P. (1993), *Phys. Rev.* **D47**, 1093–113.
[83] Carena, M., Pokorski, S. and Wagner, C. E. M. (1993), *Nucl. Phys.* **B406**, 59–89.
[84] Langacker, P. and Polonsky, N. (1994), *Phys. Rev.* **D49**, 1454–67.
[85] Olechowski, M. and Pokorski, S. (1988), *Phys. Lett.* **214B**, 393–7.
[86] Ananthanarayan, B., Lazarides, G. and Shafi, Q. (1991), *Phys. Rev.* **D44**, 1613–15.
[87] Dimopoulos, S., Hall, L. J. and Raby, S. (1992), *Phys. Rev. Lett.* **68**, 1984–7.
[88] Carena, M., *et al.* (1994), *Nucl. Phys.* **B426**, 269–300.
[89] Hall, L. J., Rattazi, R. and Sarid, U. (1994), *Phys. Rev.* **D50**, 7048–65.
[90] Hempfling, R. (1994), *Phys. Rev.* **D49**, 6168–72.
[91] Rattazi, R. and Sarid, U. (1996), *Phys. Rev.* **D53**, 1553–85.
[92] Carena, M., *et al.* (2001), *Nucl. Phys.* **B599**, 158–84.
[93] Inoue, K., Kakuto, H., Komatsu, H. and Takeshita, S. (1982), *Prog. Theor. Phys.* **67**, 1889–98.

[94] Flores, R. A. and Sher, M. (1983), *Ann. Phys.* **148**, 95–134.
[95] LEP (2003), *Phys. Lett.* **565B**, 61–75.
[96] Okada, Y., Yamaguchi, M. and Yanagida, T. (1991), *Prog. Theor. Phys.* **85**, 1–5; and (1991), *Phys. Lett.* **262B**, 54–8.
[97] Barbieri, R., Frigeni, M. and Caravaglio, F. (1991), *Phys. Lett.* **258B**, 167–70.
[98] Haber, H. E. and Hempfling, R. (1991), *Phys. Rev. Lett.* **66**, 1815–18.
[99] Ellis, J., Ridolfi, G. and Zwirner, F. (1991), *Phys. Lett.* **257B**, 83–91; and (1991), *Phys. Lett.* **262B**, 477–84.
[100] Carena, M., Quirós, M. and Wagner, C. E. M. (1996), *Nucl. Phys.* **B461**, 407–36.
[101] Haber, H. E., Hempfling, R. and Hoang, A. H. (1997), *Z. Phys.* **C75**, 539–44.
[102] Carena, M., Heinemeyer, S., Wagner, C. E. and Weiglein, G. (2003), *Eur. Phys. J.* **C26**, 601–7.
[103] Degrassi, G., *et al.* (2003), *Eur. Phys. J.* **C28**, 133–43.
[104] Drees, M. (1996), *An Introduction to Supersymmetry*, Lectures at the Asia-Pacific Centre for Theoretical Physics, Seoul, Korea APCTP-5, KEK-TH-501, November 1996 (hep-ph/9611409).
[105] Drees, M. (2005), *Phys. Rev.* **D71**, 115006-1–9.
[106] Dermíšek, R. and Gunion, J. F. (2006), *Phys. Rev.* **D73**, 111701(R)-1–5; see also Dermíšek, R. and Gunion, J. F. (2005), *Phys. Rev. Lett.* **95**, 041801-1–4.
[107] Decker, R. and Pestieau, J. (1980), *Lett. Nuovo Cim.* **29**, 560–4.
[108] Kolda, C. F. and Murayama, H. (2000), *JHEP* **07**, 035.
[109] Ibrahim, T. and Nath, P. (1998), *Phys. Rev.* **D58**, 111301-1–6.
[110] Ellis, J. R., *et al.* (1984), *Nucl. Phys.* **B238**, 453–76; see also Goldberg, H. (1983), *Phys. Rev. Lett.* **50**, 1419–22.
[111] Ellis, J. R., *et al.* (2003), *Phys. Lett.* **565B**, 176–82.
[112] Baer, H., *et al.* (2003), *JHEP* **03**, 054.
[113] Bottino, A., *et al.* (2003), *Phys. Rev.* **D68**, 043506-1–8.
[114] Harrison, P. R. and Llewellyn Smith, C. H. (1983), *Nucl. Phys.* **B213**, 223–40.
[115] Gunion, J. F. and Haber, H. E. (1986), *Nucl. Phys.* **B272**, 1–76; (1986), *ibid.* **B278**, 449–92; (1988), *ibid.* **B307**, 445–75; (1993), (E) *ibid.* **B402**, 567–8.
[116] Dawson, S., Eichten, E. and Quigg, C. (1985), *Phys. Rev.* **D31**, 1581–637.
[117] Branson, J. G., *et al.* [The ATLAS and CMS Collaborations] (2001), *EPJdirectC* **4, CN1**, 1–61.
[118] Barger, V. and Phillips, R. J. N. (1987), *Collider Physics* (Reading, MA: Addison-Wesley).
[119] Barger, V., Keung, Y. and Phillips, R. J. N. (1985), *Phys. Rev. Lett.* **55**, 166–9.
[120] Barnett, R. M., Gunion, J. F. and Haber, H. E. (1993), *Phys. Lett.* **315B**, 349–54.
[121] Baer, H., Tata, X. and Woodside, J. (1990), *Phys. Rev.* **D41**, 906–15.
[122] CDF (2001), *Phys. Rev. Lett.* **87**, 251803-1–6.
[123] CDF (1996), *Phys. Rev. Lett.* **76**, 2006–10.
[124] D0 (2001), *Phys. Rev.* **D63**, 091102-1–7.
[125] Arnowitt, R. and Nath, P. (1987), *Mod. Phys. Lett.* **A2**, 331–42.
[126] Baer, H. and Tata, X. (1993), *Phys. Rev.* **D47**, 2739–45.
[127] Baer, H., Kao, C. and Tata, X. (1993), *Phys. Rev.* **D48**, 5175–80.
[128] Kamon, T., Lopez, J., McIntyre, P. and White, J. T. (1994), *Phys. Rev.* **D50**, 5676–91.
[129] Baer, H., Chen, C.-H., Kao, C. and Tata, X. (1995), *Phys. Rev.* **D52**, 1565–76.
[130] Mrenna, S., Kane, G. L., Kribs, G. D. and Wells, J. D. (1996), *Phys. Rev.* **D53**, 1168–80.
[131] D0 (1996), *Phys. Rev. Lett.* **76**, 2228–33.

[132] CDF (1996), *Phys. Rev. Lett.* **76**, 4307–11.
[133] D0 (1998), *Phys. Rev. Lett.* **80**, 1591–6.
[134] CDF (1998). *Phys. Rev. Lett.* **80**, 5275–80.
[135] Drees, M. and Hikasa, K. (1990), *Phys. Lett.* **252B**, 127–34.
[136] D0 (1996), *Phys. Rev. Lett.* **76**, 2222–7.
[137] CDF (2000), *Phys. Rev. Lett.* **84**, 5704–9.
[138] D0 (2002), *Phys. Rev. Lett.* **88**, 171802-1–7.
[139] CDF (2001), *Phys. Rev.* **D63**, 091101-1–7.
[140] D0 (1999), *Phys. Rev.* **D60**, 031101-1–6.
[141] Hinchliffe, I., Paige, F. E., Shapiro, M. D., and Söderqvist, J. (1997), *Phys. Rev.* **D55**, 5520–40.
[142] Battaglia, M., *et al.* (2001), *Eur. Phys. J.* **C22**, 535–61.
[143] Battaglia, M., *et al.* (2004), *Eur. Phys. J.* **C33**, 273–96.
[144] Allanach, B. C., *et al.* (2002), *Eur. Phys. J.* **C25**, 113–23.
[145] Kane, G. L., *et al.* (2003), *Phys. Rev.* **D67**, 045008-1–21.
[146] Aguila-Saavedra, J. A., *et al.* (2006), *Eur. Phys. J.* **C46**, 43–60.

Index

Printed in the United States
By Bookmasters